COASTAL
OCEANOGRAPHY

NATO CONFERENCE SERIES

I   Ecology
II   Systems Science
III   Human Factors
IV   Marine Sciences
V   Air–Sea Interactions
VI   Materials Science

IV MARINE SCIENCES

COASTAL OCEANOGRAPHY

Edited by

Herman G. Gade
University of Bergen
Bergen, Norway

Anton Edwards
Scottish Marine Biological Association
Oban, United Kingdom

and

Harald Svendsen
University of Bergen
Bergen, Norway

Published in cooperation with NATO Scientific Affairs Division

Springer Science+Business Media, LLC

Library of Congress Cataloging in Publication Data

NATO Advanced Research Institute on Coastal Oceanography (1982: Os, Hordaland
fylke, Norway)
   Coastal oceanography.

   (NATO conference series. IV, Marine sciences; v. 11)
   "Proceedings of a NATO Advanced Research Institute on Coastal Oceanography,
held June 6–11, 1982, in Os, Norway"—Verso t.p.
   "Published in cooperation with NATO Scientific Affairs Division."
   Includes bibliographical references and index.
   1. Oceanography—Congresses. I. Gade, Herman G. (Herman Gerhard), 1931–
II. Edwards, Anton. III. Svendsen, Harald. IV. North Atlantic Treaty Organization. Scien-
tific Affairs Division. V. Title. VI. Series.
GC2.N33 1982   551.47   83-9724

ISBN 978-1-4615-6650-2   ISBN 978-1-4615-6648-9 (eBook)
DOI 10.1007/978-1-4615-6648-9

Proceedings of a NATO Advanced Research Institute on Coastal
Oceanography, held June 6–11, 1982, in Os, Norway

© 1983 Springer Science+Business Media New York
Originally published by Plenum Press, New York in 1983.
A Division of Plenum Publishing Corporation
233 Spring Street, New York, N.Y. 10013

PREFACE

    This volume evolved from the Coastal Oceanography Workshop held as a NATO Advanced Research Institute at Os, Norway, 6-11 June 1982. The organizing Committee, consisting of Dr. T. Carstens, Dr. M. Mork, Dr. H.G. Gade and Dr. H. Svendsen relied to a great extent on the support of an international advisory committee whose members were Dr. G. Cannon (USA), Mr. A. Edwards (UK), Dr. D.M. . Farmer (Canada), Prof. G. Kullenberg (Denmark), Prof. J.C.J. Nihoul (Belgium) and Prof. J. Sündermann (Fed. Rep. Germany).

    The aims of the Workshop were to deal with the frontiers of research on physical oceanography of coastal waters, both inshore and offshore including shelf waters and shelf seas. All relevant aspects of these topics were the concern of the conference, but stress was laid on the importance of bottom topography, particularly irregular bottoms and coastlines. Among the contributions received on conclusion of the Workshop were also a few papers aiming to review certain sections of the field.

    The papers presented in this volume range from purely descriptive to analytical studies and numerical or physical modelling. However, the papers are arranged under general subject headings. Because of the overlapping character of the various subjects, many papers could equally well have been placed in alternative sections. The majority of the contributions is in accordance with given standards of length etc., but because of the importance of recent development in certain areas, a few papers are more extensive.

    The meeting at Os provided a valuable opportunity for exchanges of information and ideas. Some of these are already reflected in the papers presented in this volume. The full impact of the Workshop is, however, yet to come and we hope that this volume will be instrumental in this purpose.

    We thank the NATO Marine Science Committee for helping to make the publication of this book possible.

Herman G. Gade
Anton Edwards
Harald Svendsen

CONTENTS

COASTAL AND SHELF SEA CIRCULATION

COASTAL UPWELLING AND OTHER LARGE SCALE COASTAL AND
SHELF PHENOMENA

CONTENTS

ON THE DYNAMICS OF COASTAL CURRENTS

Martin Mork

Geophysical Institute
Bergen, Norway

ABSTRACT

The main factors governing time dependent and steady coastal circulation are discussed. For barotropic flows the actual bottom topography is shown to play an important role while baroclinic currents have been related to buoyancy fluxes and effects of entrainment.

INTRODUCTION

The coastal boundary has a profound effect on the water motion. The constraints imposed on the motion by the coast give rise to new modes, mainly subinertial oscillations which may be excited by atmospheric forces. It has been established that the internal response to wind is intensified at the coast, mainly due to excitation of Kelvin type internal waves (Mork, 1972). Furthermore, energy from various sources is trapped in boundary waves. The bottom topography in the coastal region plays an important role in the manifestaion of the coastal circulation. This can best be demonstrated by considering the vorticity balance. Through crossisobathic flow and stretching or shrinking of vertical columns the current adjusts it's vorticity when torques are exerted by wind or frictional forces. Csanady (1981) has discussed the relationship between bottom slope and sea level adjustment for stationary coastal circulations. For time dependent responses to wind it should be noted that the bottom slope gives rise to frequency wavenumber relationships of free waves, which fits the k-σ domain of energy, containing wave components of weather systems (Mysak, 1980).

The barotropic flows along open straight coasts can usually be
related to wind conditions in a very simple way. The main forcing
factor is the wind stress component along the coast, which also
governs upwelling or downwelling. Hickey and Hamilton (1980)
have put forward the formula

$$\frac{\partial \bar{u}}{\partial t} = \frac{\tau_x}{h} - \lambda \bar{u}$$

and shown that the observed velocities compare surprisingly well with
the model results. For irregular coastlines and coasts bordering
ocean basins the situation is quite different. For example, the
Norwegian Coastal Current responds mainly to the wind induced
disturbances in the North Sea and less to the local wind, at least
in the southern part of the coastal region (Furnes, 1980; Davies
and Heaps, 1980). For a comprehensive review on coastal upwelling,
see O'Brien et al. (1977).

The trapping of tidal energy at the coast is also a factor to
be considered as a driving mechanism. The ocean tides generally
propagate along the coasts as Kelvin waves, thus giving rise to a
barotropic Stokes velocity

$$u_s = \tfrac{1}{2}c(A/H)^2 e^{-\frac{2fy}{c}}$$

where $c=\sqrt{gH}$ is the Laplacian phase velocity, H is the depth and A
is the amplitude of sea level variations. The resulting transport
is $gA^2/4f$ giving $10^5 m^3/sec$ for a tidal amplitude of 2m. The
Lagrangian transport may be higher by a factor 1.4 (Mæland, 1982a)

BAROCLINIC CURRENTS

Most coastal currents are characterised by lower density than
the adjacent water. Rivers and freshwater runoff are the main
sources of density reduction and solar heating is another cause
of density stratification. The last effect is greatly enhanced
if a top layer of light water is already present. Buoyancy fluxes
are considered as driving mechanisms for coastal currents. Heaps
(1980) has pointed out that tidal mixing as well as wind mixing
are energy sources to be considered in combination with buoyancy
fluxes. Mixing of coastal water masses leads to an increase of
potential energy which in turn promotes baroclinic currents.

Combined effects of buoyancy and wind stress have been studied
by many authors. Pietrafesa and Janowitz (1970) have also included
bottom topography in a numerical study. Diagnostic models of
baroclinic coastal currents have been developed by Backhaus (1983)
and Hamilton and Rattray (1978). The effect of buoyancy fluxes only

has been investigated in laboratory experiments by Griffith and
Linden (1981) and Vinger et al. (1981). The current structure
in such models is somewhat obscured by the presence of waves and
eddies which are almost instantly formed. The complex current
patterns and frontal mixing caused by baroclinic instability may
now frequently be observed in nature with aid of satellite imagery.
For example see Johannessen et al. (1979) and Audunson et al. (1981).

ENTRAINMENT MODELS

In order to demonstrate the effect of buoyancy fluxes from a
coast, an idealized model is presented (for details see Mork, 1981
a,b). The coast is substituted by a line source of fresh water
which is assumed to be immediately mixed through a wedge of low
salinity coastal water. Vertical entrainment from below is assumed
to be proportional to the longshore velocity.

With the lower layer at rest and the x-axis in longshore
direction the governing equation are

$$\frac{\partial}{\partial t}(uh) + \frac{\partial}{\partial y}(u\bar{v}h) - f\bar{v}h = 0$$

$$fu = -g_1\frac{\partial h}{\partial y}$$

$$\frac{\partial(uh)}{\partial x} + \frac{\partial}{\partial y}(\bar{v}h) = W_e$$

$$\frac{\partial}{\partial t}(g_1 A) = Q$$

where h is the vertical thickness of upper layer, A is the cross-
sectional area, g_1 is the reduced gravity and Q is the buoyancy
flux from the coast. In addition, the boundary conditions are h=0
at the edge and $\bar{v}h=q$ at the coast, y=0.

The main results are

$$u = u_0(1-y/L),$$

where $L\sim t^{\frac{1}{2}}$ is the width and u_0 approaches a constant value after
some initial stage of development given as

$$u_0 = \left(\frac{qf}{E}\right)^{\frac{1}{2}}$$

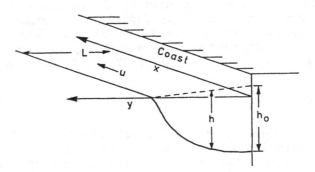

Fig. 1.   Geometry of the model.

where E is an entrainment coefficient. The width of the current
increases as the square root of time, thus

$$L = \frac{3u_0}{f} \; (t/\tau)^{\frac{1}{2}}$$

where
$$\tau = \frac{3}{2} \; \frac{u_0^3}{Qf} \; .$$

The crossectional area becomes

$$A = qt(1+(\frac{t}{\tau})^{\frac{1}{2}})$$

where the last term stems from entrainment.

    The transport is simply

$$T = \frac{3}{4} \; u_0 A.$$

The entrainment coefficient is chosen as $E=10^{-4}$ and with fluxes
from the coast

$$q = 2 \cdot 10^{-2} m^2 /sec, \; Q = 5.10^{-3} m^3 /sec^3 ,$$

representative for the Norwegian coast, the coastal current
approaches a stage in 56 days comparable with observations. For
that stage the results are

$$u_0 = .14 \text{ m/sec} \qquad \text{for } y=0$$

$$h = 72 \text{ m} \qquad \text{for } y=0$$

$$L = 102 \text{ km}$$

$$g_1 = 10^{-2} \text{m/sec}^2$$

$$T = .26 \cdot 10^6 \text{m}^3/\text{sec}.$$

The observed baroclinic transport is generally higher, up to
$10^6 \text{m}^3/\text{sec}$. Entrainment in this model has a similar effect on the
lateral velocity profile as viscous stress at the internal boundary
of immiscible fluids (Mæland, 1982b). The results of the line source
model can be related to experiments in rotating basins with circum-
ferential distribution of buoyancy fluxes, as in the experiments
by Griffith and Linden (1981).

It should be pointed out that premixing of the boundary fluxes
comparable with fjord effects would give a faster development.
Since a lateral compensation flow has to take place in the lower
layer, the volume flux, q, is increased while the buoyancy flux, Q,
is the same as before.

A STATIONARY MODEL

In this case the downstream development of a buoyant current is
studied. There are no fluxes from the coast. Thus, the buoyancy
transport remains constant while the mass transport is increased
downstream by entrainment only. With $W_e = Eu$ the equations become
in integrated form

$$\frac{\partial}{\partial x} \int u^2 h dy = -Eg_1 A,$$

$$\frac{\partial T}{\partial x} = E(2g_1 T/f)^{\frac{1}{2}},$$

$$\frac{\partial}{\partial x}(g_1 T) = 0.$$

The symbols are the same as before. It is immediately seen that
the transport increases linearly with distance downstream while the
momentum transport decreases. Obviously, the assumptions of a
confined coastal flow is violated, and we may draw the conclusion
that a coastal internal bore has a limiting distance of travelling
unless energy is supplied or some detrainment takes place. In the

present model the critical distance is

$$\frac{h_0}{2E}\left[(1+\frac{u_0^2}{g_1 h_0})^{\frac{1}{2}}-1\right].$$

where the values of thickness h_0, velocity u_0 and $g_1 = g_1(0)$ refer
to the origin, x=0. For reasonable values the distance will be of
order of a few hundred kilometers. Observations by Royer (1983) in
the Gulf of Alaska lend support to some of the foregoing arguments.
It should also be noted that the stationary model may easily be
extended to include several sources along the coast.

CONCLUDING REMARKS

        The models presented here are only meant to be the first
steps in the investigation of buoyancy effects on coastal currents.
In a more realistic model other effects than those mentioned here
must be included, but that will probably be at the expense of
analytic solutions. The variability of coastal currents has only
briefly been mentioned and frontal dynamics has not been discussed
at all. It is very likely that the meandering and eddy detachments
at the coastal front has a profound influence also on the mean
conditions of coastal currents. For example a parameterization of
eddy detachment in the way of detrainment at the current edge would
have given completely different model results.

REFERENCES

        Audunson, T. et al. 1981. Some observations of ocean fronts,
                wave and currents at the surface along the Norwegian
                coast from satellite images and drifting buoys. Proc.
                Symp. Norw. Coastal Curr., Geilo 1980.
        Backhaus, J.O. 1983. On the circulation of the Stratified
                North Sea, (this volume).
        Csanady, G.T., 1981: Shelf circulation cell. Phil. Trans.
                R. Soc. Lond. A 302, 515-530.
        Davies, A.M. and N. Heaps, 1980. Influence of the Norw. Trench
                on the circulation of the North Sea. Tellus, 32,
                164-175.
        Furnes, G.K., 1980. Wind effects in the North Sea. J. Phys.
                Oceanogr. 10, 978-984.
        Griffiths, R.W. and P.F. Linden, 1981. The stability of
                buoyancydriven coastal currents. Dyn. Atmos. Oceans,
                5: 281-306.

Hamilton, P. and M. Rattray, 1978. A numerical model of the
    depth dependent wind driven upwelling circulation on
    a continental shelf. J. Phys. Oceanogr., 8, 437-457.
Heaps, N.S. 1980. Density currents in a two-layered coastal
    system with application to the Norwegian Coastal
    Current. Geophys. J.R. Astr. Soc., 63, 289-310.
Hickey, B.M. and P. Hamilton, 1980. A spin-up model as a
    diagnostic tool for interpretation of current and
    density measurement on the continental shelf of the
    Pacific North west. J. Phys. Oceanogr., 10, 1.
Johannessen, O.M. and M. Mork et al. 1979. Remote sensing
    experiments in Norwegian coastal waters. Publ. Geoph.
    Inst. Univ. of Bergen.
Mork, M., 1972. On oceanic responses to atmospheric forces
    Rapp. et Proces-Verbaux, 162, 184-190.
Mork, M., 1981a. Experiment with theoretical models of the
    Norwegian Coastal Current. Proc. Symp. Norw. Coastal
    Current. Geilo, 1980.
Mork, M., 1981b. Circulation phenomena and frontal dynamics
    of the Norwegian Coastal Current. Phil. Trans. R. Soc.
    Lond. A, 302, 635-647.
Mæland, E. 1982a. On the steady mass transport induced by
    wave motions in a viscous rotating fluid. Geoph. Astr.
    Fluid Dyn. 19, 229-248.
Mæland, E. 1982b. A steady source-sink flow in a two-layer
    rotating fluid. Geoph. Astroph. Fluid. Dyn., 21,
    75-88.
Mysak, L. 1980. Recent advances in shelf wave dynamics. Rew.
    Geoph. and Space Phys. 18, 1, 211-241.
O'Brien, J.J. et al. 1977. Upwelling in the ocean. Modelling
    and Prediction of the upper layers of the Ocean.
    Pergamon Press.
Piatrafesa, L.J. and Janowitz. 1979. J. Phys. Oceanogr. 9,
    9, 911-918.
Royer, T.C. 1983. Observations of the Alaska coastal current
    (this volume).
Vinger, Å, T.A. McClimans and S. Tryggestad. 1981. Laboratory
    observations of instabilities in a straight coastal
    current. Proc. Norw. Coastal Curr. Geilo, 1980.

OBSERVATIONS OF THE ALASKA COASTAL CURRENT

Thomas C. Royer

Institute of Marine Science
University of Alaska
Fairbanks, Alaska 99701

ABSTRACT

The Alaska Coastal Current is a narrow, intense coastal current
bordering the southern coast of Alaska. It is characterized by
relatively low salinities and its transport is significantly
influenced by the regional freshwater discharge. The flow is
also modified by the winds which cause a downwelling condition here
throughout most of the year. In many aspects, this current is
similar to the Norwegian Coastal Current, with the important
exception that wind-driven reversals have not been observed.

THE SETTING

The rugged terrain of Alaska's southern coast contains the
northern extension of the Rocky Mountains; here called the Alaska
Range. This range has important consequences on the meteorology
over North America and this is especially true in this coastal
region. Elevations of greater than 4000 m are commonly found in
the coastal drainage area, which itself is generally less than
150 km wide. As illustrated in Fig. 1, the mountains rise abruptly
out of the sea, forming a barrier to the maritime air masses. Low
pressure systems, which propagate east-northeast over this part of
the North Pacific, frequently move into the Gulf of Alaska where
they stagnate and fill (Overland and Hiester, 1978). This contain-
ment is enhanced by cold arctic air masses over the Alaska continent.

The frequent presence of a low pressure system over the Gulf
of Alaska in all seasons, other than summer (Fig. 2), has important
consequences on the oceanographic conditions. Winds over the

9

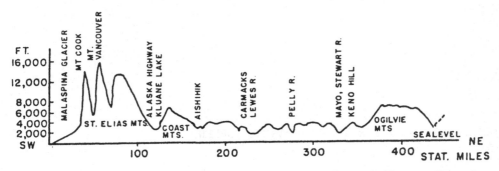

Fig. 1. Topography from ocean near Yakutat toward the northeast
(from Kendrew and Kerr, 1955).

northern Gulf of Alaska are from east to west, with the possible
exception in summer. These winds create downwelling conditions
(Fig. 3). Because of the paucity of wind measurements over the
Gulf of Alaska the winds are determined from the geostrophic winds
as calculated on a 3° surface atmospheric pressure grid (Bakun,1973).
From these calculated winds, upwelling indices are determined.
These indices are the onshore-offshore component of Ekman transport,
with downwelling being opposite to upwelling. The downwelling reaches
a maximum in January and a minimum in summer, with the annual signals
being similar for 60°N, 149°W and 60°N, 146°W. Because of the grid
scale and topographic effects from the mountainous terrain, these
indices are in slight error. They overestimate the amplitude of
the annual signal and more importantly, indicate a summer upwelling
condition while actual wind measurements do not give evidence of
eastward summer winds (Livingstone and Royer, 1980). Summer upwelling or a wind-driven current reversal has not been observed
here. This is in contrast to the annual cycle for a similar
current, the Norwegian Coastal Current, which has flow reversals
(Aure and Sætre, 1981).

The low pressure system over the Gulf of Alaska also forces
the moist, marine air to interact with the topography to cause
adiabatic cooling. This, in addition to contact with the cold
continental air masses, causes very high precipitation rates to
occur in the coastal drainage region (Fig. 4). Some areas in the
higher elevations, generally over glacial fields, receive more than
800 cm per year. Approximately 20% of the coastal drainage are
glacial fields. The precipitation rates decrease both to the west
of 150°W and in the offshore direction.

A simple hydrology model has been constructed for this area
to estimate the influx of freshwater into the ocean (Royer, 1982).
The inputs for this model are the monthly means of the divisional
precipitation rates and air temperatures. (The U.S. Weather

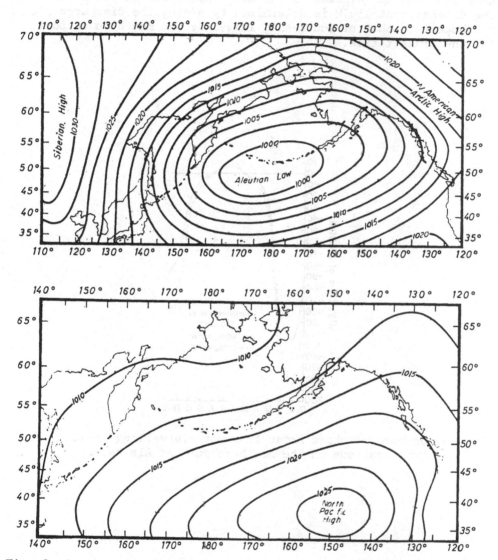

Fig. 2 Average sea level atmospheric pressure for winter (upper
        panel and summer (lower panel). After Dodimead et al., 1963.

Service divides this coast from Canada to 155°W into two divisions
for their averaging. These are Southeast (SE) and Southcoast (SC).)
The runoff is obtained from the precipitation rate and drainage
area with storage allowed if the air temperature is below freezing.
The discharge at 150°W is determined by adding the discharge from
Southcoast to that of Southeast, lagged by one month to account
for the transit time.

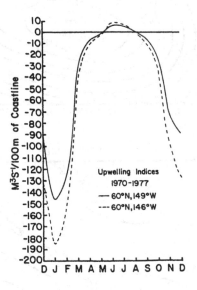

Fig. 3 Onshore-offshore Ekman transport (upwelling indices) for
       two locations in the northern gulf of Alaska.

        The maximum annual discharge occurs in October (Fig. 5), though
the precipitation rate has a maximum slightly later in the fall.
An early cutoff of discharge is caused by freezing air temperatures
and the storage of precipitation as snow. There is also a small

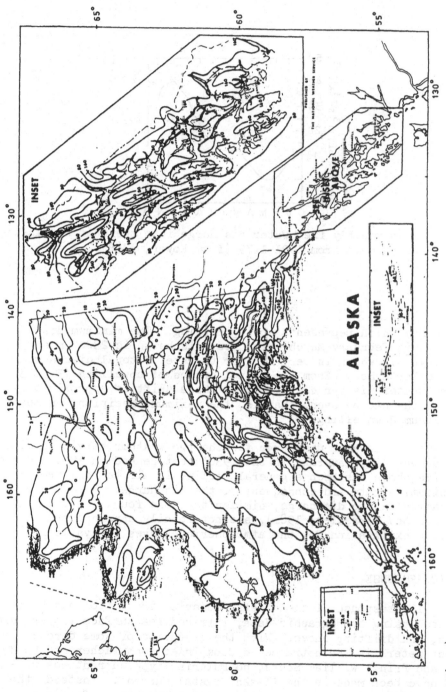

Fig. 4. Average distribution of precipitation for Alaska in inches per year (1 inch = 2.54 cm) (Courtesy of Larry Mayo).

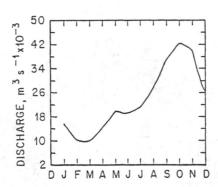

Fig. 5. Mean monthly freshwater discharge into the Gulf of Alaska
using data from 1931-1979 (from Royer, 1982).

peak in May which represents spring melting. The minimum discharge is in February-March. The mean discharge is 23000 m s
(730 km^{3}yr^{-1}), which is very similar to rate of total discharge
into the North Sea. From the annual cycles of upwelling (Fig. 3)
and freshwater discharge (Fig. 5), a phase shift of approximately
three months can be seen, that is, the maximum precipitation leads
the maximum downwelling.

The freshwater discharge for the two weather service divisions
(Fig. 6) was below average in the 1930's but peaked in 1940. Since
that time there has been a general decline in air temperature,
precipitation rate and consequently, the discharge. A periodic
fluctuation in the discharge, similar to those found for the
Norwegian Sea (Skreslet, 1981), are not readily apparent and no
immediate explanation for the atmospheric changes is available.

OCEANIC RESPONSE

The response of the coastal oceanic circulation has been
described using hydrographic data, infrared sea surface temperature
images, and drifting buoys. With the exception of some direct
current meters in the entrance to Cook Inlet (Schumacher and Reed,
1980) and Prince William Sound, no Eulerian-type current measurements have been made in the Alaska Coastal Current. Instead, the
response of the coastal circulation to the freshwater forcing

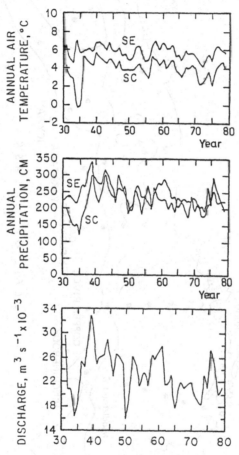

Fig. 6 Annual mean air temperature (top panel) for
        southeast Alaska (SE) and southcoast Alaska
        (SC), precipitation (middle panel) and mean
        annual freshwater discharge for both regions
        (from Royer, 1982).

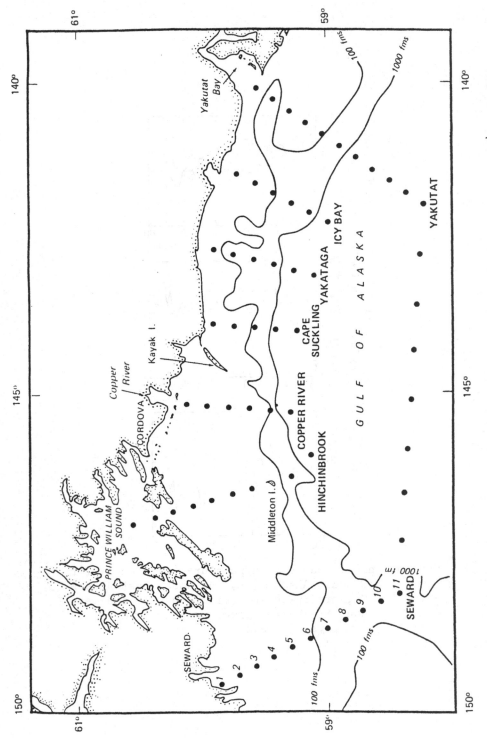

Fig. 7 Station positions for hydrographic surveys conducted between 1974 and 1979.

has been estimated indirectly through baroclinic, geostrophic
current computations. A hydrographic section on the western
boundary of the grid (Fig. 7) has been occupied at irregular intervals since 1974. The comparison is quite good between the 0/100 db
transport between stations 1 and 7 and the estimate of the coastal
freshwater discharge from Southwest and Southcoast(Fig. 8). (In
Fig. 8 the range of the baroclinic transport is 0 to $1.5 \times 10^6 m^3 s^{-1}$.)
The correlation coefficient, r, for transport discharge is 0.763,
which is in the confidence interval of 99.9%. However, both the
transport and discharge have large amplitude annual cycles which
influence this correlation. A better test of their interdependence
is one using their anomalies, that is, the signals after the
removal of the annual cycles. Under these conditions, the correlation between transport and discharge is reduced slightly to 0.603,
which is in the 99.5% confidence interval.

Multiple linear regression analysis on the discharge, winds
and transport allows an estimation of the transport to these forcing
functions with possible phase shifts. For this analysis, the
transport was considered for both a nearshore band (Stations 1-2)
and entire inner half of the shelf (Station 1-7). Phase shifts
up to 12 months were considered. The best fits were for; 1) the
discharge, wind and transport in phase for the nearshore flow and
2) the wind lagging the discharge and transport by one month for the

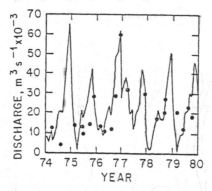

Fig. 8. Monthly freshwater discharge for 1974-1979 (line) with baro-
    clinic transport (data) for stations 1-7 (see Fig. 7)
    0/100 db superimposed. The range of the baroclinic trans-
    port is $1 - 1.5 \times 10^6 m^3 s^{-1}$ (from Royer, 1982).

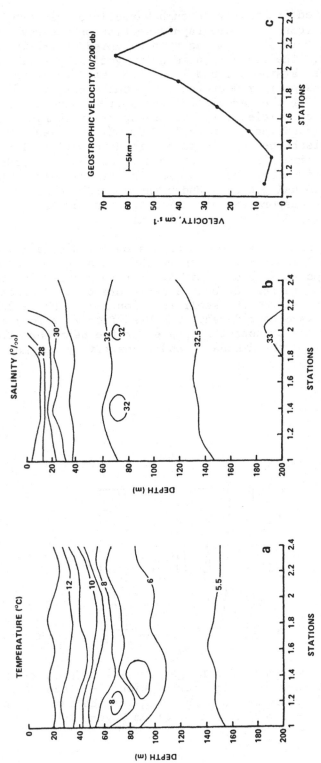

Fig. 9. Detailed cross-section of a) temperature and b) salinity with a c) geostrophic speed profile for September 1979, for stations between Stations 1 and 2 (see Fig. 7) (from Royer,
1981).

inner shelf. The F-statistic for the fitting of discharge to
transport is 18.13 for nearshore and 7.17 for inner shelf, while
the wind to transport fits are 0.85 and 2.15, respectively. For
this analysis, it must be assumed that wind and discharge are
independent. However, this is not necessarily the case, since
their correlation coefficient r, is 0.140 when in phase and 0.542
when the winds lead the discharge by one month. It is conceivable
that high winds accompany a subsequent increased discharge.

The Alaska Coastal Current, as observed near Seward, in
September 1979 is evident in both the temperature and salinity
structure. There is a seasonal thermocline in the upper eighty
meters where temperatures decrease from 12° to less than 7°C, while
the salinity increases from 26 ‰ to 32 ‰ (Fig. 9). The width of
the current here is 15-20 km. The width of the current in this
section is greater than elsewhere along the coast because the coast
turns southward immediately downstream, causing an offshore (southward) deflection of the current. The low water temperatures and
high salinity gradients cause the density structure to follow the
salinity structure quite closely. While the maximum baroclinic
current for this section is about 70 cm s^{-1}, baroclinic speeds in
excess of 100 cm s^{-1} relative to 100 db have been calculated for
other sections in this current.

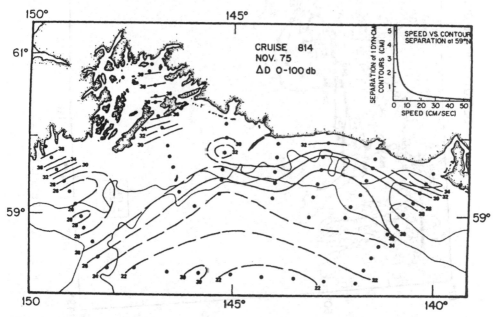

Fig. 10. Dynamic topography for the northern Gulf of Alaska, 0/100 db,
September 1975.

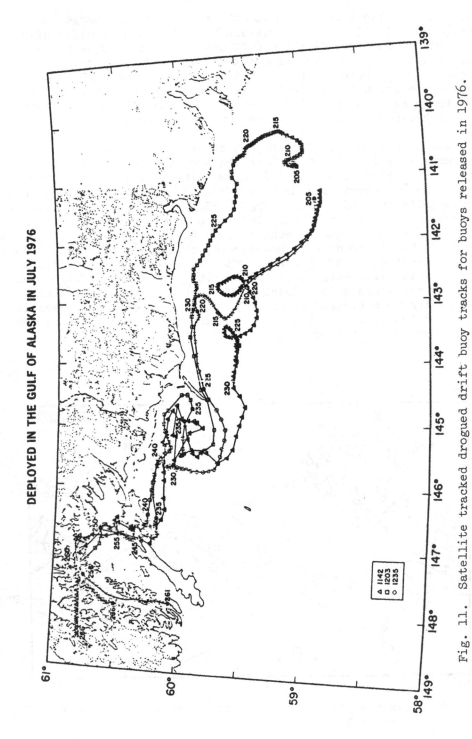

Fig. 11. Satellite tracked drogued drift buoy tracks for buoys released in 1976.

Fig. 12 Satellite IR image near Prince William Sound, 17 July 1974 (dark tones indicate warm water).

T. C. ROYER

Fig. 13. Satellite IR-image near Kodiak Island, 28 January 1980.

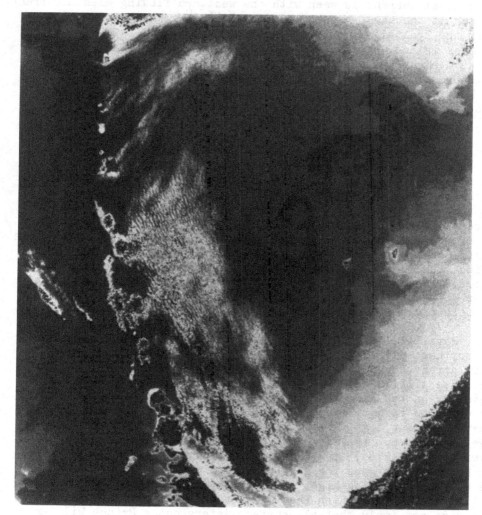

Fig. 14 Satellite IR image near Aleutian Islands, 11 January 1980.

The alongshore distribution of the Alaska Coastal Current is not well known. While hydrographic sections have been taken frequently since 1974, the coastal current was not recognized until 1979, with a result being that often sections were terminated outside of the coastal flow. However, as can be seen in dynamic topography from September 1975 (Fig. 10), there are traces of the Alaska Coastal Current. Progressing just offshore at about 143°W the coastal current is seen with the westward flowing Alaska Current located at the shelf break. At about 144.5°W, Kayak Island interrupts the coastal current, deflecting it southward. To the west of Kayak Island, a permanent eddy is created. The Alaska Current (located along the shelf break) west of Kayak Island along the shelf break carries low salinity, coastal current water with it. For hundreds of kilometers westward, the Alaska Current can be identified by this fresher surface water.

The coastal current can be observed by the offshore, horizontal gradients at the mouth of Prince William Sound (located at 147°W) and near Seward (149°W) (Fig. 7). The separation of the Alaska Coastal Current and the Alaska Current can be seen best on the western limit of this dynamic topography, because the stations are located in both currents. The separation of the two currents is distinct here because of the discharge of freshwater out of the western side of Prince William Sound. While most of the influx of freshwater into the Gulf of Alaska should be treated as a line source, this outflow from Prince William Sound can be regarded as a point source. Mork (1981) suggests that a freshwater driven coastal current will be narrow and well-defined near to their source, as in this case.

Satellite-tracked drifting buoys (Fig. 11) were released in the northern Gulf of Alaska in July 1976 to describe the shelf circulation (Royer et al., 1979). These buoys followed paths that are quite similar to those given by the dynamic topography. Beginning on the eastern part of the shelf, the buoys drifted to the west and inshore (cross-shelf) until they encountered the coastal current. At that point they were held on an interface and move alongshore. (For a full explanation of the process, see Royer et al., 1979.) The buoys moved around Kayak Island, entered the Kayak Island gyre for a brief period and eventually moved into the Copper River discharge. From there, most entered Prince William Sound, from which they did not exit while transmitting. This does not imply that all surface waters enter Prince William Sound since the buoys were drogued at 30 m. However, because these drogues move along a front (see Royer et al., 1979), it implies that the coastal current, as indicated by the front, does not maintain itself within Prince William Sound. However, as can be seen in Fig. 10, the current is again evident to the west of Prince William Sound.

Satellite IR images are presently the only technique to obtain
a synoptic view of the Alaska Coastal Current. Many of the features
discussed up to this point, such as the cold (light tone) water in
the coastal band, the Kayak Island gyre, the Alaska Current and flow
out of Prince William Sound can be seen in the IR image for July
1974 (Fig. 12). To the west, along the shelf beyond Kodiak Island,
the separation of the Alaska Current and Alaska Coastal Current can
be seen clearly (Fig. 13). Here, the Alaska Current is identified
by a warm surface layer (dark tone) along the shelf break. Progressing further westward through the Aleutian Island chain and into the
Bering Sea, the influence of the cold coastal current into the
Bering Sea can be seen. The passage of the Alaska Coastal Current
into the Bering Sea has been verified by current meter measurements
(Schumacher et al., 1982). Therefore, this coastal current acts as
a source of relatively low salinity water for the Bering Sea. The
discharge entering the Alaska Coastal Current is somewhat greater
than the contribution of rainfall over the Northwest Pacific Ocean
east of 180° and north of 50°N (Royer, 1982).

The theoretical treatment of freshwater driven coastal flow
indicates that a cross-shelf component of the velocity should exist
(Heaps, 1980; Kao, 1981; and Mork, 1981). This cross-shelf circulation is depicted in Fig. 15, as an offshore surface flow with
approcimately 10% of the magnitude of the alongshore velocity.
Beneath the upper offshore flow is an onshore lower layer, which
is consistent with the drifter observations if the 30 m drogue was
within the lower layer when deployed. Moving offshore (the middle

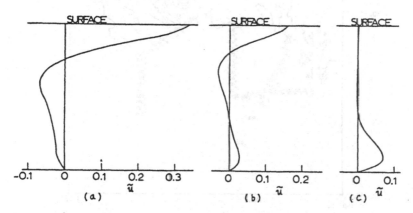

Fig. 15. Vertical profiles of the nondimensional cross-shelf
         velocity at the a) coast b) mid-shelf and c) shelf break;
         offshore velocities are to the right at the vertical axis
         (from Kao, 1981).

and right panels in Fig. 15), a third layer, at the bottom, forms
in the flow which is directed offshore over the middle and outer
shelf. This bottom offshore current remains at the shelf break.
Given this cross-shelf flow, the maintenance of the Alaska Coastal
Current along several hundred kilometers of coast is not straightforward since a 50 cm s^{-1} alongshore current might have a cross-
shelf component of 5 cm s^{-1} (4.3 km day^{-1}). The current will spread

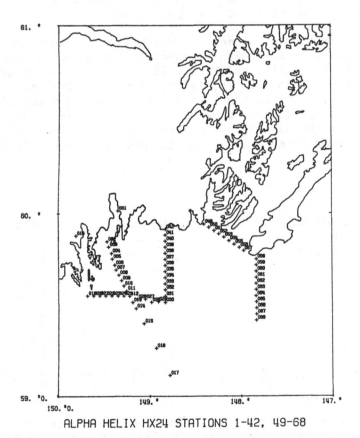

ALPHA HELIX HX24 STATIONS 1-42, 49-68

Fig. 16. Station positions for November 1981; Cape Fairfield Line,
          Stations 30-42.

horizontally with an angle of about $6°$. Such a spreading is not
observed in this current. Bathymetric control of the alongshore
flow could be the cause of this lack of alongshore cross-shelf
spreading, but control of these surface layers in depth of several
hundred meters is questioned. A more likely possibility is that
the onshore component of the Ekman transport from the alongshore
(easterly) winds balances this offshore spreading.

Evidence for this balance between cross-shelf flow and wind
stress transport was observed in November 1981 in the hydrographic
section data at Cape Fairfield (Fig. 16). This section was taken
on three occasions with the times between transects being 8 and
29 hours respectively. Prior to and during the initial transect
the alongshore wind component was 5.7 m s^{-1} westward. This was
greater than the average November component of 3.7 m s^{-1} during
the last transect. Precise location of the front associated with
the coastal current is difficult, but if $\sigma_t = 24.5$ is used for
transects I and II and 24 is used for transects II and III, there
is an offshore displacement of about 10 km in each case. These
yield offshore frontal speeds of 35 cm s^{-1} and 10 cm s^{-1}. The
first speed is somewhat high, possibly indicating a transient
relaxation. The second one agrees with the theory of an alongshore current of 100 cm s^{-1} without winds. Therefore, easterly
winds (downwelling) could be important to the configuration of
this current as a narrow flow. In the western Gulf of Alaska,
storms have a greater tendency to pass over the shelf and cause
an increased variability in the wind direction. Hence, downwelling is not steady in the western gulf and there is a greater
spreading of the coastal current there, as observed from satellite
IR images.

The distribution of wind stress over the Gulf of Alaska is
the major way in which the conditions in the North Pacific differ
from the North Atlantic. In the northern Gulf of Alaska high winds
are in the direction of the ocean currents, while for the Norwegian
coast they can oppose the flow. Thus, wind stress over the Norwegian
shelf can reverse the flow (Mork, 1982), whereas this has never
been observed for the coastal current in the Gulf of Alaska. Otherwise, the flows are similar; each with a freshwater influx of
$2-6 \times 10^{-2}$ m^{3}s^{-1} per length of coastline and a high sensitivity of
water density to salinity because of the low temperature of water.

CONCLUSIONS

The Alaska Coastal Current is the major avenue for the influx
of freshwater into the North Pacific. Its transport can exceed
10^6 m^{3}s^{-1} and can have speeds greater than 100 cm s^{-1}. It borders
the Gulf of Alaska from Southeast Alaska to beyond Kodiak Island

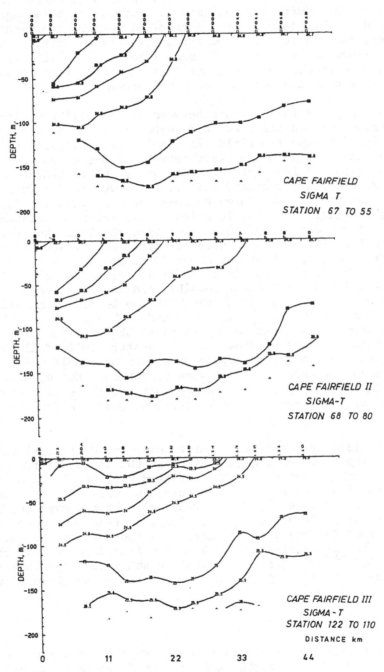

Fig. 17   Density cross-section for three occupations of Cape
          Fairfield line (see Fig. 16).   Coast is on the left.

and into the Bering Sea, serving as an important source of low
salinity water for the Bering Sea. While quite similar to another
subarctic coastal current, the Norwegian Coastal Current, it differs
in that the winds have never been observed to reverse the flow.

The implications of the Alaska Coastal Current on the regional
biological, chemistry and geology have not yet been explored. However, the response of fish stocks to this current should be similar
to those found for the Norwegian Coastal Current as described by
Skreslet (1981). In that paper, it was speculated that the coastal
marine food webs were connected to freshwater outflow. A study of
the fish stocks in the Gulf of Alaska as related to the historical
freshwater discharges (Fig. 6) appears to be appropriate.

ACKNOWLEDGEMENTS

I appreciate the reviewing of this manuscript by D.L. Nebert
and W.R. Johnson and the discussions with H.G. Gade concerning this
work. The field work was supported by the U.S. Department of
Commerce and Bureau of Land Management under Contract 03-5-022-56.
The later analysis and writing of this manuscript was financed by
the State of Alaska, Institute of Marine Science Contribution
No. XXX.

REFERENCES

Aure, J. and R. Sætre, 1981, Wind effects on the Skagerak out-
    flow. In The Norwegian Coastal Current, Vol. I, Sætre
    and Mork (eds.), Univ. of Bergen, Norway, pp. 263-293.
Bakun, A., 1973, Coastal upwelling indices, west coast of North
    America, 1946-71. U.S. Dept. of Commerce, NOAA Tech.
    Rep. NMFS SSRF-671, 103 pp.
Dodimead, A.J., Favorite F. and Hirano, T., 1963, Review of
    the oceanography of the subarctic Pacific Ocean. In
    Salmon of the North Pacific Ocean. Intern. N. Pac. Fish.
    Comm. Bull. No. 13, 195 pp.
Heaps, N.S., 1980, Density currents in a two-layered coastal
    system, with application to the Norwegian Coastal
    Current. Geophys. J.R. Astron. Soc. 63:289-310.
Kao, T., 1981, The dynamics of ocean fronts. Part II: Shelf
    water structure due to freshwater discharge. J. Phys.
    Oceanogr., 11:1215-1223.
Kendrew, W.G. and Kerr, D., 1955, The Climate of British
    Columbia and Yukon Territory, Edmond Cloutier, Ottawa.
Livingstone, D. and Royer, T.C., 1980, Observed surface winds
    at Middleton Island, Gulf of Alaska and their influence
    on the ocean circulation. J. Phys. Oceanogr. 10:753-764.

Mork, M., 1981, Experiments with theoretical models of the
    Norwegian Coastal Current. In the Norwegian Coastal
    Current, Vol. II, Sætre and Mork (eds.), Univ. of Bergen,
    Norway, pp. 518-530.

Mork, M., 1982, On the dynamics of coastal currents. This
    volume.

Overland, J.E. and Hiester, T.R., 1978, A synoptic climatology
    of surface winds along the southern coast of Alaska.
    NOAA/OCSEAP, RU 140, Final Rep.

Royer, T.C., 1981, Baroclinic transport in the Gulf of Alaska,
    Part II, Fresh water driven coastal current. J. Mar. Res.,
    39:251-766.

Royer, T.C., 1982, Coastal fresh water discharge in the North-
    east Pacific. J. Geoph. Res., 87:2017-2021.

Royer, T.C., Hansen, D.V. and Pashinski, D.J., 1979, Coastal
    flow in the northern Gulf of Alaska as observed by
    dynamic topography and satellite-tracked drogued drift
    buoys. J. Phys. Oceanogr., 9:785-801.

Schumacher, J.D., Pearson, C.A. and Overland, J.E., 1982, An
    exchange of water between the Gulf of Alaska and the
    Bering Sea through Unimak Pass. J. Geoph. Res., 87:5785-
    5795.

Schumacher, J.D. and Reed, R.K., 1980, Coastal flow in the
    Northwest Gulf of Alaska: The Kenai Current. J.Geoph.
    Res., 85:6680-6688.

Skreslet, S., 1981, Information and opinions on how freshwater
    outflow to the Norwegian Coastal Current influences
    biological production and recruitment of fish stocks
    in adjacent seas. In The Norwegian Coastal Current, Vol.
    II, Sætre and Mork (eds.), Univ. of Bergen, Norway,
    pp. 712-748.

ASPECTS OF THE DYNAMICS OF THE RESIDUAL
CIRCULATION OF THE ARABIAN GULF

J. R. Hunter

Unit for Coastal and Estuarine Studies
Marine Science Laboratories
Menai Bridge
Anglesey, United Kingdom

INTRODUCTION

Until recently, estimates of the residual circulation of the
Arabian Gulf have been based mainly on observations of the drifts
of ships underway and on arguments based on the distribution of
salinity. Both of these methods are prone to errors, and in an
effort to improve our knowledge in this area, two analyses have
been undertaken:

(1) All the data on ships drifts collected by the British
Meteorological Office have been subjected to a simple statistical
analysis in order to reject observations that may be deemed to be
in error, or to be not contributing to a "long term" velocity
structure. Rather surprisingly, the results show a systematic
circulation pattern that is not obviously related to the local
wind velocity (we would expect a simple relationship between wind
and a ship's drift if the ship's windage had not been adequately
corrected for). It hence appears that the observations of the
drift of ships at sea are not simply a measure of the local wind,
but rather that they tell us something about the underlying water
circulation.

(2) An attempt was made to describe this "observed" residual
flow in terms of the dynamical forcing involved (primarily the wind
and horizontal density gradients). It appears that density forcing
is the most probable generator of the major characteristics of
the flow, involving a geostrophic balance across the axis of the
Arabian Gulf, and a frictional balance along the axis. A three-
dimensional sigma-coordinate model has been used to predict the
Gulf circulation from a prescribed density field (derived from
observations) and an applied wind stress.

31

The physical oceanography of the Arabian Gulf has been reviewed
by Grasshoff (1976), Hughes and Hunter (1979) and Hunter (1982).
The present paper summarises the major aspects of these reviews and
investigates the important mechanisms driving the long-term circulation pattern.

The main topographic features of the Gulf are shown in Fig. 1,
indicating a channel near the Iranian coast, and shallow areas at
the north-west end and off the coasts of Saudi Arabia, Qatar and
the United Arab Emirates.

The oceanographic conditions in the Gulf differ markedly from
winter to summer, and unfortunately few data exist for the summer
regime, when the temperature structure is dominated by existence
of a seasonal thermocline (Emery, 1956). The Gulf is a region of
strong evaporation, leading to salinity values frequently exceeding
40 ppt, under both summer and winter conditions. At the north-west
end of the Gulf, the temperature and salinity distribution is
generally dominated by the river inflows of the Tigris, the
Euphrates and the Karun, all of which enter the Gulf in the vicinity
of the Shatt al Arab.

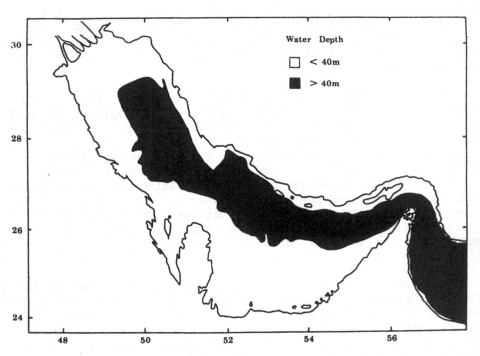

Fig. 1. Topography of the Arabian Gulf. Source: Purser and
        Seibold (1973).

The most recent set of observational data for winter conditions was given by Brewer et al. (1978). Observations of temperature, salinity and chemical properties were made over the whole Gulf in February and March of 1977. Points of note were:

(a) The influence of river inflow at the north-west end of the Gulf was evident.
(b) Surface temperatures in the Gulf were lower than those outside the Strait of Hormuz in the East.
(c) High surface salinity was evident along the south-west and south coasts of the Gulf.
(d) In the portion of the Gulf east of Qatar, isopycnals sloped downwards towards the north.
(e) In the portion of the Gulf east of Qatar, the vertical density stratification was relatively low (the difference in sigma-t from surface to bottom in the deeper areas was roughly 1.5 units).
(f) In the portion of the Gulf west of Qatar, the vertical density stratification was very low, except in a region in the north-west corner that was influenced by the river inflow.

These observations are indicative of high evaporation rates with consequent high salinities and densities in the shallower waters of the south west and south coasts of the Gulf. The conventional description of conditions has been that this water sinks, is deflected to the right by Coriolis force and passes out of the Gulf through the Strait of Hormuz. This flow must be balanced by a surface inflow through the Strait, also deflected to the right, hence passing up the Iranian coast of the Gulf and giving rise to an anti-clockwise circulation (Grasshoff, 1976).

The only published account of a comprehensive survey of the physical oceanography during the summer has been due to Emery (1956), based on observations made in August, 1948. Most of the data presented relate to temperature, but some salinity and density data were given. Points of note were:

(a) The influence of river inflow at the north-west end of the Gulf was evident.
(b) Surface temperatures in the Gulf were generally higher than those outside the Strait of Hormuz in the East. Coastal surface temperatures were also generally higher than temperatures away from the coasts.
(c) The salinity distribution was very similar to that of winter conditions, as reported by Brewer et al. (1978).
(d) Over most of a section taken from north to south in the portion of the Gulf east of Quatar, isotherms sloped downwards towards the North.

(e)   A seasonal thermocline was evident over most of the Gulf at a
       depth of about 100 ft. (30 metres). The surface to bottom
       temperature difference was around 20 degrees F. (11 degrees
       C.).
(f)   The difference in sigma-t from surface to bottom was around
       5 units in the deeper areas (about three times as large as
       the difference for the winter data of Brewer et al. (1978)).

       Unfortunately, Emery gave little density information and
hence it is not easy to see whether the "winter" mechanism, of
high evaporation in the south west and south coastal areas, consequent sinking of dense water and an anticlockwise water circulation,
holds for the summer as well. However, from inspection of the data
available, it appears that the horizontal density variations are
roughly the same in both winter and summer off the Saudi coast,
but rather smaller in summer in the region east of Qatar. Hence
the density forcing needed to drive an anti-clockwise circulation
is possibly reduced in summer. This effect is however probably
counteracted by the presence of strong vertical density stratification in summer, which would reduce vertical "friction".

SHIP DRIFT OBSERVATIONS

       The simple density driven anti-clockwise circulation described
above would undoubtedly be modified by the effects of meteorological
forcing. However, observations of ship drifts published by the
Hydrographic Department of the British Admiralty (chart C6120),
(based on data supplied by the British Meteorological Office)
generally indicate this type of circulation. As the data used in
chart C6120 was based on information collected earlier than 1941,
contained only about 600 observed ship drifts and apparently
consisted of all current estimates (no matter how low their
statistical significance), it was decided to analyse all the ship
drifts collected by the British Meteorological Office up to the
present time.

       These data consisted of 1806 observations distributed among
28 "one degree" (latitude and longitude) squares. The data were
presented as vector averages for the whole year, for each of four
seasons, and for each month. For each vector average, the number
of observations contributing to the average, and the standard
deviations of the velocity components in the east-west and north-
south directions were given. For the present work, a current
vector was accepted or rejected using the following criteria:

(a)   If the number of observations was less than 5, the vector
       was rejected.
(b)   Of the remainder, a vector was accepted if:
       (i) Using a "Student's t" test, either the east-west or

the north-south velocity was significant at the 5% confident
level or
(ii) The magnitude of the standard vector error (the estimated
error in the vector average) was less than 0.05 m/s.

    It was hoped that the resultant selected data would be repre-
sentative of a "long-term" circulation pattern. Fig. 2 shows the
data for the whole year, after selection using the above criteria;
 vector "tails" are centred on each "one degree" square, and "•"
indicates the centre of a "one degree" square for which there is
no vector. The vectors show considerably less variability than
the raw data and generally indicate a surface flow westwards into
the Gulf along the Iranian coast, of magnitude about 1.1 m/s.
There is also the suggestion of an anti-clockwise circulation.

    The analysis was also carried out for the four seasons. The
results did not differ markedly from those for the whole year,
except that less points were accepted because there were on
average less observations per point, the westward flow into the
Gulf appeared strongest in summer (about 0.2 m/s) and weakest in
spring and autumn (about 0.1 m/s), and there was little evidence
of an anti-clockwise circulation (no eastward return flow present).

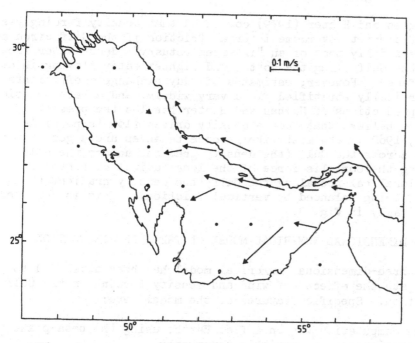

Fig. 2.   Selected ship drift data for whole year.

Current estimates based on ship drift observations may be
subject to a number of errors (HMSO, 1977), one of the main ones
being due to the ship's windage, tending to bias estimates in the
direction of the prevailing wind if sufficient allowance is not
made for this factor. However, the prevailing wind in the Gulf is
from the north west and west (IMCOS, 1974), in opposition to the
dominant inferred flow. It would hence appear that the ship drift
observations are an indicator of the "long term" circulation
pattern, namely a surface inflow towards the west along the Iranian
coast of strength between 0.1 and 0.2 m/s (strongest in summer).
There are, unfortunately, little data on the currents in the south
and south-west regions of the Gulf, due to the absence of ships'
observations in these areas.

MODELS OF THE RESIDUAL CIRCULATION

The observed residual flow could be explained by wind or
density forcing. Vertically averaged models of wind-driven circu-
ation (eg. Csanady, 1973) do not appear to be appropriate in this
instance, as estimates of the Ekman depth are of the same order
as the water depth. Models of circulation forced by surface
Ekman flows also appear not to fit the observed velocity field.
It therefore seems that density forcing is the most likely candi-
date for the generation of the "long-term" circulation.

Hughes and Hunter (1979) concluded that density forcing was
not sufficient to overcome internal friction if the Gulf circulation
were essentially that of an "inverted estuary" (dense water flowing
out of the Gulf along the bottom and lighter water flowing in at
the surface). However, estimates of eddy exchange coefficients
in a vertically stratified fluid vary widely, and a recomputation
of the predictions of Hughes and Hunter has led the present
author to believe that such a density driven flow is possible
(Hunter, 1982). The most likely force balances are a geostrophic
balance across the Gulf (the density gradient across the Gulf
balancing the Coriolis force of the longitudinal currents) and
a frictional balance along the Gulf (the density gradient along
the Gulf being balanced by vertical friction). This is depicted
schematically in Fig. 3.

A THREE-DIMENSIONAL NUMERICAL MODEL OF THE GULF CIRCULATION

A three-dimensional numerical model has been developed to
investigate the effects of wind and density forcing on the Gulf
circulation. Specific features of the model were:

(a) The computation was on a flat Earth, using the beta-plane
     approximation and variable topography.
(b) The equations were time-dependent and linearised, with

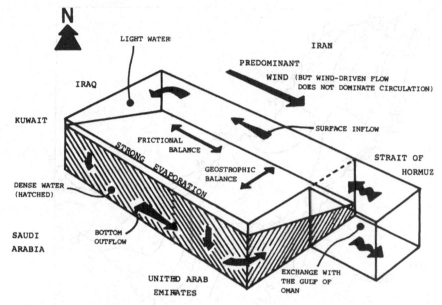

Fig. 3. The probable circulation pattern in the Arabian Gulf.

the damping effect of tides included through appropriate
bottom friction.

(c) "Sigma coordinates" were used in the vertical (the vertical
axis was scaled by the local depth).

(d) A quadratic law was used to relate wind stress to wind
velocity.

(e) A fixed sea level was prescribed at the open boundary.

(f) The density field was prescribed from observational data
and was not subject to modification by the model.

(g) The vertical eddy viscosity and linear bottom friction
coefficient were chosen to be .005 m^2/sec and .0005 m^2/sec
respectively. These were best guesses from available data –
they were not "tuned" to obtain optimum model results.

(h) The model mesh consisted of 331 cells in the horizontal,
each divided into 5 levels.

(i) The model was run to "steady state" (for the internal
velocity field), which took about 23 days.

The prescribed three-dimensional density field was obtained
by objective analysis and a little smoothing of the observations
of Brewer et al. (1978) (see Fig. 4). The model predictions are
therefore relevant to winter conditions. Two model runs are
presented here:

(a) With no wind forcing (Fig. 5).

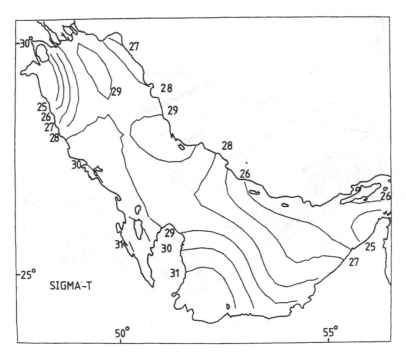

Fig. 4. The density data used in the numerical model – surface values.

(b) With an imposed wind stress equivalent to a 5 m/sec wind
     from the north-west, a typical wind for the region (the
     eastward and southward components of the kinematic stress
     were .00003 SI units) (Fig. 6).

     Both sets of predictions show some "noise" which is inevitable
with the use of observational density data collected over a period
of a month. The density field also is subject to error in coastal
regions especially in the southern parts of the Gulf where the
data have been extrapolated from the station positions. The
predictions under zero wind conditions indicate a surface inflow
of strength around 0.1 m/sec along the Iranian coast, some evidence
of river inflow at the north-western end of the Gulf, and outflow
of water along the bottom from the coastal areas off Saudi Arabia,
Qatar and the United Arab Emirates. The predictions are somewhat
modified by the imposed wind stress, which generates clear Ekman
rotation and a surface inflow into the region north of the
United Arab Emirates. The predictions are generally in accord
with the schematic representation of Fig. 3.

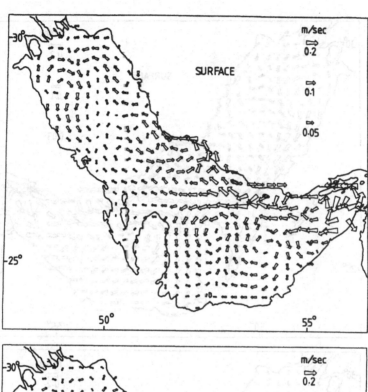

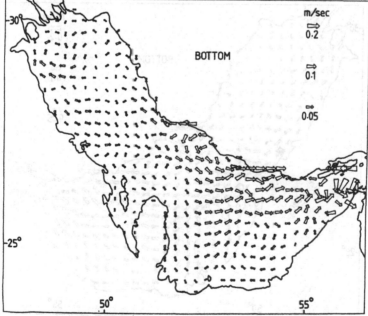

Fig. 5. Predicted model velocities for surface and bottom cells
         with no wind stress (vector lengths proportional to cube
         root of velocity).

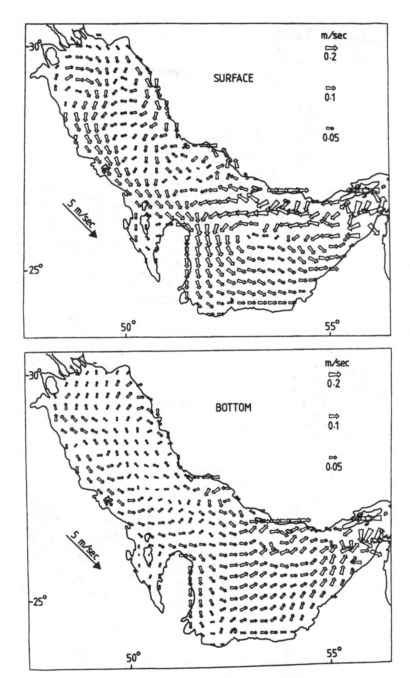

Fig. 6.   Predicted model velocities for surface and bottom cells
          with prescribed wind stress (vector lengths proportional
          to cube root velocity).

CONCLUSIONS

The major features of the observed "long-term" circulation of
the Arabian Gulf may be accounted for by a geostrophic balance
across the channel axis and a frictional balance along it. A
three-dimensional numerical model with appropriate density and wind
forcing can also reproduce these features. The next stage is to
introduce transport of the density field into the model, such that
it no longer depends on observed density data but rather on buoyancy input such as evaporation and surface runoff. Unfortunately
such models are far less efficient to run than the present one,
since the flushing time of the Gulf is from 2 to 5 years (Hunter,
1982). Any satisfactory model would have to cover this period of
time at least, compared with the 23 days used for the present
model.

ACKNOWLEDGEMENTS

The author wishes to thank James J. O'Brien and Alan Davies for
helpful comments on the numerical model during the workshop and for
indicating an omission in the density forcing terms in a preliminary
version of that model.

REFERENCES

Brewer, P.G., Fleer, A.P., Kadar, S., Shafer, D.K., and Smith,
        C.L., 1978, Report A, chemical oceanographic data from the
        Persian Gulf and Gulf of Oman, Woods Hole Oceanographic
        Institution, Report WHO-78-37.
Csanady, G.T., 1973, Wind-induced barotropic motions in long
        lakes, J. Phys. Oceanogr., 3:429-438.
Emery, K.O., 1956, Sediments and water of Persian Gulf,
        Bull.Amer.Ass.Petrol. Geol., 40:10:2354-2383.
Grasshoff, K., 1976, Review of hydrographical and productivity
        conditions in the Gulf region, in: "Marine Sciences in
        the Gulf Area", UNESCO Technical Papers in Marine Sciences,
        No. 26.
HMSO, 1977, "The Marine Observer's Handbook", Meteorological
        Office Report MO522.
Hughes, P., and Hunter, J.R., 1979, Physical oceanography and
        numerical modelling of the Kuwait Action Plan region,
        UNESCO, Division of Marine Sciences, Report MARINEF/27.
Hunter, J.R., 1982, The physical oceanography of the Arabian
        Gulf: a review and theoretical interpretation of pre-
        vious observations, presented at the First Gulf Conference
        on Environment and Pollution, Kuwait, February 1982.
IMCOS, 1974, "Handbook of the Weather in the Gulf", IMCOS
        Marine Ltd., London, IM 102.

Purser, B.H., and Seibold, E., 1973, The principal environmen-
    tal factors influencing Holocene sedimentation and diage-
    nesis in the Persian Gulf, in: "The Persian Gulf", Purser,
    B.H., ed., Springer-Verlag.

HYDRODYNAMIC MODEL OF A STRATIFIED SEA

N.S. Heaps

Institute of Oceanographic Sciences
Bidston Observatory, Birkenhead, U.K.

INTRODUCTION AND BASIC EQUATIONS

Figure 1 delineates the areas of summer stratification in the seas around the British Isles (Robinson 1973). Surprisingly perhaps, little has been done to model the dynamics of the region shown, taking stratification into account. Notable progress has been made with a three-dimensional shelf model for tides and surges, assuming homogeneous water (Davies 1981). However, to predict the motion of the water during the summer season, a three-dimensional shelf model is needed allowing for the presence of homogeneous and stratified areas and the fronts between them (James 1981).

Bearing in mind the latter requirement, a start has been made with a three-dimensional model for the motion of a sea stratified in three layers. The notation is illustrated in Figure 2. The depths of the surface, intermediate and bottom layers are h_1, h_2, h_3 respectively : h_1 and h_2 are constants while h_3 varies with horizontal position (x, y). During the motion the sea surface is elevated by ζ_1, the upper interface by ζ_2 and the lower interface by ζ_3. The x- and y-directed components of current are u_1, v_1 (surface layer), u_2, v_2 (intermediate layer) and u_3, v_3 (bottom layer). Density and vertical eddy viscosity in the respective layers are P_1, N_1; P_2, N_2; P_3, N_3. These are prescribed constants.

Considering a rectangular area ABCD of the Celtic Sea shown in Figure 3, the following values were taken (Heaps and Jones 1983a):

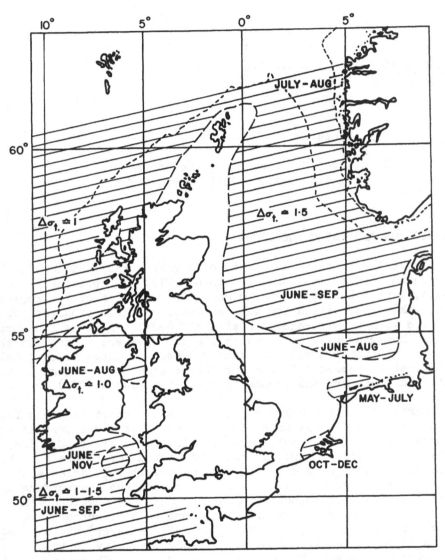

Figure 1. Areas of summer stratification in the sea areas surrounding the British Isles.

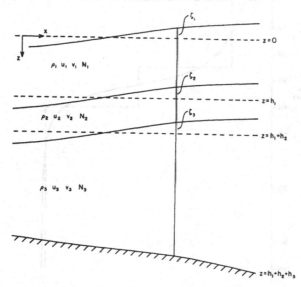

Fig. 2. Three-layered model: a vertical section showing the notation.

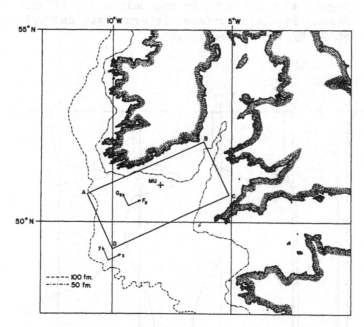

Fig. 3. Celtic Sea showing the rectangular area ABCD modelled.

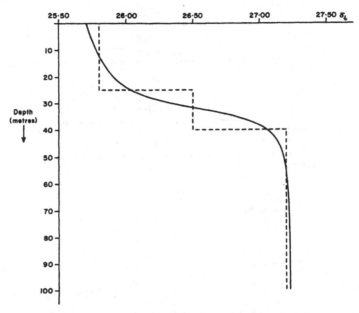

Figure 4. Mean vertical distribution of density for the month of
August at station MU in the Celtic Sea (figure 3).
Resolution into surface, intermediate and bottom layers
shown by the dashed lines.

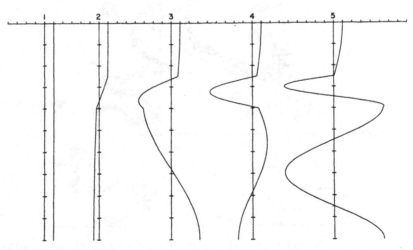

Figure 5. Current modes 1-5 through the vertical, plotted against
fractional depth.

$$h_1 = 25 \quad , \qquad h_2 = 15 \quad , \qquad h_3 = 60 \qquad m$$

$$\rho_1 = 1.0258 \, , \qquad \bar{\rho}_2 = 1.0265 \, , \qquad \rho_3 = 1.0272 \; gcm^{-3} \qquad (1)$$

$$N_1 = 300 \quad , \qquad N_2 = 10 \quad , \qquad N_3 = 100 \qquad cm^2 s^{-1}$$

The above depths and densities come from fitting a step-like variation to an average vertical density profile (Figure 4). For simplicity, the total depth is assumed to be constant at 100 m: a reasonably good first approximation to the actual depth topography within the rectangular area. The values of eddy viscosity are tentative, reflecting wind-driven turbulence in the surface layer, high vertical stability in the intermediate thermocline layer, and tidal mixing in the bottom layer.

The basic equations of the model (Heaps 1983) are, for motion in the x-direction :

$$\frac{\partial u_1}{\partial t} - \gamma v_1 = -g \frac{\partial \zeta_1}{\partial x} + g \frac{\partial \bar{\zeta}}{\partial x} - \frac{1}{\rho_1} \frac{\partial p_a}{\partial x} + \frac{\partial}{\partial z} \left(N_1 \frac{\partial u_1}{\partial z} \right) \qquad (2)$$

$$\frac{\partial u_2}{\partial t} - \gamma v_2 = -g \frac{\rho_1}{\rho_2} \frac{\partial \zeta_1}{\partial x} - g \left(\frac{\bar{\rho}_2 - \rho_1}{\rho_2} \right) \frac{\partial \zeta_2}{\partial x} + g \frac{\partial \bar{\zeta}}{\partial x}$$

$$- \frac{1}{\rho_2} \frac{\partial p_a}{\partial x} + \frac{\partial}{\partial z} \left(N_2 \frac{\partial u_2}{\partial z} \right) \qquad (3)$$

$$\frac{\partial u_3}{\partial t} - \gamma v_3 = -g \frac{\rho_1}{\rho_3} \frac{\partial \zeta_1}{\partial x} - g \left(\frac{\rho_2 - \rho_1}{\rho_3} \right) \frac{\partial \zeta_2}{\partial x} - g \left(\frac{\rho_3 - \rho_2}{\rho_3} \right) \frac{\partial \zeta_3}{\partial x}$$

$$+ g \frac{\partial \bar{\zeta}}{\partial x} - \frac{1}{\rho_3} \frac{\partial p_a}{\partial x} + \frac{\partial}{\partial z} \left(N_3 \frac{\partial u_3}{\partial z} \right) \qquad (4)$$

for motion in the y-direction :

$$\frac{\partial v_1}{\partial t} + \gamma u_1 = -g \frac{\partial \zeta_1}{\partial y} + g \frac{\partial \bar{\zeta}}{\partial y} - \frac{1}{\rho_1} \frac{\partial p_a}{\partial y} + \frac{\partial}{\partial z} \left(N_1 \frac{\partial v_1}{\partial z} \right) \qquad (5)$$

$$\frac{\partial v_2}{\partial t} + \gamma u_2 = -g \, \frac{\rho_1}{\rho_2} \, \frac{\partial \zeta_1}{\partial y} - g \left(\frac{\rho_2 - \rho_1}{\rho_2} \right) \frac{\partial \zeta_2}{\partial y} + g \, \frac{\partial \bar{\zeta}}{\partial y}$$

$$- \frac{1}{\rho_2} \frac{\partial p_a}{\partial y} + \frac{\partial}{\partial z} \left(N_2 \frac{\partial v_2}{\partial z} \right) \tag{6}$$

$$\frac{\partial v_3}{\partial t} + \gamma u_3 = -g \, \frac{\rho_1}{\rho_3} \, \frac{\partial \zeta_1}{\partial y} - g \left(\frac{\rho_2 - \rho_1}{\rho_3} \right) \frac{\partial \zeta_2}{\partial y} - g \left(\frac{\rho_3 - \rho_2}{\rho_3} \right) \frac{\partial \zeta_3}{\partial y}$$

$$+ g \, \frac{\partial \bar{\zeta}}{\partial y} - \frac{1}{\rho_3} \frac{\partial p_a}{\partial y} + \frac{\partial}{\partial z} \left(N_3 \frac{\partial v_3}{\partial z} \right) \tag{7}$$

and for continuity

$$\frac{\partial \zeta_1}{\partial t} - \frac{\partial \zeta_2}{\partial t} + \frac{\partial}{\partial x} \int_0^{h_1} u_1 \, dz + \frac{\partial}{\partial y} \int_0^{h_1} v_1 \, dz = 0 \tag{8}$$

$$\frac{\partial \zeta_2}{\partial t} - \frac{\partial \zeta_3}{\partial t} + \frac{\partial}{\partial x} \int_{h_1}^{h_1 + h_2} u_2 \, dz + \frac{\partial}{\partial y} \int_{h_1}^{h_1 + h_2} v_2 \, dz = 0 \tag{9}$$

$$\frac{\partial \zeta_3}{\partial t} + \frac{\partial}{\partial x} \int_{h_1 + h_2}^{h} u_3 \, dz + \frac{\partial}{\partial y} \int_{h_1 + h_2}^{h} v_3 \, dz = 0 \tag{10}$$

These constitute nine equations in the nine variables

$$\zeta_s = \zeta_s(x, y, t) \; ; \quad u_s, v_s = u_s, v_s \, (x, y, z, t) \; ; \quad s = 1, 2, 3$$

and are solved in three-dimensional space x,y,z through time t subject to surface conditions specifying wind stress:

$$- \rho_1 N_1 \frac{\partial u_1}{\partial z} = F_s \, , \qquad - \rho_1 N_1 \frac{\partial v_1}{\partial z} = G_s \quad \text{at } z = 0 \tag{11}$$

interfacial conditions satisfying continuity of current and stress:

$$u_1 = u_2 \quad , \quad v_1 = v_2 \quad \text{at} \quad z = h_1 \tag{12}$$

$$\rho_1 N_1 \frac{\partial u_1}{\partial z} = \rho_2 N_2 \frac{\partial u_2}{\partial z} \quad , \quad \rho_1 N_1 \frac{\partial v_1}{\partial z} = \rho_2 N_2 \frac{\partial v_2}{\partial z}$$

$$\text{at} \quad z = h_1 \tag{13}$$

$$u_2 = u_3 \quad , \quad v_2 = v_3 \quad \text{at} \quad z = h_1 + h_2 \tag{14}$$

$$\rho_2 N_2 \frac{\partial u_2}{\partial z} = \rho_3 N_3 \frac{\partial u_3}{\partial z} \quad , \quad \rho_2 N_2 \frac{\partial v_2}{\partial z} = \rho_3 N_3 \frac{\partial v_3}{\partial z}$$

$$\text{at} \quad z = h_1 + h_2 \tag{15}$$

and sea-bed conditions specifying bottom stress:

$$- \rho_3 N_3 \frac{\partial u_3}{\partial z} = F_B \quad , \quad - \rho_3 N_3 \frac{\partial v_3}{\partial z} = G_B \text{ at } z = h \tag{16}$$

The coordinates x, y, z form a left-handed set in which x, y
are measured in the undisturbed sea surface and z is depth below
that surface. The x and y axes are directed parallel to the
sides of the Celtic Sea rectangle, as shown in Figure 3. Notation
introduced above, so far undefined, is:

F_S , G_S components of wind stress in the x , y directions

F_B , G_B components of bottom stress

$\bar{\zeta}$ the equilibrium tide

p_a atmospheric pressure on the sea surface

γ geostrophic coefficient .

With F_S, G_S, $\bar{\zeta}$, p_a as prescribed forcing functions, solutions of (2) - (10) are developed from an initial condition of no
motion:

$$\zeta_1 = \zeta_2 = \zeta_3 = u_s = v_s = 0 \quad \text{everywhere at} \quad t = 0$$

$$(s = 1,2,3). \tag{17}$$

Radiation boundary conditions are applied along the edges of the
model rectangle. Thus along the y-directed side BC :

$$u_j = \sum_{s=1}^{3} u_{j,s} \zeta_s \quad (j = 1.2.3) \tag{18}$$

where the coefficients $u_{j,s}$ depend, in turn, on certain radiation
coefficients ε_1, ε_2, ε_3. Condition (18) is based on one-
dimensional gravity-wave propagation in an inviscid three-layered
system corresponding to that shown in Figure 2 (Heaps and Jones
1983a). That propagation consists of a surface and two internal
modes (modes 1,2,3 say) and the degree to which those modes are
radiated across the boundary is determined by ε_1, ε_2, ε_3
respectively. Specific cases governing the transmission of modes
are

$\varepsilon_s = 0$: total reflection, zero flow

$\varepsilon_s = 1$: no reflection

$\varepsilon_s \rightarrow \infty$: total reflection, zero elevation.

Conditions of the type (18) are applied to all four sides of
the rectangular basin. Thus, for outward radiation across all four
sides, with no reflection,

$\varepsilon_1, \varepsilon_2, \varepsilon_3 = 1, 1, 1$ on side BC

$= -1,-1,-1$ on side AD

$= 1, 1, 1$ on side AB

$= -1,-1,-1$ on side DC

giving a radiation matrix

$$R_1 = \begin{pmatrix} 1 & 1 & 1 \\ -1 & -1 & -1 \\ 1 & 1 & 1 \\ -1 & -1 & -1 \end{pmatrix} \qquad (19)$$

For the bottom stress, use is made of the quadratic law

$$F_B = k\rho_3 u_3(h) \left[u_3^2(h) + v_3^2(h) \right]^{\frac{1}{2}}$$

$$G_B = k\rho_3 v_3(h) \left[u_3^2(h) + v_3^2(h) \right]^{\frac{1}{2}} \qquad (20)$$

and also the linear law

$$F_B = k\rho_3 u_3(h) \quad , \quad G_B = k\rho_3 v_3(h) \qquad (21)$$

where $u_3(h)$, $v_3(h)$ denote u_3, v_3 evaluated at the sea bed $z = h$.

TRANSFORMATION OF THE EQUATIONS

The hydrodynamic equations formulated in the preceding section are transformed, by vertical integration, to a two-dimensional set in the horizontal. These are then solved numerically using well-known finite-difference techniques. Subsequently, an inverse transformation recovers the vertical structure of current. The procedure has been described in detail by Heaps (1983) and only the main idea can be discussed here.

Basically, equations (2) – (4) are combined, involving integrations through the depth in each layer, to give

$$\frac{\partial \hat{u}}{\partial t} - \gamma \hat{v} = - g\, a_1 \frac{\partial \zeta_1}{\partial x} - g\, a_2 \left(\frac{\rho_2 - \rho_1}{\rho_1} \right) \frac{\partial \zeta_2}{\partial x} - g\, a_3 \left(\frac{\rho_3 - \rho_2}{\rho_1} \right) \frac{\partial \zeta_3}{\partial x}$$

$$+ g\, a_4 \frac{\partial \bar{\zeta}}{\partial x} - \frac{a_1}{\rho_1} \frac{\partial p_a}{\partial x} + I_x \qquad (22)$$

where

$$\hat{u} = \frac{1}{h} \left(\int_0^{h_1} f_1 u_1 \, dz + \frac{\rho_2}{\rho_1} \int_{h_1}^{h_1+h_2} f_2 u_2 \, dz + \frac{\rho_3}{\rho_1} \int_{h_1+h_2}^{h} f_3 u_3 \, dz \right) \tag{23}$$

$$\hat{v} = \frac{1}{h} \left(\int_0^{h_1} f_1 v_1 \, dz + \frac{\rho_2}{\rho_1} \int_{h_1}^{h_1+h_2} f_2 v_2 \, dz + \frac{\rho_3}{\rho_1} \int_{h_1+h_2}^{h} f_3 v_3 \, dz \right) \tag{24}$$

$$a_1 = \frac{1}{h} \left(\int_0^{h_1} f_1 \, dz + \int_{h_1}^{h_1+h_2} f_2 \, dz + \int_{h_1+h_2}^{h} f_3 \, dz \right) \tag{25}$$

$$a_2 = \frac{1}{h} \left(\int_{h_1}^{h_1+h_2} f_2 \, dz + \int_{h_1+h_2}^{h} f_3 \, dz \right) \tag{26}$$

$$a_3 = \frac{1}{h} \left(\int_{h_1+h_2}^{h} f_3 \, dz \right) \tag{27}$$

$$a_4 = \frac{1}{h} \left(\int_0^{h_1} f_1 \, dz + \frac{\rho_2}{\rho_1} \int_{h_1}^{h_1+h_2} f \, dz + \frac{\rho_3}{\rho_1} \int_{h_1+h_2}^{h} f_3 \, dz \right) \tag{28}$$

$$I_x = \frac{1}{\rho_1 h} \left(\int_0^{h_1} f_1 \frac{\partial}{\partial z} \left(\rho_1 N_1 \frac{\partial u_1}{\partial z} \right) \, dz + \int_{h_1}^{h_1+h_2} f_2 \frac{\partial}{\partial z} \left(\rho_2 N_2 \frac{\partial u_2}{\partial z} \right) \, dz \right.$$

$$\left. + \int_{h_1+h_2}^{h} f_3 \frac{\partial}{\partial z} \left(\rho_3 N_3 \frac{\partial u_3}{\partial z} \right) \, dz \right) \tag{29}$$

For each x, y and t, the f_1, f_2, f_3 are functions of z defined in the surface, intermediate and bottom layers respectively. Their choice is important in the reduction of I_x. Thus, taking

$$\frac{\partial}{\partial z}\left(N_i \frac{\partial f_1}{\partial z}\right) = -\lambda f_1 \quad , \quad 0 \leqslant z \leqslant h_1$$

$$\frac{\partial}{\partial z}\left(N_2 \frac{\partial f_2}{\partial z}\right) = -\lambda f_2 \quad , \quad h_1 \leqslant z \leqslant h_1 + h_2 \qquad (30)$$

$$\frac{\partial}{\partial z}\left(N_3 \frac{\partial f_3}{\partial z}\right) = -\lambda f_3 \quad , \quad h_1 + h_2 \leqslant z \leqslant h$$

where λ is independent of z, and also satisfying the conditions

$$f_1 = 1 \quad , \quad \frac{\partial f_1}{\partial z} = 0 \text{ at } z = 0 \qquad (31)$$

$$f_1 = f_2 \quad , \quad \rho_1 N_1 \frac{\partial f_1}{\partial z} = \rho_2 N_2 \frac{\partial f_2}{\partial z} \quad \text{at } z = h_1 \qquad (32)$$

$$f_2 = f_3 \quad , \quad \rho_2 N_2 \frac{\partial f_2}{\partial z} = \rho_3 N_3 \frac{\partial f_3}{\partial z} \quad \text{at } z = h_1 + h_2 \qquad (33)$$

$$\frac{\partial f_3}{\partial z} = 0 \text{ at } z = h \qquad (34)$$

it turns out that

$$I_x = \left(\frac{F_S - f_3(h) F_B}{\rho_1 h}\right) - \lambda \hat{u}$$

where $f_3(h)$ denotes f_3 evaluated at $z = h$. Therefore (22) reduces to

$$\frac{\partial \hat{u}}{\partial t} + \lambda \hat{u} - \gamma \hat{v} = -g\, a_1 \frac{\partial \zeta_1}{\partial x} - g\, a_2 \left(\frac{\rho_2 - \rho_1}{\rho_1}\right)\frac{\partial \zeta_2}{\partial x}$$

$$- g\, a_3 \left(\frac{\rho_3 - \rho_2}{\rho_1}\right)\frac{\partial \zeta_3}{\partial x} + g\, a_4 \frac{\partial \zeta}{\partial x} - \frac{a_1}{\rho_1}\frac{\partial p_a}{\partial x} - \frac{f_3(h)}{\rho_1 h} F_B + \frac{F_S}{\rho_1 h} \qquad (35)$$

Similarly, working from (5) - (7), it may be shown that

$$\frac{\partial \hat{v}}{\partial t} + \lambda \hat{v} + \gamma \hat{u} = - g\, a_1 \frac{\partial \zeta_1}{\partial y} - g\, a_2 \left(\frac{\rho_2 - \rho_1}{\rho_1} \right) \frac{\partial \zeta_2}{\partial y}$$

$$- g\, a_3 \left(\frac{\rho_3 - \rho_2}{\rho_1} \right) \frac{\partial \zeta_3}{\partial y} + g\, a_4 \frac{\partial \bar{\zeta}}{\partial y} - \frac{a_1}{\rho_1} \frac{\partial p_a}{\partial y} - \frac{f_3(h)}{\rho_1 h} G_B$$

$$+ \frac{G_S}{\rho_1 h} \qquad (36)$$

EIGENVALUES AND EIGENFUNCTIONS

Solution of the eigenvalue problem (30) - (34) yields a set of ascending eigenvalues and corresponding eigenfunctions given formally by

$$\lambda = \lambda_r \quad , \quad f_1 = f_{1,r} \quad , \quad f_2 = f_{2,r} \quad , \quad f_3 = f_{3,r}$$

$$(r = 1,2 \ldots, \infty). \qquad (37)$$

With constant values of N_1, N_2 and N_3 it transpires that

$$\lambda_r = N_1 \alpha_r^2 / h^2 \qquad (38)$$

where α_r ($r = 1,2,\ldots, \infty$) denote the non-negative roots, arranged in ascending order, of the trigonometric equation

$$(q+e+d+1) \sin \left((a+b+c)\alpha \right) + (q+e-d-1) \sin \left((a+b-c)\alpha \right)$$

$$+ (q-e-d+1) \sin \left((a-b+c)\alpha \right) + (q-e+d-1) \sin \left((a-b-c)\alpha \right)$$

$$= 0 \qquad (39)$$

where

$$a = h_1/h \qquad , \quad d = (\rho_1/\rho_2)(N_1/N_2)^{\frac{1}{2}}$$

$$b = (N_1/N_2)^{\frac{1}{2}} h_2/h \quad , \quad e = (\rho_2/\rho_3)(N_2/N_3)^{\frac{1}{2}} \qquad (40)$$

$$c = (N_1/N_3)^{\frac{1}{2}} h_3/h \quad , \quad q = (\rho_1/\rho_3)(N_1/N_3)^{\frac{1}{2}}.$$

Solutions of (39) are easily determined numerically. Corresponding to λ_r deduced from (38)-(40), the eigenfunctions $f_{1,r}$, $f_{2,r}$, $f_{3,r}$ have simple trigonometric forms. Also, from (35) and (36):

$$\frac{\partial u_r}{\partial t} = -\lambda_r \hat{u}_r + \gamma \hat{v}_r - g \, a_{1,r} \frac{\partial \zeta_1}{\partial x} - g \, a_{2,r} \left(\frac{\rho_2 - \rho_1}{\rho_1} \right) \frac{\partial \zeta_2}{\partial x}$$

$$- g \, a_{3,r} \left(\frac{\rho_3 - \rho_2}{\rho_1} \right) \frac{\partial \zeta_3}{\partial x} + g \, a_{4,r} \frac{\partial \overline{\zeta}}{\partial x} - \frac{a_{1,r}}{\rho_1} \frac{\partial p_a}{\partial x}$$

$$- \frac{f_{3,r}(h)}{\rho_1 h} F_B + \frac{F_S}{\rho_1 h} \qquad (r = 1, 2, \ldots, \infty) \qquad\qquad (41)$$

$$\frac{\partial \hat{v}_r}{\partial t} = - \lambda_r \hat{v}_r - \gamma \hat{u}_r - g \, a_{1,r} \frac{\partial \zeta_1}{\partial y} - g \, a_{2,r} \left(\frac{\rho_2 - \rho_1}{\rho_1} \right) \frac{\partial \zeta_2}{\partial y}$$

$$- g \, a_{3,r} \left(\frac{\rho_3 - \rho_2}{\rho_1} \right) \frac{\partial \zeta_3}{\partial y} + g \, a_{4,r} \frac{\partial \overline{\zeta}}{\partial y} - \frac{a_{1,r}}{\rho_1} \frac{\partial p_a}{\partial y}$$

$$- \frac{f_{3,r}(h)}{\rho_1 h} G_B + \frac{G_S}{\rho_1 h} \qquad (r = 1, 2, \ldots, \infty) \qquad\qquad (42)$$

where $\hat{u}_r$, $\hat{v}_r$, $a_{1,r}$, $a_{2,r}$, $a_{3,r}$, $a_{4,r}$ are given by (23)-(28) with f_1, f_2, f_3 replaced respectively by $f_{1,r}$, $f_{2,r}$, $f_{3,r}$. Equations (41),(42) constitute a doubly-infinite set of equations, two-dimensional in the horizontal : transforms of the three-dimensional set (2)-(7).

INVERSE TRANSFORMATION

Invoking an orthogonal property of the eigenfunctions, the current components u_s, v_s ($s = 1, 2, 3$) are expressed in terms of their transforms $\hat{u}_r$, $\hat{v}_r$. Thus,

$$u_1 = \sum_{r=1}^{\infty} \phi_r \hat{u}_r f_{1,r} \quad , \quad v_1 = \sum_{r=1}^{\infty} \phi_r \hat{v}_r f_{1,r} \quad (0 \leqslant z \leqslant h_1)$$

$$u_2 = \sum_{r=1}^{\infty} \phi_r \hat{u}_r f_{2,r} \quad , \quad v_2 = \sum_{r=1}^{\infty} \phi_r \hat{v}_r f_{2,r} \quad (h_1 \leqslant z \leqslant h_1 + h_2)$$

$$u_3 = \sum_{r=1}^{\infty} \phi_r \hat{u}_r f_{3,r} \quad , \quad v_3 = \sum_{r=1}^{\infty} \phi_r \hat{v}_r f_{3,r} \quad (h_1 + h_2 \leqslant z \leqslant h$$

$$(43)$$

where

$$\phi_r = h \left[\int_0^{h_1} f_{1,r}^2 \, dz + \frac{\rho_1}{\rho_2} \int_{h_1}^{h_1+h_2} f_{2,r}^2 \, dz + \frac{\rho_3}{\rho_1} \int_{h_1+h_2}^{h} f_{3,r}^2 \, dz \right]^{-1} .$$

$$(44)$$

The form of (43) indicates that each horizontal current component is expressed in terms of a series of current modes through the vertical corresponding to r=1,2,3,.... The shape of the r th mode is defined by $f_{1,r}(z)$, $0 \leqslant z \leqslant h_1$: $f_{2,r}(z)$, $h_1 \leqslant z \leqslant h_1 + h_2$; $f_{3,r}(z)$, $h_1 + h_2 \leqslant z \leqslant h$. The first five modes, calculated on the basis of the values of depth, density and eddy viscosity in equation (1) are shown in Figure 5. The first mode (r = 1) is barotropic and the higher modes (r > 1) baroclinic.

Substituting (43) into the equations of continuity (8)-(10) yields

$$\frac{\partial \zeta_1}{\partial t} = - \sum_{r=1}^{\infty} \left[\frac{\partial}{\partial x} (h \, a_{1,r} \phi_r \hat{u}_r) + \frac{\partial}{\partial y} (h \, a_{1,r} \phi_r \hat{v}_r) \right] \qquad (45)$$

$$\frac{\partial \zeta_2}{\partial t} = - \sum_{r=1}^{\infty} \left[\frac{\partial}{\partial x} (h \, a_{2,r} \phi_r \hat{u}_r) + \frac{\partial}{\partial y} (h \, a_{2,r} \phi_r \hat{v}_r) \right] \qquad (46)$$

$$\frac{\partial \zeta_3}{\partial t} = - \sum_{r=1}^{\infty} \left[\frac{\partial}{\partial x} (h \, a_{3,r} \phi_r \hat{u}_r) + \frac{\partial}{\partial y} (h \, a_{3,r} \phi_r \hat{v}_r) \right] \qquad (47)$$

The vertically-integrated equations (45)-(47),(41),(42) are solved numerically for ζ_1, ζ_2, ζ_3, $\hat{u}_r$, $\hat{v}_r$ on a horizontal finite-difference grid. Then the three-dimensional current structure is determined from (43). In practice the series for r is taken to M terms so that there are 2M + 3 equations for ζ_1, ζ_2, ζ_3, $\hat{u}_r$, $\hat{v}_r$ $(r = 1,2,\ldots, M)$. The F_B, G_B in (41),(42) are given by either (20) or (21) with $u_3(h)$ and $\hat{v}_3(h^B)$ written in terms of $\hat{u}_r$, $\hat{v}_r$ using (43). The initial conditions (17) and the boundary conditions (18) may easily be expressed in terms of $\hat{u}_r$, $\hat{v}_r$ using (23) and (24).

COMPUTATIONAL PROBLEMS

     First experiments with the model determined the motion produced in the Celtic Sea rectangle (Figure 3) by a uniform, longitudinally-directed wind stress pulse of 1 dyncm^{-2}, acting over the water surface for 10 hours (Figure 6). Values of depth, density and eddy viscosity given by (1) were assumed along with open boundary radiation given by (19). Ten vertical modes were included, i.e. M = 10. In formulating finite-difference forms of (45)-(47), (41), (42) it was found necessary to employ two calculation grids (Figure 7) : a coarse one 20 km square (G_1) on which the fields of ζ_1, $\hat{u}_1$, $\hat{v}_1$ were generated through time and a fine one 20/3 km square (G_3) on which ζ_2, ζ_3, $\hat{u}_r$, $\hat{v}_r$ $(r = 2,3\ldots, M)$ were generated. A semi-implicit finite-difference scheme was used in which elevation is evaluated at the centre of each grid box, $\hat{u}_r$ at the mid-point of each y-directed box side and $\hat{v}_r$ at the mid-point of each x-directed box side. The fine grid was required to properly resolve the relatively small horizontal space scales of the internal waves. On the other hand, the coarse grid was sufficient to resolve the longer surface waves. In the computations on G_1 a time-step Δt = 360s achieved computational stability, this step being related to the speed of propagation of the surface gravity waves. However, on G_3 a time-step $\Delta t'$ = 5 Δt was found to be adequate, because the internal waves have lower propagation speeds.

     The combination of time and space splitting described above, based on a separation of the barotropic and baroclinic motions, only works if ζ_1, $\hat{u}_1$, $\hat{v}_1$ are generated independently of:

          ζ_2, ζ_3, $\hat{u}_r$, $\hat{v}_r$ $(r = 2,3\ldots, M)$.
This entails neglecting the influence of the latter variables on the former. However, the values of ζ_1, $\hat{u}_1$, $\hat{v}_1$ on the coarse grid are inperpolated on to the fine grid in the updating of ζ_2, ζ_3, $\hat{u}_r$, $\hat{v}_r$ $(r = 2,3,\ldots, M)$. To avoid aliasing, $\Delta t'$ must be small enough to resolve the time variations of the barotropic motion. The complete set of variables ζ_1, ζ_2, ζ_3, $\hat{u}_r$, $\hat{v}_r$ $(r = 1,2,\ldots M)$ is evaluated at intervals of $\Delta t'$ when the computations on G_1 are followed directly by those on G_3. At such times, the three-dimensional current structure may be evaluated using (43).

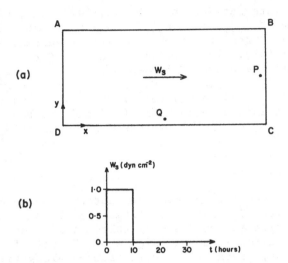

Figure 6. Wind stress pulse W_s acting longitudinally over the
rectangular basin.

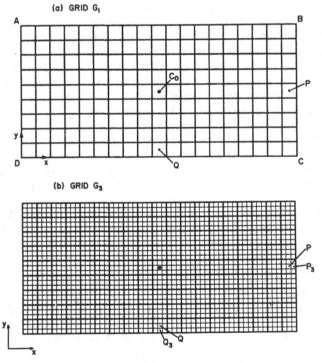

Figure 7. Finite-difference grids: G_1 (20 km square mesh) and
G_3 (20/3 km square mesh).

The computational split between the barotropic and baroclinic motions, described above, was found to be necessary in order to limit the size of the calculations and thereby make them practically possible.   A limited number of comparisons between results obtained from the model with and without the splitting showed good agreement, justifying the procedure.   However, the physical assumptions implied by the procedure need not necessarily be valid under all circumstances and therefore care needs to be taken in its general application.

FIRST RESULTS FROM THE MODEL

Figure 8 shows inertial currents set up in the rectangular basin by the wind pulse, at the central position C_0.   The horizontal current (u, v) is plotted in polar form at hourly intervals from t = 0 to t = 100 hours, at five levels through the vertical corresponding to the fractional depths : ξ = 0.0 (surface), 0.25,

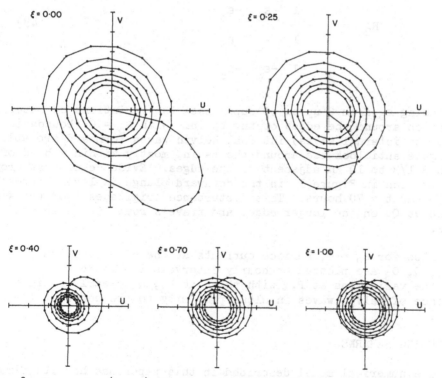

Fig. 8. Currents (u, v) at the central position C_0 of the rectangular sea area, plotted at hourly intervals from t = 0 to t = 100 hours, at five levels through the vertical water column including the sea surface ξ = 0 and the sea bed ξ = 1.

0.40, 0.70, 1.00 (bottom). In the surface layer an initial current surge is followed by damped inertial oscillations. Smaller inertial currents, set up mainly by pressure forces associated with the surface gradients produced by the wind, occur in the bottom layer (below the thermocline) and are out of phase with the inertial currents near the surface.

The inertial elevations ζ_2, ζ_3 also exhibits inertial oscillations as shown in Figure 9. These oscillations are concentrated around the edges of the basin during the early stages of the motion and spread slowly into the interior (Figure 10). They emmanate from the boundaries, as opposed to being directly generated by the wind.

Effects on ζ_2 of changing the radiation coefficients are shown in Figure 11. The Figure illustrates that, with a radiation matrix R_2 instead of R_1 (Equation 19), where

$$R_2 = \begin{matrix} 1 & \varepsilon_o & \varepsilon_o \\ -1 & -\varepsilon_o & -\varepsilon_o \\ 1 & \varepsilon_o & \varepsilon_o \\ -1 & -\varepsilon_o & -\varepsilon_o \end{matrix} \qquad (48)$$

increasing ε_o from 0 up to 50 suppresses the variations of internal elevation around the edges of the basin. With $\varepsilon_o = 0$ (boundaries closed to internal waves) internal Kelvin waves form and are able to propagate anticlockwise around the basin, mostly within a band of width 3 1/3 to 10 km adjacent to the edges. Evidence of that propagation is seen in Figure 11 in the downward plunge in level between t = 50 and t = 70 hours. This disturbance originates from the wind effect at Q_3 on the longer edge, and travels round to P_3 in about 50 hours.

Also for $\varepsilon_o = 0$, surface currents at the edge positions P, F_3, Q, Q_3 are plotted at hourly intervals in Figure 12. Comparing the variations at P,Q with those at P_3, Q_3 clearly indicates the presence of Kelvin waves in close proximity to the edges of the basin.

CONCLUDING REMARKS

The numerical model described in this paper has had its first test in an application to wind-induced motion in a flat-bottomed basin representing the northern Celtic Sea. Much research and development lies ahead and, in particular, detailed comparisons of the model's results with observational data are envisaged. The

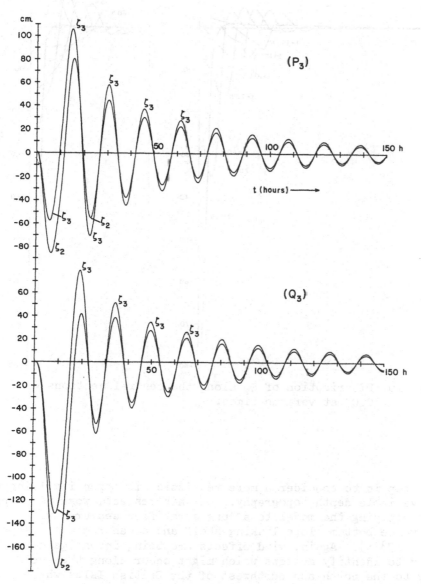

Figure 9. Variations of ζ_2 and ζ_3 at the edge positions P_3, Q_3 of the rectangular sea area (see Figure 7).

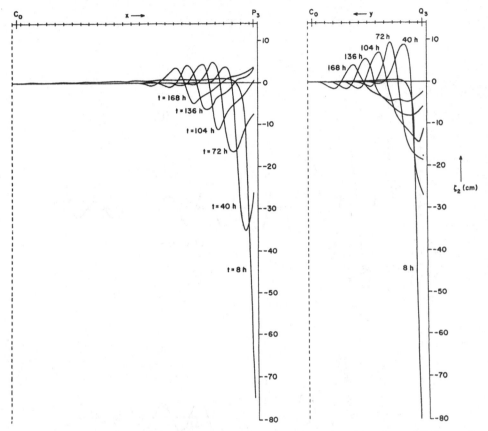

Figure 10. Distribution of ζ_2 along the central sections C_oP_3 and C_oQ_3 at various times.

next step is to consider a more realistic situation involving a sea with variable depth topography. In this respect, work is in progress applying the model to a long stratified sea area with a transverse bottom slope linking shelf and ocean regions (Heaps and Jones, 1983b). Again, wind effects are being investigated. The aim is to identify motions which might occur along the shelf edge lying to the north and northwest of the British Isles where a large observational programme is presently under way.

ACKNOWLEDGEMENT

    The work described in this paper was funded by the Department of Energy in the United Kingdom.

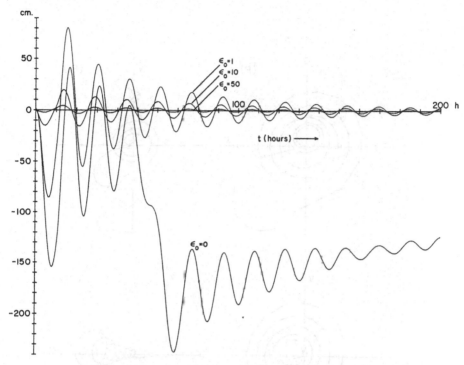

Figure 11. Variation of ζ_2 at P_3, obtained using the radiation
matrix R_2 (equation (48)) with $\varepsilon_0 = 0,1,10,50$.

REFERENCES

Davies, A.M., 1981, Three-dimensional hydrodynamic numerical
    models. Part 1: "A homogeneous ocean-shelf model".
    Part 2: "A stratified model of the northern North Sea".
    pp. 370-426 in: Vol.2, The Norwegian Coastal Current,
    (ed., R. Saetre & M. Mork), Bergen University, 795 pp.
Heaps, N.S., 1983, Development of a three-layered spectral
    model for the motion of a stratified sea. Part I: Basic
    equations. Chapter 9 in:"Coastal and Shelf Dynamical
    Oceanography", ed. B. Johns. Amsterdam: Elsevier Scien-
    tific Publishing Company. (In press).
Heaps, N.S., and Jones, J.E., 1983a, Development of a three-
    layered spectral model for the motion of a stratified
    sea.   Part II: Experiments with a rectangular basin
    representing the Celtic Sea.   Chapter 10 in:"Coastal
    and Shelf Dynamical Oceanography", ed. B. Johns.
    Amsterdam: Elsevier Scientific Publishing Company.(In
    press).

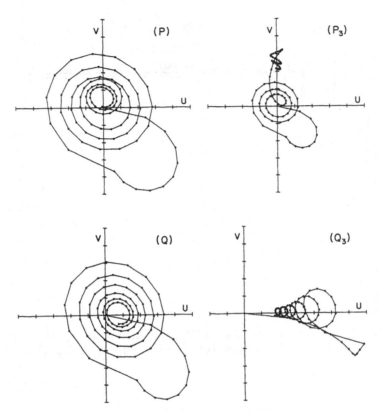

Figure 12. Surface current (u, v) at positions P, P₃, Q, Q₃
          (marked in figure 7), plotted at hourly intervals from
          t = 0 to t = 100 hours. Obtained using the radiation
          matrix R_2 with ε_o = 0.

Heaps, N.S., and Jones, J.E., 1983b, A three-layered spectral
     model with application to wind-induced motion in the
     presence of stratification and a bottom slope (In pre-
     paration).
James, I.D., 1981, Fronts and shelf-circulation models,
     Phil. Trans. R. Soc., A, 302:597-604.
Robinson, I.S., 1973, Internal tides in the British shelf
     seas, Institute of Coastal Oceanography and Tides,
     Internal Report No. ICOT/IR/28, 28 pp. (Unpublished
     manuscript).

THREE DIMENSIONAL MODELS OF NORTH SEA CIRCULATION

Alan M. Davies

Institute of Oceanographic Sciences
Bidston Observatory
Birkenhead, Merseyside, U.K.

INTRODUCTION

The JONSDAP '76 Oceanographic Exercise which took place during the Spring of 1976 in the northern North Sea, provided the stimulus for a significant program of three dimensional North Sea modelling.

The initial aims of both the experiment and the modelling were to try to isolate the mechanisms which produced significant northern North Sea currents and the factors which influenced the temporal and spatial variability of these currents.

Previous work using limited sets of observations (Dooley 1974) had hypothesised on the circulation of the northern North Sea. The JONSDAP '76 experiment however provided a large synoptic data set of measurements in the northern North Sea.

In this paper a very brief description of the numerical methods employed in the North Sea models is given. A decomposition into external and internal modes is developed which together with a time splitting algorithm, yields a computationally economic method to integrate the three dimensional equations.

A finite difference grid is used in the horizontal and the Galerkin method through the vertical. By this means a physically realistic continuous vertical variation of density and eddy viscosity can be used. The models yield continuous current profiles from sea surface to sea bed.

Current patterns from various North Sea models used in connection with JONSDAP '76 are examined in detail. Particular

attention is focused on the northern North Sea especially the FLEX
area. An area in the central northern North Sea where an intense
biological experiment took place. In this region large spatial
variations of current direction occurred.

By using a range of numerical models, the mechanisms producing
and influencing the spatial variability of northern North Sea
currents during JONSDAP '76 can be identified. The problem of the
summer circulation of the northern North Sea is also examined.

MODEL FORMULATION AND SOLUTION FOR A HOMOGENEOUS SEA REGION

In this section only the essential features of the model formulation are presented for illustrative purposes. Linear hydrodynamic
equations in Cartesian coordinates are used throughout in order to
simplify the analysis, although in practice in many of the models
described later the fully non-linear equations in polar coordinates
have been used (see Davies 1980a, b, and 1981a for details and
examples).

The linear three dimensional hydrodynamic equations describing
flow in a homogeneous sea in Cartesian coordinates are,

$$\frac{\partial \zeta}{\partial t} = -\frac{\partial}{\partial x} \int_0^h u\,dz - \frac{\partial}{\partial y} \int_0^h v\,dz \qquad (1)$$

$$\frac{\partial u}{\partial t} = \gamma v - g\frac{\partial \zeta}{\partial x} + \frac{\partial}{\partial z}\left(\mu \frac{\partial u}{\partial z}\right) \qquad (2)$$

$$\frac{\partial v}{\partial t} = -\gamma u - g\frac{\partial \zeta}{\partial y} + \frac{\partial}{\partial z}\left(\mu \frac{\partial v}{\partial z}\right) \qquad (3)$$

where: t denotes time, x, y, z a left handed set of coordinates,
with z the depth below the undisturbed surface, h the undisturbed
depth of water, ζ elevation, u, v the x and y components of current,
γ geostrophic coefficient, g acceleration due to gravity and μ
vertical eddy viscosity.

To complete the formulation of the model, sea surface and sea
bed boundary conditions have to be specified. Surface boundary conditions evaluated at z = 0 are

$$-\rho\left(\mu\frac{\partial u}{\partial z}\right)_0 = F_s \quad , \quad -\rho\left(\mu \frac{\partial v}{\partial z}\right)_0 = G_s$$

where F_s, G_s denote components of wind stress acting on the water
surface in the x and y directions; ρ density of sea water assumed
constant.

Similarly at the sea bed

$$-\rho(\mu \frac{\partial u}{\partial z})_h = F_B \quad, \quad -\rho(\mu \frac{\partial v}{\partial z})_h = G_B \tag{5}$$

where F_B, G_B denote the components of bottom friction.

In the models described later a quadratic law of bottom friction was used, with

$$F_B = k\rho u_h (u_h^2 + v_h^2)^{\frac{1}{2}} \quad, \quad G_B = k\rho v_h (u_h^2 + v_h^2)^{\frac{1}{2}} \tag{6}$$

where k is the coefficient of bottom friction.

The solution of equations (1),(2) and (3) is now developed using the Galerkin method. The two components of velocity u and v are expanded in terms of m depth-dependent functions $f_r(z)$ (the basis functions) and coefficients $A_r(z,y,t)$ and $B_r(x,y,t)$, thus,

$$u = \sum_{r=1}^{m} A_r f_r \tag{7}$$

$$v = \sum_{r=1}^{m} B_r f_r$$

Substituting (7) and (8) into (1) and transforming from the interval $0 \leqslant z \leqslant h$, which varies with horizontal position, into a depth following set of coordinates from b to a, using,

$$\sigma = z(a-b)/h + b$$

gives

$$\frac{\partial \zeta}{\partial t} = \frac{-1}{(a-b)} \sum_{r=1}^{m} \left[\left(\frac{\partial}{\partial x} (A_r h) + \frac{\partial}{\partial y} (B_r h) \right) \cdot \int_b^a f_r d\sigma \right] \tag{10}$$

Applying the Galerkin method to equation (2), it is multiplied by f_r and integrated from sea surface to sea bed. By integrating by parts the term involving μ surface and bottom boundary conditions (4) and (5) can be incorporated into the solution (see Davies 1980b, for details). Substituting (7) and (8) into the resulting equations, and transforming using (9) gives,

$$\sum_{r=1}^{m} \frac{\partial A_r}{\partial t} \int_{b}^{a} f_r f_k d\sigma = \gamma \sum_{r=1}^{m} B_r \int_{b}^{a} f_r f_k d\sigma - g \frac{\partial \zeta}{\partial x} \int_{b}^{a} f_k d\sigma$$

$$- \frac{(a-b)}{\rho h} f_k(a) F_B + \frac{(a-b)}{\rho h} f_k(b) F_s - \left(\frac{a-b}{h}\right)^2 \sum_{r=1}^{m} A_r$$

$$\int_{b}^{a} \mu \frac{df_r}{d\sigma} \frac{df_k}{d\sigma} d\sigma \qquad \text{where } k = 1,2,\ldots,m \qquad (11)$$

The Galerkin form of the v equation of motion can be obtained in a similar manner.

In theory the choice of basis function and interval a to b to use with the Galerkin method is arbitrary. However as has been shown by Davies (1982a) by choosing certain basis functions, a computationally efficient algorithm can be developed in which the solution can be split into external and internal modes. It is then possible to integrate each solution with a time step appropriate to the time evolution of that mode (see Davies 1982a for details). This method is briefly illustrated here.

Separation into External and Internal Modes

If the basis function f_r are taken as cosines, given by,

$$f_r = \cos \alpha_r \sigma \quad \text{where } \alpha_r = (r-1)\pi , \quad r = 1,2,\ldots,m \qquad (12)$$

with a = 1 and b = 0, then it is evident from (12) that

$$\int_{0}^{1} f_1 d\sigma = 1 , \quad \int_{0}^{1} f_1^2 d\sigma = 1, \quad \int_{0}^{1} f_r d\sigma = 0 \quad r = 2,3,\ldots,m$$

also $\quad f_1(0) = 1, \; f_1(1) = 1, \; df_1/d\sigma = 0$ $\qquad\qquad\qquad$ (13)

Since the cosine functions are orthogonal, then,

$$\int_{0}^{1} f_r f_k d\sigma = 0 \qquad r \neq k \qquad\qquad (14)$$

A consequence of (13) and the orthogonality condition (14) is that equation (11) can be separated into an equation describing the mean flow (the external mode) given by

$$\frac{\partial A_1}{\partial t} = \gamma B_1 - g \frac{\partial \zeta}{\partial x} - \frac{F_s}{\rho h} + \frac{F_B}{\rho h} \qquad\qquad (15)$$

and the set of equations

$$\frac{\partial A_k}{\partial t} \int_0^1 f_k f_k d\sigma = \gamma B_k \int_0^1 f_k f_k d\sigma$$

$$- \frac{F_B}{\rho h} f_k(1) + \frac{F_s}{\rho h} f_k(0) - \frac{1}{h^2} \sum_{r=1}^m A_r \int_0^1 \mu \frac{df_r}{d\sigma} \frac{df_k}{d\sigma} d\sigma \qquad (16)$$

$$k = 2.3, \ldots, m$$

The set of equations (16) describe the deviations from the mean flow (the internal modes). The external and internal modes are coupled by frictional effects.

It is apparent from the above that the free surface elevation only enters into equation (15) and not into (16), and advantage can be taken of this when an explicit time stepping algorithm is used to integrate these equations forward through time. With such an algorithm, the time step τ is restricted by the speed of propagation of the free surface waves. Since equation (16), unlike equation (15), does not involve the free surface it can be integrated with a much longer time step Δt, where for convenience $\Delta t = n\tau$, with n an integer. By this means a computationally efficient time splitting algorithm can be developed, and calculations show that Δt can be significantly larger than τ (see Davies 1982a for details).

This separation into internal and external modes is not restricted to the basis set of cosine functions described above. A set of Legendre polynomials $P_r(\sigma)$, r=1,2,..,m over the interval $-1 \leqslant \sigma \leqslant +1$ can also be used.

The Legendre polynomials $P_r(\sigma)$ are defined by the recurrence formula:

$$P_{r+1}(\sigma) = \left(\frac{2r+1}{r+1} \right) \sigma P_r(\sigma) - \left(\frac{r}{r+1} \right) P_{r-1}(\sigma) \qquad (17)$$

with

$$P_0(\sigma) = 1, \quad P_1(\sigma) = \sigma \qquad (18)$$

From (17) and (18) we have,

$$\int_{-1}^{+1} P_0(\sigma) d\sigma = 2, \quad \int_{-1}^{+1} P_0^2(\sigma) d\sigma = 2 \quad \text{and} \quad \int_{-1}^{+1} P_r(\sigma) d\sigma = 0$$

$$r = 1, 2, \ldots, m$$

also $P_o(+1) = 1$, $P_o(-1) = 1$, $\dfrac{dP_o}{d\sigma} = 0$ $\hspace{2cm}$ (19)

Since the Legendre polynomials are orthogonal,

$$\int_{-1}^{+1} P_r P_k d\sigma = 0 \qquad r \neq k \hspace{2cm} (20)$$

It is evident from (19) and (20) that when a basis set of Legendre polynomials is employed in the calculation, a separation into external and internal modes is again possible. In practice the modal separation can be further extended to any orthogonal basis set which can be generated by Gram–Schmidt orthogonalization (Davies 1980b).

It is evident from equation (16) that although a separation into external and internal modes can be readily achieved by using various orthogonal basis sets, the internal modes are still coupled through the eddy viscosity term in (16). For the general case in which μ varies with x, y, z and t, it is advantageous to expand it in terms of coefficients E_j (x, y, t) and functions $\Phi_j(\sigma)$, thus

$$\mu = \sum_{j=1}^{m'} E_j \; \Phi_j(\sigma) \hspace{2cm} (21)$$

with m' in general not equal to m.

For the restricted case, when the vertical variation of μ is fixed, then we can write

$$\mu = \psi(x, y, t) \; \Phi(\sigma) \hspace{2cm} (22)$$

where Φ is a fixed function representing the vertical variation of μ.

In this particular case the internal modes in (16) can be decoupled by using as a basis set functions which are eigenfunctions of an eigenvalue problem of the form

$$\frac{d}{d\sigma} \left[\Phi \frac{df}{d\sigma} \right] = -\varepsilon f \hspace{2cm} (23)$$

Davies (1983) solved this eigenfunction problem for an arbitrary Φ, by expanding the eigenfunctions f in (23) in terms of B-spline functions (Cox 1972). By this means eigenfunctions f and associated eigenvalues ε for arbitrary vertical variations of Φ can be computed and used as basis functions in the Galerkin method (for details see Davies 1983). These functions are such that all

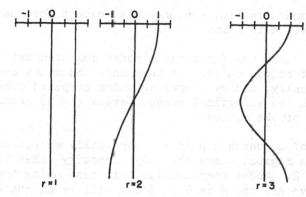

Fig. 1. First three eigenfunctions computed with $\mu = 650$ cm^2/s

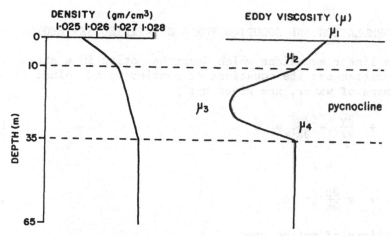

Fig. 2. Schematic variation of density and viscosity

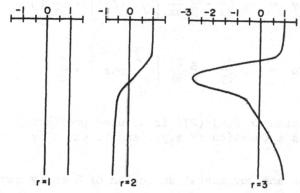

Fig. 3. First three eigenfunctions computed with $\mu_2 = 500$ cm^2/s,
$\mu_2 = 200$ cm^2/s, $\mu_3 = 10$ cm^2/s and $\mu_4 = 100$ cm^2/s

the internal modes in (16) are uncoupled except for the coupling
produced by bottom friction.

Figure 1 shows the first three modes computed using this method
for μ constant at 650 cm^2/s. In this case the modes can be determined analytically, and by comparing modes computed using spline
functions with the analytical modes, Davies (1983) demonstrated the
high accuracy of the method.

The use of a constant μ is not physically acceptable in a
stratified sea region, where the eddy viscosity takes the form
given in Fig. 2. Modes computed for this case using the method
described above are shown in Fig. 3 and will be discussed in detail
in the next section.

MODEL FORMULATION AND SOLUTION FOR A STRATIFIED SEA REGION

The linear equations which describe motion in a stratified sea
may be written as: the equations of continuity of volume of water
and of mass of water, are given by:

$$\frac{\partial u}{\partial x} + \frac{\partial v}{\partial y} + \frac{\partial w}{\partial z} = 0 \tag{24}$$

and

$$\frac{\partial \rho}{\partial t} + w \frac{\partial \rho}{\partial z} = 0 \tag{25}$$

The equations of motion are

$$\frac{\partial u}{\partial t} - \gamma v = -g \frac{\partial \zeta}{\partial x} - \frac{g}{\rho} \frac{\partial}{\partial x} \left[\int_o^z \rho dz' \right] + \frac{\partial}{\partial z} \left(\mu \frac{\partial u}{\partial z} \right) \tag{26}$$

$$\frac{\partial v}{\partial t} + \gamma u = - g \frac{\partial \zeta}{\partial y} - \frac{g}{\rho} \frac{\partial}{\partial y} \left[\int_o^z \rho dz' \right] + \frac{\partial}{\partial z} \left(\mu \frac{\partial v}{\partial z} \right) \tag{27}$$

The notation in (24)-(27) is as used previously, w is vertical
velocity, ρ is a function of x,y,z and t, and ρ̄ is the mean
density.

Diffusion and horizontal advection of density have been omitted
from the model.

Separation into Barotropic and Baroclinic Modes.

The solution of equations (24)–(27) using the Galerkin method has been described by Davies (1981b, 1982b) and details will not be given here. Rather we will briefly consider the separation of equations (24)–(27) into external and internal modes. In stratified seas, the external mode is associated with barotropic flow, and the internal modes with baroclinic flow and this terminology will be used here.

One method of separating into external and internal modes would be to expand the currents in terms of orthogonal polynomials (possibly Legendre polynomials) as described in the previous section. An alternative would be to compute the eigenfunctions of the associated depth variation of eddy viscosity. Fig. 2 shows schematically the depth variation of density and eddy viscosity used by Davies (1982a) in a stratified North Sea model. The first three modes computed with this distribution of viscosity are shown in Fig. 3. The first mode here represents the barotropic flow, the other modes are orthogonal to it, being the 1st and 2nd baroclinic modes. When the velocity is expanded in terms of these modes, the equations separate naturally into a barotropic and baroclinic set. It is interesting to compare Figs 1 and 3, for r = 2,3. The r = 2,3 modes in Fig. 3 exhibit a nearly constant vertical variation above the pycnocline, corresponding to the mixed layer, a rapid variation through the pycnocline – the region of high shear – and a smooth variation below it. However a more gradual variation throughout the whole water column is evident in Fig. 1, corresponding to the homogeneous case.

An alternative to using the orthogonal properties of the basis set to separate into internal and external modes, is to perform the separation by writing the two components of velocity as,

$$u = \bar{u} + u', \quad v = \bar{v} + v' \qquad \bar{u} = \frac{1}{h} \int_0^h u\,d\sigma, \quad \bar{v} = \frac{1}{h} \int_0^h v\,d\sigma \qquad (28)$$

where $\bar{u}$, $\bar{v}$ are the mean flows and u', v' are deviations from them.

Considering for illustrative purposes the u equation of motion, vertically integrating (26) and incorporating sea surface and sea bed boundary conditions, gives

$$\frac{\partial \bar{u}}{\partial t} - \gamma \bar{v} = -g\frac{\partial \zeta}{\partial x} - \frac{1}{h}\frac{g}{\rho} \int_0^h (\frac{\partial}{\partial x} \int_0^z \rho\,dz')dz - \frac{F_B}{\rho h} + \frac{F_S}{\rho h} \qquad (29)$$

The deviation from the mean flow μ' is computed from

$$\frac{\partial u'}{\partial t} - \gamma v' = - \frac{g}{\rho} \frac{\partial}{\partial x} \int_0^z \rho dz' + \frac{1}{h} \frac{g}{\rho} \int_0^h (\frac{\partial}{\partial x} \int_0^z \rho dz') dz$$

$$+ \frac{\partial}{\partial z} (\mu \frac{\partial u}{\partial z}) + \frac{F_B}{\rho h} - \frac{F_S}{\rho h} \qquad\qquad (30)$$

Equation (29) describes the mean flow and involves the gradient of sea surface elevation $\partial \zeta / \partial x$ and a term involving the density gradient contribution to the mean flow. The gradient of sea surface elevation is not involved in (30) which can consequently be integrated together with (25) using a much longer time step than that used in (29). By inspection, adding (29) and (30) we obtain equation (26).

Equation (30) and the corresponding v equation of motion together with (25) can be solved using the Galerkin method, with u', v', ρ expanded in terms of a set of basin functions. Vertical integration of (24) from sea surface to sea bed gives the standard form of the continuity equation (1) which can be used to advance ζ through time. Equation (24) can also be used in a diagnostic form to compute w. The application of the Galerkin method to the calculation of flow in a stratified sea is given in detail in Davies (1981b, 1982b) and will not be described here.

HOMOGENEOUS NORTH SEA MODELS

Current meter observations taken during JONSDAP '76 in the Norwegian Trench (Fig. 4) by Furnes and Saelen (1977) revealed that on average during the time of the JONSDAP experiment there was a strong northerly transport of water out of the North Sea within the Norwegian Trench. However measurements taken off the east coast of Scotland showed (Riepma, 1980) that there was on average net inflow into the North Sea during this period. Currents in the central northern North Sea in general were found to vary significantly in direction over quite small distances (of order 30 to 50 km).

To try to identify the mechanisms producing and controlling these major features of northern North Sea circulation a number of idealized models were developed (Furnes (1980), Davies and Heaps (1980)).

Idealized Models

In Furnes' two dimensional model, the North Sea was represented by a rectangular basin, open at its northern end. The depth of the basin increased linearly from south to north and from west to east, representing in an idealized manner, known changes in the bottom topography of the North Sea from the shallow German Bight to the deep Norwegian Trench (Fig.4).

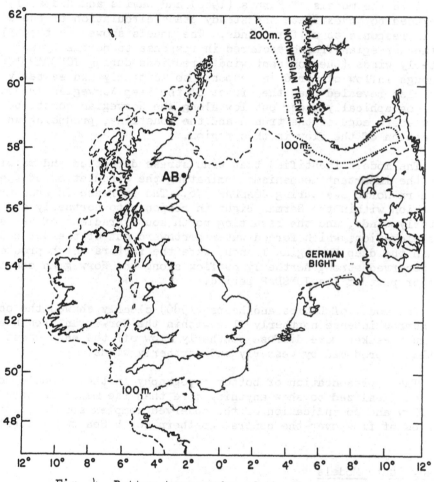

Fig. 4. Bottom topography of the North Sea

In the three dimensional model of Davies and Heaps (1980) the
North Sea was also represented by a rectangular basin, open at its
northern end. The basin had a constant depth of 65 m (a typical
average North Sea depth), except for a much deeper region (depth
260 m) extending from the open boundary to a point half way along
the eastern side of the basin. The width and length of this deeper
region being chosen to represent the Norwegian Trench.

Both the models of Furnes (1980) and Davies and Heaps (1980),
were used to investigate the steady state circulation of the North
Sea in response to uniform winds. The models showed that outflow
in the Norwegian Trench occurred in response to northerly and
westerly winds (the dominant wind directions during JONSDAP '76)
although inflow occurred in response to southerly and easterly winds.
They also revealed that the effect of the deep Norwegian Trench was
to topographically steer outflow along the Norwegian coast, between
the western edge of the trench and the coastline, producing an intensification of the flow in this region.

The models identified that wind stress direction and magnitude
were the important mechanisms dominating the circulation of the
northern North Sea during JONSDAP '76. The increase in sea surface
elevation within the German Bight in response to northerly and
westerly winds, and the resulting north south gradient of sea surface elevation, which forced water northward out of the North Sea
within the deep Norwegian Trench, were the factors which produced
the observed strong northerly outflow along the Norwegian coast for
a major part of the JONSDAP period.

The model of Davies and Heaps (1980) clearly showed the observed narrow intense northerly flow within the Norwegian Trench and
the much weaker more diffuse southerly flow off the east coast of
Scotland produced by westerly and northerly winds.

The representation of bottom topography in these models however
was too idealized to show anything more than the major features of
the flow and no indication of the observed complex spatial distributions of flow over the central northern North Sea was evident.

Steady State Models

In order to investigate in detail the observed spatial variability of current in the northern North Sea (Riepma 1980, Dooley
and Furnes 1981) further modelling work was performed using models
in which an accurate physically realistic representation of bottom
topography was included. Again, time variations of the currents
were removed by running the models to a steady state. However,

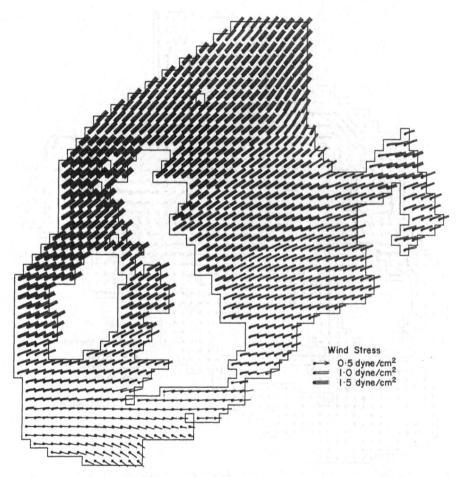

Fig. 5. Distribution of mean-wind stress over the continental
shelf, for the period 15 March to 15 April 1976.

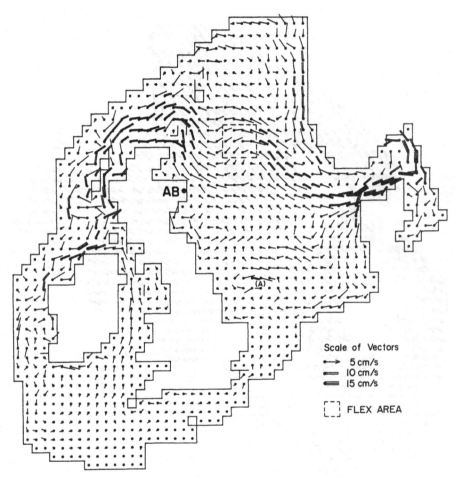

Fig. 6. Meteorologically-induced depth averaged currents, means
for the period 15 March to 15 April 1976, computed with
the three-dimensional shelf model.

currents generated using a monthly mean wind stress distribution
for the period 15 March to 15 April 1986 (the monthly period during
JONSDAP '76 when the winds were strongest) were employed rather than
the uniform winds used previously (Furnes 1980, Davies and Heaps
1980). The wind stress distribution used in the calculation is
shown in Fig. 5. (The convention used in this figure and all other
figures in this paper is that the grid point in the model is represented by a circle and the wind or current direction is away from
this point, with magnitude and direction indicated by the vector
lines).

    The three dimensional numerical model used to compute the
circulation of the North Sea in response to this imposed wind stress
was based upon the Galerkin method described previously. Spherical
polar rather than Cartesian coordinates were employed to allow for
the curvature of the earth. A radiation open boundary condition
was applied at the shelf edge. For details of the open boundary
condition and the numerical model, see Davies (1980a, 1981a).

    The steady state meteorologically induced depth-averaged cur-
rents computed with the model are shown in Fig. 6. It is evident
from this figure that although the wind stress distribution over
the North Sea is to first order uniform, corresponding to a wind
from the south-west (Fig.5), the meteorologically induced currents
show a high degree of spatial variability.

    The major features found with the steady state idealized models
(Furnes 1980, Davies and Heaps 1980), namely the northerly outflow
in the Norwegian Trench and the weaker southerly inflow down the east
coast of Scotland are clearly evident in Fig. 6. The existence of
these flows within this model confirms the findings of the earlier
idealized models, namely the importance of wind direction and bottom
topography.

    The influence of coastal and bottom topography in determining
the directions of flow can be readily appreciated from Figs.4 and 6.
It is apparent from Fig.5 that the region of maximum wind stress
during this period is situated off the west coast of Scotland.
This wind stress induces the essentially northerly flow of water
along the west coast of Scotland, which is evident in Fig.6. This
flow subsequently moves eastward along the north coast of Scotland
and then flows southward. Just to the north of Aberdeen (denoted
by AB in Fig.6) a significant proportion of the flow turns eastward and is topographically steered along the 100 m depth contour
(see Figs.4 and 6). The persistence of this easterly flow has been
reported by Dooley (1974). The reason for its persistence in winter
time is probably the dominance of northerly and westerly winds at
this time of year. They induce a southerly flow into the North
Sea, which is topographically steered off shore in this region
along the 100 m contour (Fig.4).

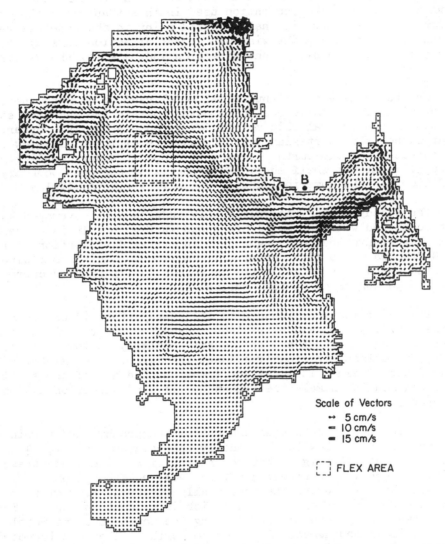

Fig. 7. Meteorologically-induced depth-averaged currents, means
for the period 15 March to 15 April 1976, computed with
the enlarged two-dimensional North Sea model.

It is evident from Figure 4, that the 100 m depth contour has
a west-east alignment off shore from Aberdeen. However in the
central northern North Sea (the region of the FLEX area, Fig.6) it
runs northward and to the east of the FLEX area it is aligned in a
south easterly direction forming the western edge of the Norwegian
Trench (Fig.4). It is apparent that the easterly flow to the north
of Aberdeen is subsequently steered northward along the 100 m depth
contour (Figs. 4 and 6), passing through the FLEX area (Fig.6) and
then flowing to the south-east along the 100 m depth contour into
the Skagerrak.

The flow of water from the central northern North Sea is
enhanced as it enters the Skagerrak by water flowing northward from
the German Bight. This enhancement in the volume of water, together
with the shallow depths along the north coast of Denmark, produces
the high currents in this area that are evident in Fig.6.

It is clear from this figure that very little of the water
flowing from the FLEX area into the Skagerrak crosses the 100 m and
200 m depth contours (Fig.4) into the Norwegian Trench, until it
reaches the eastern end of the Skagerrak. Here coastal boundaries
constrain the flow and force the water westward out of the Skagerrak
within the deep Norwegian Trench. This water subsequently flows
northward (Fig.6), channeled between the 200 m depth contour (Fig.4)
and the Norwegian coast.

It is interesting to note that at its northern end the Norwegian
Trench, defined here as the region between the 200 m depth contour
and the Norwegian coast, widens (see Fig.4). This widening is apparent in the broadening and decreasing intensity of the flow as it
moves northward along the west coast of Norway (see Fig.6).

From Fig.6 it is evident that current patterns in the central
and southern parts of the North Sea show complex spatial variations.
In particular in the vicinity of point A (Fig.6) where there is a
small grid scale gyre. To examine in detail the fine scale features
of the flow, such as this gyre, a higher resolution numerical grid
was used.

Depth averaged currents for the same period computed using a
1/3rd finer grid are shown in Fig.7. It is apparent from this
figure that there is a gyre in the region of point (A), and complex
onshore-offshore flows in this region have been observed by Lee and
Ramster (1968) (see Davies (1980a) for a full discussion of model
results).

It is interesting to note that Figure 7 shows an easterly
flow into the Skagerrak along the south coast of Norway (south of
point B in Figure 7) at a time when off shore flows were to the
west. This change in current direction near shore has been observed

a number of times (Furnes personal communication). However there
is no evidence of such a return flow in the larger model (Fig.6),
presumably due to its coarser grid.

   A close comparison of Figs. 6 and 7 does reveal some slight
differences in the shallow coastal regions, suggesting that the
grid in Fig.6 is too coarse to resolve near shore spatial variability. However,there are no distinct differences in the major features of the northern North Sea circulation computed with the two
models. Comparisons between computed and observed flows for this
period (Davies 1980a, 1981a) showed that the models could reproduce
the major features of North Sea circulation), a result which tends
to confirm that during this period meteorological effects were the
primary driving mechanism producing North Sea currents.

   These computed circulation patterns highlight and explain the
spatial variability of flow within the central northern North Sea
(the FLEX area and the region to the south of it). Within the FLEX
box currents are topographically steered along the 100 m depth contour. The alignment of this contour changes rapidly over distances
of the order of 50 km, producing the spatial variability of the
currents found in the FLEX area (Dooley and Furnes 1981). Because
the eastward flow off shore from Aberdeen is channeled northward
through the FLEX area and subsequently southward along the western
edge of the Norwegian Trench, a weak flow field, possibly influenced
by local winds,is maintained to the south east of the FLEX area
(Fig.6).

Time Dependent Models

   In these calculations the meteorologically induced steady state
circulation of the North Sea in response to a steady wind field was
investigated. However,in order to gain some insight into temporal
and vertical variations of North Sea currents, a three dimensional
numerical model was used to reproduce the period 1-10 April 1976.

   This model was based upon the numerical methods developed
earlier in this paper. Real bottom topography was incorporated in
the model and realistic hourly values of wind stress and pressure
gradients were used to drive the model. Model details and an analysis of the meteorological conditions and computed flow patterns for
the period are given in Davies (1981a). Only a brief outline of
the computed circulation patterns is presented here.

   During the 4-5 April 1976 intense north-westerly winds over
the continental shelf had developed, driving water into the North
Sea and raising water levels in the German Bight by approximately
80 cms. By 0600 GMT 7 April these intense northerly winds over
the North Sea had decreased, although light northerly winds per-
sisted over the northern North Sea. Figures 8 and 9 show surface

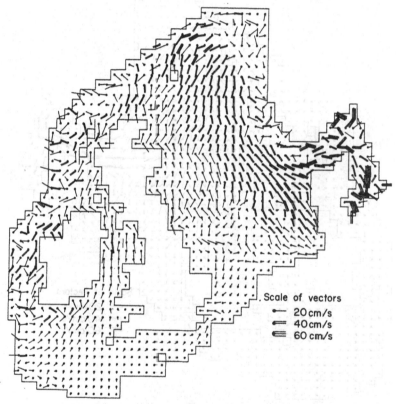

Fig. 8. Computed surface current vectors at 0600 GMT April 7, 1976.

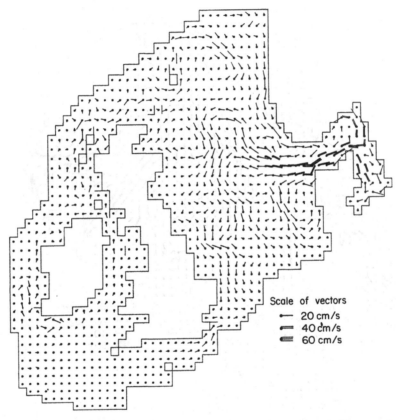

Fig. 9. Computed depth-mean current vectors at 0600 April 7,
        1976.

and depth mean currents at this time.

It is evident from Fig.8 that in the central northern North
Sea surface currents are directly wind driven, flowing in essence
in the same direction as the wind field (Davies 1981a). Away from
the influence of the coast, the surface surrent field in the central
northern North Sea is to first order uniform and not significantly
affected by bottom topography. Davies (1981a) found that this
southerly flowing wind driven surface layer was in many areas confined to the upper 5 to 10 m of the water column and that below this
layer water in some cases was flowing in the opposite direction.
An example of this can be seen in Fig.9, where off the west coast
of Denmark a northerly transport is evident, although the surface
currents are to the south-east.

It is apparent from Fig.9 that the depth mean flow is significantly influenced by bottom topography. Many of the strong spatial
features found in the steady state circulation patterns (Fig.7),
namely the easterly flow to the north east of Aberdeen, the south
easterly flow into the Skagerrak, the strong currents along the
north coast of Denmark and the northerly flow along the west coast
of Norway, are also present in the instantaneous flow field (see
Fig.9. Note difference in scales between Figs. 7 and 9). It is
evident from Fig.9, that currents in the central northern North Sea
are weaker and more variable in direction, a feature which was
also found in the steady state case.

A comparison of Figs. 9 and 10 shows that the spatial features
observed in the numerical model in the northern North Sea are confirmed by current meter observations made in the area at this time
(Riepma 1980).

A detailed analysis (Davies 1981a) of the time variations of
currents for the period 1-10 April, showed significantly hourly
variations of current magnitude in response to the rapidly changing
wind field at this time.

In general, circulation patterns computed with the numerical
models showed that wind direction and topographic steering were of
major importance in controlling the flow of the northern North Sea
during JONSDAP '76.

STRATIFIED NORTH SEA MODELS

In the previous section computed northern North Sea circulation
patterns determined using three dimensional homogeneous models were
described. These models are adequate for winter and spring conditions in the northern North Sea, although not for summer conditions
when the region is stratified.

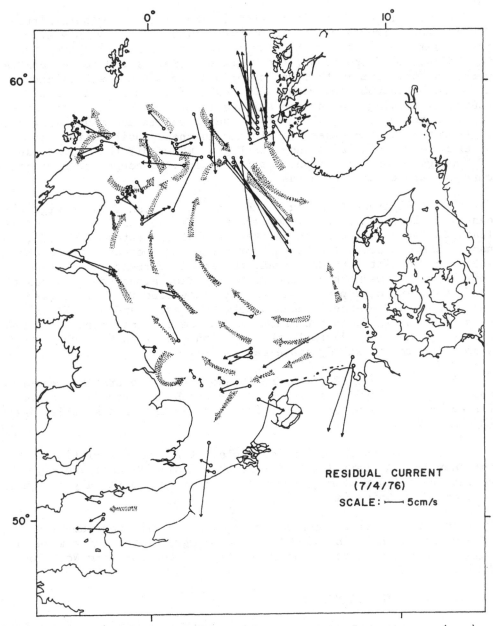

Fig. 10. Observed current vectors, and postulated currents (→),
         from Riepma (1980). The upper circle is located at the
         station position. The circles below this represent the
         measured residual currents at the various depths at that
         position.

Both models and observations showed that during JONSDAP '76 meteorological forcing was the primary mechanism producing currents in the northern North Sea. However, a comparison of seasonal steady state meteorologically induced currents (Davies 1982c) has shown that in the summer time northern North Sea currents arising from this mechanism are small compared with those in the winter period.

Recent observations by Dooley (1983) do however show significant currents in the summer time which cannot be explained using idealized models of density driven flow (Heaps 1980). These findings suggest that there is a need to develop three dimensional stratified North Sea models, and further North Sea observations are required in order to understand the mechanisms and time scales of the summer circulation of the region.

Using the methods described in this paper, two idealized models have been developed to examine the response of a stratified North Sea to meteorological forcing. In the first, an x-z model, (Davies 1981b), the internal displacements of pycnoclines at the Norwegian coast in response to a 12 hour wind impulse were examined. This model showed that the barotropic response of the North Sea took place on a time scale of two to three days. The baroclinic response however was associated with a much longer time scale of the order of ten days, and internal oscillations of the pycnocline were still evident twenty days after the wind impulse had ceased.

Data collected by I.O.S. (Bidston) in the North Sea (Sauvel 1982) during summer time revealed strong inertial currents following periods of wind activity (Sauvel, personal communication). An idealized three dimensional model of depth h = 65 m has been developed using the method described previously to examine the response of a stratified North Sea to wind (Davies 1982a).

A continuous vertical variation of eddy viscosity and density were used in this model (see Fig.3). A linear vertical variation of eddy viscosity with μ_1 = 500 cm^2/s and μ_2 = 200 cm^2/s, corresponding to a wind generated turbulent layer, was employed. Below this layer within the pycnocline turbulence was suppressed and μ_3 = 10 cm^2/s. In the bottom layer turbulence was assumed to be of tidal origin with μ_4 = 100 cm^2/s. The vertical variation of the modes computed with this viscosity distribution is shown in Fig.3. Motion in the basin was generated by a 12 hour wind impulse.

The time dependent response of a constant depth, stratified basin to wind impulses shorter than the inertial period (this case) is well known and numerous papers are available in the literature (for a review see Simons (1980)).

The initial response of the North Sea basin close to the land boundaries takes the form of internal Kelvin waves. The length scale of these waves is determined by the internal Rossby radius of deformation $Ro_i = (g_v h \, \Delta\rho/\rho)^{\frac{1}{2}}/\gamma$ with $\Delta\rho$ the density difference over the depth h and γ the Coriolis parameter. For typical North Sea values of these parameters Ro_i is the order of kilometres. Consequently initially the internal Kelvin waves are confined against the lateral boundaries and do not have a marked effect upon the interior solution, although they subsequently do propagate into the interior. In the numerical model it is necessary to have a sufficiently fine horizontal grid to resolve these waves. If too coarse a grid is used they appear as artificial grid scale oscillations away from the coast.

Away from lateral boundaries internal Poincaré waves, which are essentially equivalent to inertial motions, dominate the currents. Due to viscous and some boundary effects, surface currents computed with the model (see Davies 1982a for details) do not show purely circular motion but exhibit and elliptic spiral motion decaying rapidly with time (see Fig.11).

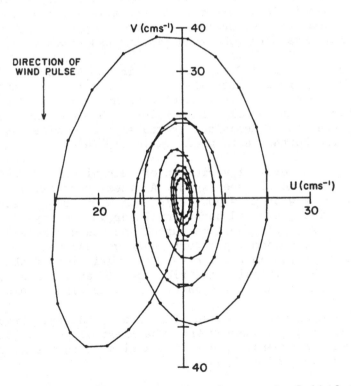

Fig. 11. Surface u and v components of current plotted at hourly intervals in polar form.

After the wind impulse had decayed, a 180° phase difference in currents above and below the pycnocline was evident, a feature characteristic of inertial currents (Pollard 1970). The vertical variation of current was characterized by a surface layer through which the velocity changed very little; a rapid nearly linear variation within the pycnocline, through which the flow reversed. Below the pycnocline the current varied only slightly through the vertical. The main characteristics of the flow could be adequately described using the second mode (1st baroclinic mode) shown in Fig. 3.

The scale difference between the internal Rossby radius of deformation Ro_i (order kilometres) and its external equivalent $Ro_e = (gh)^{\frac{1}{2}}/\gamma$ (order up to hundred kilometres) means that previous numerical mesh sizes used to model barotropic North Sea circulation must be reduced significantly in order to model internal flows.

These idealized calculations indicate that future stratified northern North Sea models will be concerned with time scales considerably longer than those previously associated with homogeneous models. Space scales however will be much smaller.

CONCLUDING REMARKS

In this paper the solution of the three dimensional hydrodynamic equations, which describe flow in homogeneous and stratified sea regions, has been developed. The approach used involves the application of the Galerkin method through the vertical and a finite difference grid in the horizontal. By this means physically realistic continuous vertical variations of density and eddy viscosity can be included in the model, which computes a continuous current profile from sea surface to sea bed.

By separating the flows into external and internal modes a computationally efficient time splitting algorithm is developed which allows the modes to be integrated using different time steps.

Computed circulation patterns from a range of North Sea models have been analysed. Both model results and observations confirm that during JONSDAP '76 the major driving mechanism producing currents in the northern North Sea was meteorological forcing. The circulation of the North Sea was influenced in part by wind direction, although below the surface wind driven layer bottom topography also played an important role in determining the direction and spatial variability of currents over the northern North Sea.

Models are presently being developed to examine the circulation of the North Sea under stratified conditions. Initial calculations indicate that longer time scales and shorter space scales are involved in these models than have previously been used in the homogeneous models.

ACKNOWLEDGEMENTS

The author is indebted to Dr. N.S. Heaps for various comments.
Thanks are due to Mr. R. Smith for drawing the diagrams and
Mrs. E. Bell for typing the paper.

This work was funded by the Natural Environment Research Council
and the Department of Energy.

REFERENCES

Cox, M., 1972, The numerical evaluation of B-splines,
    J.Inst.Math.Appl., 10:134-149.
Davies, A.M., 1980a, Application of numerical models to the
    computation of the wind-induced circulation of the North
    Sea during JONSDAP '76. Meteor.Forschungsergebn Reihe A,
    22:53-68.
Davies, A.M., 1980b, Application of the Galerkin method to the
    formulation of a three-dimensional non-linear hydro-
    dynamic numerical sea model, Applied Mathematical Model-
    ling, 4:245-256.
Davies, A.M., 1981a, Three-dimensional modelling of surges,
    in:"Floods due to High Winds and Tides", ed. D.H.Pere-
    grine, published Academic Press (London), pp. 45-74.
Davies, A.M., 1981b, Three dimensional hydrodynamic numerical
    models. Part 1. A homogeneous ocean-shelf model. Part 2.
    A stratified model of the northern Sea.Proc.Symp.Norwegian
    Coastal Current, Geilo September 1980, edited R.Saetre
    and M.Mork, published Bergen University, pp. 370-426.
Davies, A.M., 1982a, Formulating a three dimensional hydro-
    dynamic sea model using a mixed Galerkin-Finite diffe-
    rence method, in: "Finite Elements in Water Resoruces",
    ed. K.P. Holz, Proceedings of the 4th International
    Conference on Finite Elements in Water Resources,
    pp. 5-27 to 5-41.
Davies, A.M., 1982b, On computing the three dimensional flow
    in a stratified sea using the Galerkin method,
    Applied Mathematical Modelling, 6:347-362.
Davies, A.M., 1982c, Meteorologically-induced circulation on
    the North-West European continental shelf: from a three-
    dimensional numerical model. Oceanologica Acta,
    5:269-280.
Davies, A.M., 1983, Formulation of a linear three-dimensional
    hydrodynamic sea modelusing a Galerkin-Eigenfunction
    method (in press), International Journal of Numerical
    Methods in Fluids.
Davies, A.M. and Heaps, N.S., 1980, Influence of the Norwegian
    Trench on the wind-driven circulation of the North Sea,
    Tellus, 32:164-175.

Dooley, H.D., 1974, Hypotheses concerning the circulation of
    the northern North Sea, J.Cons.Int.Explor.Mer.,
    36:54-61.

Dooley, H.D., and Furnes, G.K., 1981, Influence of the wind
    field on the transport of the northern North Sea,
    Proc.Symp.Norwegian Coastal Current, Geilo September
    1980, edited R. Saetre and M. Mork, published Bergen
    University, pp. 57-71.

Dooley, H.D., 1983, Seasonal Variability in the Position and
    Strength of the Fair Isle Current, in: "Proceedings of
    the Symposium on North Sea Dynamics", Hamburg September
    1981, ed., J.Sündermann and W.Lenz, published Springer
    Verlag, pp. 108-119.

Furnes, G.K., and Saelen, O.H., 1977, Currents and hydrography
    in: "the Norwegian coastal current off Utsira during
    JONSDAP '76", Report No. 2/77, University of Bergen.

Furnes, G.K., 1980, Wind effects in the North Sea,
    J. Phys. Oceanogr., 10:978-984.

Heaps, N.S., 1980, Density currents in a two-layer coastal
    system with application to the Norwegian coastal current,
    Geophys. J. R. Astron. Soc., 63:289-310.

Lee, A.J., and Ramster, J.W., 1968, The hydrography of the
    North Sea. A review of our knowledge in relation to
    pollution problems. Helgolander Wiss. Meeresunters,
    17:44-63.

Pollard, R.T., 1970, On the generation by wind of inertial
    waves in the ocean, Deep-Sea Res.17:785-812.

Riepma, H.W., 1980, Residual currents in the North Sea during
    the INOUT phase of JONSDAP '76, Meteor Forschungsergeb
    Reihe A, 22:19-32.

Simons, T.J., 1980, Circulation models of lakes and inland
    seas, Can. Bull. Fish. Aquat. Sci, 203, 146 pg,
    published Dept, of Fisheries and Oceans Ottawa.

Sauvel, J., 1982, The tidal dynamics of the Western North Sea,
    I.O.S. Report 138: 94.

ON THE CIRCULATION OF THE STRATIFIED NORTH SEA

Jan O. Backhaus

Institut für Meereskunde
der Universität Hamburg
D 2000 Hamburg 13
F.R.G.

ABSTRACT

The circulation of the North Sea for the summer and the winter
season is simulated by means of a diagnostic three-dimensional model.
The simulations are compared with the sparse observational information available on currents with long time scales. The influence
of stratification and the consequences of its seasonal changes on
the circulation in a wind-driven tidal shelf sea are illustrated
by a number of model experiments.

INTRODUCTION

In the past, shelf sea models mainly focused on the simulation
of sea surface elevations due to tides and surges (Heaps, 1965;
Sündermann, 1966; Flather, 1976; Davies and Flather, 1978; Dolata
and Engel, 1979). Usually, with some few exceptions (Heaps and
Jones, 1977), the assumption of a well mixed, homogeneous sea was
made for these so-called "storm surge" models. The assumption of
homogeneity may be justified when events of short time scale, like
storm surges, are to be simulated or, especially, when only water
levels are the main interest.

In recent publications (Maier-Reimer, 1979; Davies, 1980;
Backhaus, 1980; Backhaus and Maier-Reimer, 1983; Davies, 1983) the
focus seems to shift towards the simulation of currents and current
fluctuations with longer time scales, involving a wider range of
physical processes but often still neglecting stratification. The
present study shows that the assumption of a "well mixed sea" may
not be appropriate when a quantitative estimate of the circulation
of a shelf sea is desired.

   The development and validation of a baroclinic shelf sea circulation model requires high quality data. Observations of sufficient
quality and density are available only for a very few shelf seas.
Far more than 50 years of research and routine observations, carried
out in the North Sea, qualify this region of the European shelf to
be a test case for the development of a prognostic circulation
model.

   In this study, however, the nascent stage of a three dimensional circulation model is used to determine diagnostically the circulation for summer and winter. The simulations are compared with
known features of the circulation. Some model experiments were
carried out in order to illustrate the effects of winds, stratification and tide, and their contribution to the circulation of the
North Sea. The above cited authors were mainly concerned with tide-
and wind-induced currents with time scales longer than the tide,
also called - without definition - "residual currents", whereas
this study will focus on the effects of stratification. So far
there is practically no information on the influence of density
currents on the circulation of the North Sea, except for the Norwegian Trench and Skagerrak (The Norwegian Coastal Current, Saetre
and Mork, editors, 1981). Since baroclinicity is most pronounced
here, special attention will be paid to the flow patterns in these
regions.

THE MODEL

   The model is based on the primitive shallow water equations
for a rotating earth, incorporating advection and horizontal and
vertical diffusion of momentum. Hydrostatic pressure distribution
and the Boussinesq approximation are assumed. At the sea surface
and bottom, kinematic boundary conditions are applied. Energy
dissipation at the sea bed is parameterized by nonlinear bottom
friction. At open boundaries tidal forcing is prescribed by sea
surface elevations, whereas at closed lateral boundaries conservation of mass requires no flux through the coast line. At coastal
boundaries a semi-slip condition is assumed for the horizontal
diffusion of momentum. At the open boundaries the normal derivatives of the mass flux are assumed to vanish. Because the model
is diagnostic, no boundary conditions are needed for the incorporation of stratification. Thus, at present the open boundary problems arising in connection with a prognostic baroclinic model are
avoided.

   A detailed description of the numerical scheme of the model
will be published elsewhere (Backhaus, 1983). Therefore, only a
brief outline is given here. The space domain is approximated
horizontally by a lattice of the "Richardson" or "Arakawa C" type
(Arakawa and Lamb, 1977). A number of horizontal computational
planes, discretized by the C-grid, are arranged in the vertical to

allow for an approximation of the vertical structure of the flow
and mass. The primitive equations are vertically integrated over
the depth range of a model layer, which is defined by the space
between two adjacent computational planes. Thus, the model belongs
to the class of "grid-box" models, also used in numerical weather
prediction. Within one grid-box vertically integrated quantities
are defined. Boxes next to the bottom may have different sizes due
to the approximation of topography.

    In the time domain the model involves two time levels. For
the sake of computing economy, favourable for long term simulations,
an implicit scheme is formulated for the external gravity waves and
the vertical diffusion of momentum. A similar approach was proposed by Kurihara (1965). Thus the time step can be chosen independently of the "Courant-Friedrichs-Lewy" stability criterion, strin-
gent for explicit schemes. However, it cannot be chosen arbitrarily
since implicit schemes tend to distort the wave characteristics
with increasing time step. Nevertheless, a saving of at least 50%
in computing time, compared with commonly used explicit schemes,
can be achieved without negative effects by means of the above
sketched semi-implicit scheme. In order to avoid confusion: the
numerical scheme is prognostic; however, the model must be regarded
as diagnostic, because all input (wind, stratification, topography
etc.) is kept constant in time, as is the periodic tide.

    Tidal oscillations are removed from the simulations, retaining
nonlinear effects of the tide, by an integration in time over one
tidal cycle, after a (periodic) steady state is reached. For each
layer, the resulting flow patterns are displayed by means of
Lagrangian trajectories. A reasonable reduction of the model output trajectories allows for an estimate of the displacements of
water masses, valuable information for both the physical oceano-
grapher and the marine biologist. The displacements of virtual
particles are computed for an interval of one month because the
model input will be monthly means of stratification and wind stress.
The trajectories start at locations which are each 4 grid sizes
apart in longitudinal and meridional directions. The vertical
velocity component of the flow was not taken into account for the
following reason: a considerable overestimation of the vertical
velocity in some portions of the model area occurred due to an
imbalance of the flow and the density field. The density distribution prescribed is fixed in time, according to the diagnostic
approach; this prohibits an adjustment of the flow and the density
field.

OBSERVATIONS

    The stratification of the North Sea shows a pronounced sea-
sonal signal. In winter a fairly homogeneous water body is formed
by convection and tide- and wind-induced stirring. During summer

however, wide areas are stratified. The Norwegian Trench and the
Skagerrak, the deepest portions of the North Sea, are stratified
throughout the year. An important feature, commonly observed in
shelf seas, is a pronounced horizontal density gradient in coastal
regions, present at all times of the year. Fresh water discharges
from the coast maintain this density gradient. They reach peak
values in spring because of enhanced precipitation and melting
snow. This may illustrate that the assumption of a "well-mixed"
sea, equivalent to homogeneity, frequently made for numerous shelf
sea "current" models, is questionable.

Monthly mean values of temperature and salinity distributions
at oceanographic standard depths were compiled by Tomczak and
Goedecke (1962), and by Goedecke et al.(1967). These data were
digitized and interpolated on the three dimensional grid to provide
model input. The spatial distribution of density was computed
from the T,S data by means of the equation of state for sea water.

To approximate the main features of the T,S distribution, as
well as the topography, the horizontal resolution of the model grid
was chosen to be 12 and 20 minutes in latitude and longitude
respectively. The vertical resolution was set to match the oceanographic standard depths of observation, resulting in a maximum
number of 10 layers for the deepest portion, the Skagerrak.

Tides and topography in the North Sea, both important model
inputs, are known with sufficient accuracy. For this study the
semi-diurnal lunar tide M2, which contributes more than 80% to the
total tidal signal, was chosen to represent tidal dynamics.

In order to contrast summer and winter conditions, model simulations for March and August will be presented. Figures 1a, b
show wind stress patterns for the months under consideration, derived from monthly mean values (Hellerman 1967, 1968). These figures
also provide information about the area covered by the model, and
the horizontal resolution of the grid.

The input data described here must be regarded as the best
information presently available for the purpose of this study.

Whereas sufficient data is available for model input, there
is only a little information on flow patterns for a comparison
with the model simulations. Many authors, amongst them Böhnecke
(1922) and Lee (1980), have already published charts of the surface
circulation of the North Sea, deduced from observations. Since
Böhnecke, the flow pattern regarded as the commonly accepted
sketch of the circulation has not changed very much. In Fig. 2
the most recent version is shown, proposed by Lee (1980). In contrast to this fairly well established pattern there is very little
information about the circulation at depth. In Lee's chart however,

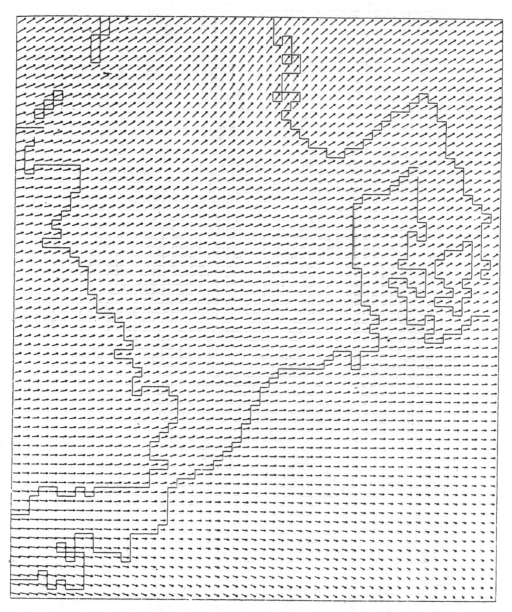

Fig. la. Mean wind stress distribution for March (Hellerman, 1967, 1968) interpolated on model grid, arbitrary units.

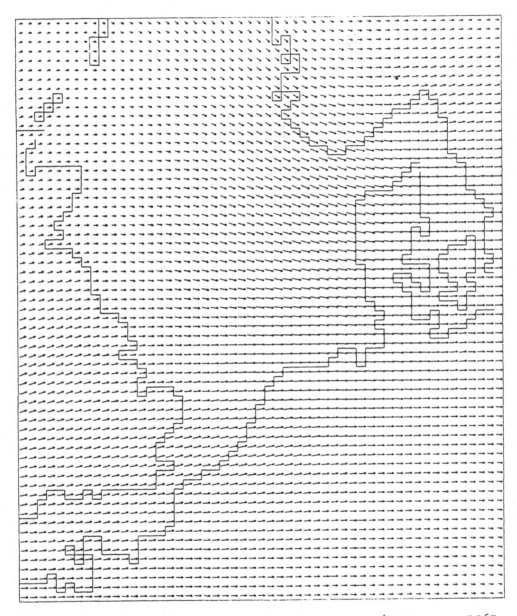

Fig. 1b. Mean wind stress distribution for August (Hellerman, 1967, 1968) interpolated on model grid, arbitrary units.

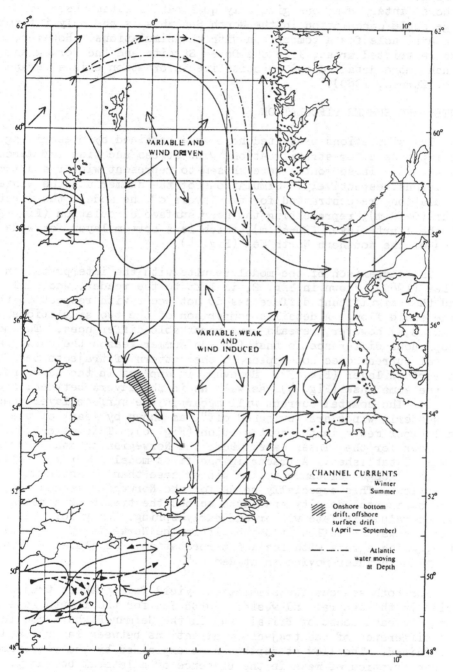

Fig. 2. North Sea circulation, adopted from Lee (1980).

there are some remarks about the interior flow. His and the other
authors' interpretations give only qualitative estimates. For the
large scale circulation in the North Sea this is the only information available for a comparison with the simulations. However, for
some restricted areas, like the Dover Straits and the Norwegian
Trench, more detailed and quantitative information exists (Prandle,
1978; Furnes, 1980).

WINTER AND SUMMER CIRCULATION

The simulations presented here were obtained by prescribing
the input data for stratification, wind stress and tide for March
and August. These months were chosen to represent winter and summer
conditions respectively. Simulations of the summer and the winter
circulation are contrasted for two layers of the model: the first
layer (0-10 m), representing the near surface circulation (Fig. 3);
and the fourth layer (30-60 m), which is a bottom layer for wide
areas of the southern North Sea (Fig. 4).

The comparison of the model results with the interpretation
of Lee (1980), shown in Fig. 2, is left to the reader, who will
find that significant differences do not occur with respect to the
large scale flow. A detailed comparison of the two simulations
(Figs. 3,4) however, reveals some remarkable differences. The most
conspicuous difference is that for the summer season the Norwegian
coastal current does not appear in the pattern of trajectories for
the surface layer (Fig. 3). However, it appears in the layer no.4
for the same case (Fig. 4), as well as in the layers between, see
Fig. 8. The coastal current will apear in the surface layer when
the westerly winds are "switched off", as shown by means of the
model experiments in the next section (Fig. 6). This effect does
not occur for the winter case. Here the Norwegian coastal current
is well established. A tracer in the first model layer generally
travels a longer distance in the winter case than in the summer
simulation. This especially holds for the Norwegian Trench area,
where horizontal density gradients across the trench axis reach
maximum values in late winter and early spring. Only for this
situation was an inflow at depth within the Norwegian Trench simulated, which agrees with Lee's interpretation. He states in Fig.
2: "Atlantic water moving in at depth."

For both seasons the simulations yield fairly short trajectories in the central and western North Sea for the fourth layer.
Along the east coast of Britain and in the German Bight a considerable difference of the trajectory directions between layer 1 and 4
is obtained. The flow at depth can be explained by the requirement
for conservation of mass in the presence of a lateral boundary.
The wind forcing in the semi-enclosed North Sea basin generates a
large scale barotropic pressure gradient (wind set up), which is
compensated by the near bottom currents. For the German Bight these

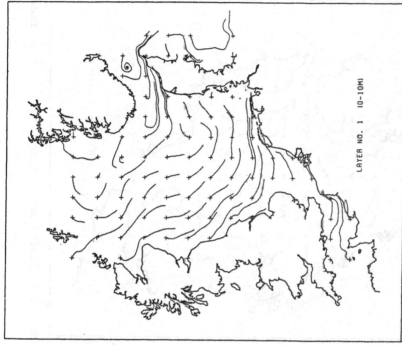

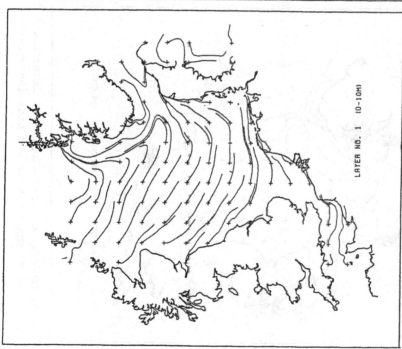

Fig. 3. Lagrangian trajectories in layer no. 1 for winter (left) and summer (right); stratification, wind stress and tide incorporated.

J. O. BACKHAUS

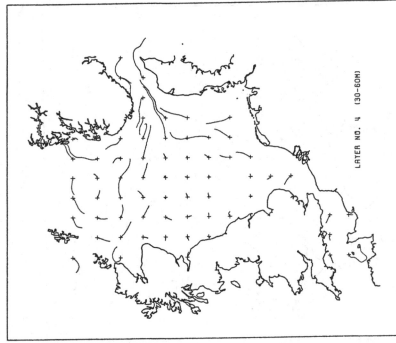

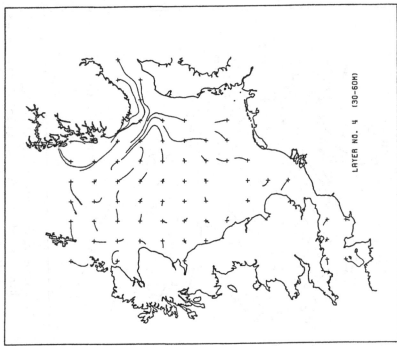

Fig. 4. Lagrangian trajectories in layer no. 4 for winter (left) and summer (right); stratification, wind stress and tide incorporated.

slope currents are discussed in more detail in Backhaus (1980).

For the August simulation (Fig. 4) the Fair Isle current (Dooley 1974), setting eastwards in the northern North Sea guided by topography and joining the Norwegian coastal current, is slightly more pronounced in comparison to the winter case. The Jutland current at the north-west coast of the Danish mainland (Jutland) is guided by topography (into the Skagerrak) for the summer case, whereas it close-circuits south of Norway into the coastal current in the winter simulation. The simulations clearly reveal that the Norwegian coastal current is an integral part of the general circulation of the North Sea.

The magnitude of the flow through a section off Utsira, on the west coast of Norway, was compared with observations obtained during March and April as part of the JONSDAP '76 field program. In the data presented by Furnes (1980), fluctuations of the coastal current range from -0.8 (inflow) up to 3.0 sverdrup (outflow). The mean value of this record agrees well with the winter simulation, which yields for the same section and outflow of 1.6 sverdrup. Good agreement is also found for the inflow through the Dover Straits, if compared with values given by Prandle (1978).

On the basis of the input data presently available the simulations must be regarded to be the most realistic estimates of the circulation to be obtained by the diagnostic model described above. However, it is known that wind action in a shallow sea like the North Sea plays a dominant role. Therefore it is questionable whether a constant mean wind forcing will lead to realistic estimates of the wind induced flow.

ON THE EFFECTS OF STRATIFICATION

In general the diagnostic simulations presented in the previous section reflect our present knowledge about the circulation of the North Sea. Quantitative agreements were found. By contrasting summer and winter circulations, seasonal changes were found with respect to the Norwegian coastal current especially but also to the circulation at depth.

Thus, gained with the confidence in the model, it will be applied to illustrate the effects of stratification on the circulation. This is done by a qualitative comparison of simulations carried out for a stratified and a non-stratified North Sea, again contrasting summer and winter conditions. Incorporating tidal forcing because it is always present in nature, model experiments are presented, which first exclude, then include, the wind-forcing. This approach was chosen to determine qualitatively the effects of combinations of the flow components due to stratification, wind and tide.

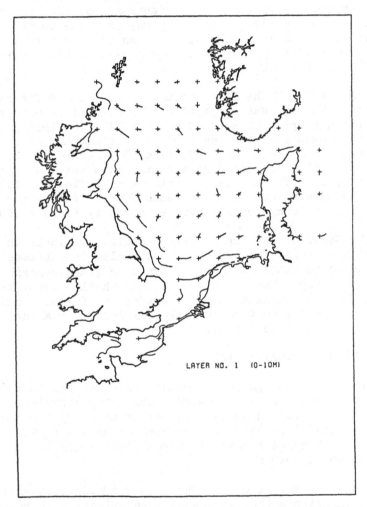

Fig. 5. Lagrangian trajectories in layer no. 1 for a homogeneous
        North Sea, tide only.

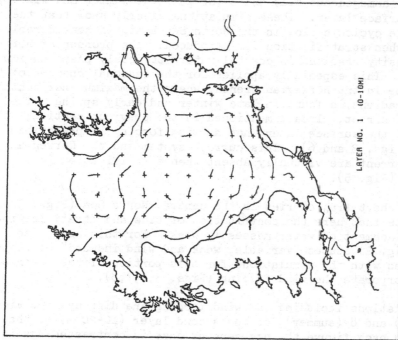

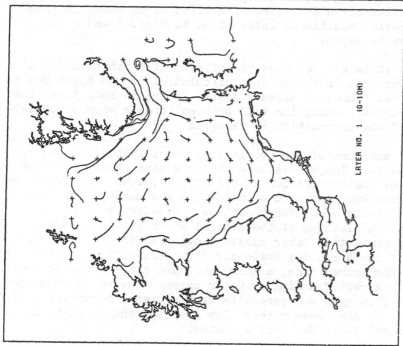

Fig. 6. Lagrangian trajectories in layer no. 1 for a stratified North Sea: winter, left: summer, right: tide included.

Tidal simulations, excluding the wind forcing, are presented in Figs.5 (homogeneous sea) and 6 (stratified, summer and winter) for the surface layer. These simulations clearly show that the large scale cyclonic flow in the North Sea basin is considerably enhanced when stratification is introduced. The pronounced horizontal density gradient in coastal waters obviously plays an important role. This especially applies for the seasonal changes of the flow regime in the Norwegian Trench, where the maximum horizontal density gradient is found in late winter and early spring. The Fair Isle current, already mentioned in the previous section, only appears in the surface layer when no wind forcing is prescribed (compare Figs. 3 and 6). This current system and the Norwegian coastal current are virtually absent when stratification is neglected (Fig. 5).

Very short trajectories in the central North Sea (Figs. 5 and 6) indicate that here the contribution of tide and stratification to a displacement of water masses is weak. For this area, Lee (1980) (Fig. 2) states "variable, weak and wind induced" currents. This agrees with the simulations for this portion of the North Sea, which incorporate the wind forcing (Figs. 3,4,7,8).

Simulations including the wind forcing are displayed in Figs. 7 (winter) and 8 (summer) for the second layer (10-20 m) of the model. In each figure the homogeneous case is contrasted against the stratified North Sea. The stratified cases shown in these figures provide additional information to Figs. 3 and 4 with respect to the flow at depth.

Again it is evident that the effects of stratification do play an important role in the general circulation of the North Sea for both seasons. Thus, as already stated in the introduction, the "well-mixed sea" assumption will be questionable when a realistic estimate of the circulation is required.

The comparison of the two seasons shows that on average a tracer travels a longer distance in the winter case. This can be explained by the combination of two effects, both significant for winter conditions: enhanced wind forcing and maximum horizontal density gradients along the coastline. A seasonal change with respect to the flushing of the central North Sea is obtained. In the winter simulation water masses flowing in through the northern entrance of the North Sea obviously flush the entire central region where, in the summer case, a large cyclonic gyre is formed, which prohibits a direct flushing by water masses flowing in from the Atlantic. This gyre and generally weak currents occurring during calms (Fig. 6) may cause a very slow displacement of water masses in the central North Sea during summer.

A possible interaction of wind- and density-induced currents

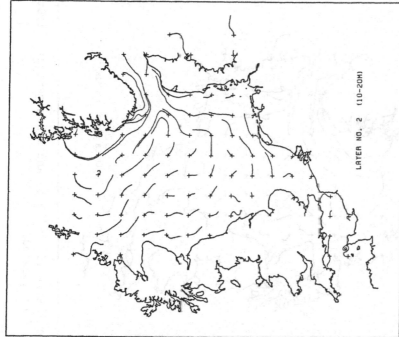

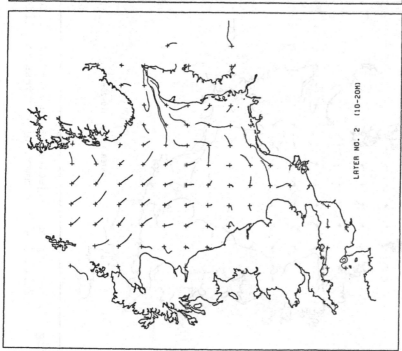

Fig. 7. Lagrangian trajectories in layer no. 2 for a stratified (left) and a homogeneous (right) North Sea in winter; wind stress and tide incorporated.

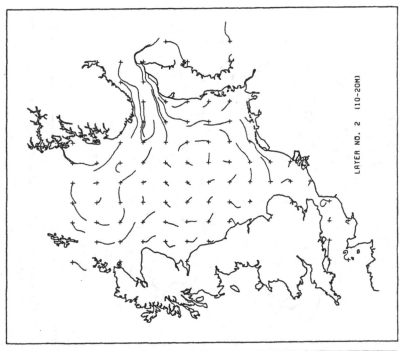

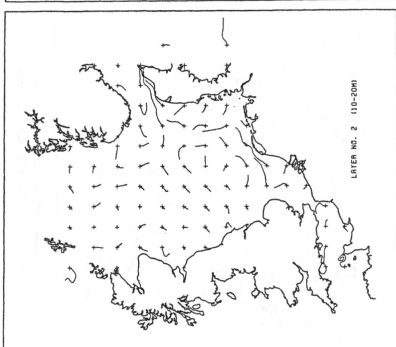

Fig. 8. Lagrangian trajectories in layer no. 2 for a stratified (left) and a homogeneous (right) North Sea in summer; wind stress and tide incorporated.

may be inferred from Fig. 8 (summer). The above mentioned gyre is obviously formed by the combined effect of wind forcing and stratification, as can be seen from a comparison of the unstratified and the stratified case in this figure. Furthermore, a considerable shift of the trajectory directions along the east coast of Britain is simulated for the same model layer. The direction of the flow obtained for the unstratified case is nearly normal to the coast line. This is mainly a wind induced phenomenon, also simulated by Davies (1983) for a similar wind forcing (see also slope currents, discussed in the previous section). The introduction of stratification results in a flow which is almost parallel to the coast line. It does not need much imagination to deduce that an increased wind forcing will weaken or even overcome this effect of stratification. Thus, changing westerly winds may lead to considerable changes of the flow direction in this region, resulting from the combined effect of stratification and wind forcing.

FINAL REMARKS

In the course of the development of a prognostic baroclinic shelf sea circulation model, a diagnostic simulation of the circulation of the North Sea was carried out. In general a good agreement between the model results and the few observations of the circulation was found.

The strong influence of stratification on the circulation of the North Sea, found in this study, leads to the conclusion that, for a realistic and quantitative simulation of the circulation, stratification must be taken into consideration. For the entire region of the North Sea the large scale flow will generally be underestimated when stratification is neglected, as illustrated by model experiments. A qualitative comparison of summer and winter simulations revealed remarkable seasonal changes of the circulation, especially with respect to the flushing of the central North Sea.

Shelf seas cover less than 10% of the world ocean area; their contribution to the total volume of water masses on our globe is even smaller. Nevertheless, shelf seas have always been important for mankind. They are a source of valuable nutrients; a considerable amount of the world fish catch is found in shelf seas and adjacent ocean areas. Mineral resources beneath the sea bed are accessible in these shallow regions. Coastal seas are often used as dumping sites for industrial and municipal wastes all over the world. Offshore drilling, the dumping of waste, and heavy ship traffic in coastal waters constitute a menace to the marine ecosystem. For a rapidly increasing world population, shelf seas will become even more important in the near future, because of these uses and their implications. Oceanographers should be prepared for a future demand of an enlarged use of shelf seas for the benefit of

mankind. One urgent question arising from this demand will certainly concern the quantitative estimate of the general circulation of
shelf seas.

ACKNOWLEDGEMENT

For fruitful discussions on the subject of this study the
author expresses his thanks to his colleagues H. Friedrich,
D. Quadfasel and E. Maier-Reimer.
Thanks also to Prof. Gade and Dr. Svendsen for the arrangement
of an excellent workshop at Os in Norway.

REFERENCES

Arakawa, A., Lamb, V.R., 1977, Computational design of the
        basic dynamical processes of the UCLA general circula-
        tion model, Methods of comp. physics,16:173-263.
Backhaus, J.O., 1980, Simulation von Bewegungsvorgängen in der
        Deutschen Bucht, Dtsch. Hydrogr. Z. Erg.-H. B 15.
Backhaus, J.O., Maier-Reimer, E., 1983, On seasonal circula-
        tion patterns in the North Sea. In:"North Sea Dynamics",
        Sündermann, Lenz,editors, Springer Verlag, 63-84.
Backhaus, J.O., 1983, A semi-implicit primitive equation model
        for shelf seas; outline and practical aspects of the
        scheme. Continental Shelf Research, special issue on
        JONSMOD '82.
Böhnecke, G., 1922, Salzgehalt und Strömungen der Nordsee.
        Veröff. d. Inst. f. Meereskd. Univ. Berlin, N.F.A.10,34.
Davies, A.M., Flather, R.A., 1978, Application of numerical
        models of the North West European Continental Shelf
        and the North Sea to the computation of the Storm
        surges of November to December 1973, Dtsch. Hydrogr.Z.
        Erg.-H. A, 14.
Davies, A.M., 1980, Application of numerical models to the
        computation of the wind-induced circulation of the
        North Sea during JONSDAP '76. Meteor Forsch.-
        Ergebnisse, A, 22:54-68.
Davies, A.M., 1983, Application of a three dimensional shelf
        sea model to the calculation of North Sea currents.
        In: "North Sea Dynamics", Sündermann, Lenz, editors,
        Springer Verlag.
Dolata, L.F., Engel, M., 1979, Sturmflutvorhersagen mit
        mathematisch- physikalischen Modellen, Die Küste,
        34:203-225.
Dooley, H.D., 1974, Hypotheses concerning the circulation of
        the northern North Sea, J. Cons. Int. Explor. Mer.,
        36:54-61.
Flather, R.A., 1976, Tidal model of the north-west European
        continental shelf, Memoir. Soc. R. Sci. Liege, 6, ser.
        10:141-164.

Furnes, G.K., 1980, Wind effects in the North Sea,
    J. Phys. Oceanography, 10:978-984.
Goedecke, E., Smed, J., Tomczak, G., 1967, Monatskarten des
    Saltzgehaltes der Nordsee, dargestellt für verschiedene
    Tiefenhorizonte, Dtsch. Hydrogr. Z., Erg.-H. B,9.
Heaps, N.S., 1965, Storm surges on a continental shelf.
    Phil. Trans. R. Soc. London, (A), 1082, Vol. 257:351-383.
Heaps, N.S., Jones, J.E., 1977, Density currents in the Irish
    Sea, Geophys. J.R. Astron. Soc. 51:393-429.
Hellerman, S., 1967, 1968, An updated estimate of the wind
    stress on the world ocean, Monthly Weather Rev.,
    95:607-626; and 96:62-74.
Kurihara, Y., 1965, On the use of implicit and iterative
    methods for the time integration of the wave equation.
    Monthly Weather Rev., 93:33-46.
Lee, A.J., 1980, North Sea: Physical Oceanography. In: "The
    North- West European Shelf Seas: the sea bed and the
    sea in motion; II. Physical and chemical oceanography
    and physical resources", Banner, Collins, Massie,
    editors, Elsevier Oceanographic Series 24 B.
Maier-Reimer, E., 1979, Some effects of the Atlantic circula-
    tion and of river-discharges on the residual circulation
    of the North Sea, Dtsch. Hydrogr. Z., 32:126-130.
Prandle, D., 1978, Residual flows and elevations in the
    southern North Sea, Proc. R. Soc. London, (A) 359:189-288.
Saetre, R., Mork, M., (editors), 1981, The Norwegian Coastal
    Current. Proc. from the Norwegian coastal current
    symposium. Vol. I, II.
Sündermann, J., 1966, Ein Vergleich zwischen der analytischen
    und der numerischen Berechnung winderzeugter Strömungen
    und Wasserstände mit Anwendungen auf die Nordsee.
    Univ. Hamburg, 4.
Tomczak, G., Goedecke, E., 1962, Monatskarten der Temperatur
    der Nordsee, dargestellt für verschiedene Tiefen-
    horizonte, Dtsch. Hydrogr. Z.,Erg.-H. B, 7.

THE STATE-OF-THE-ART IN COASTAL OCEAN MODELLING:
A NUMERICAL MODEL OF COASTAL UPWELLING OFF PERU-INCLUDING MIXED
LAYER DYNAMICS

James J. O'Brien and George W. Heburn*

Mesoscale Air Sea Interaction Group
Florida State University
Tallahassee, Florida 32306

Abstract

This is a study of three-dimensional, time dependent
upwelling off Peru. The model is the most complete study of ocean
coastal circulation ever done. It is an excellent example of our
actual capability to simulate ocean circulation and variables on a
timescale of days to weeks.

The principal new theoretical result is the detection of a
signature of continental shelf waves in the sea surface
temperature pattern. When the model results are compared with
observations, we learn that we can predict the horizontal flow,
vertical velocity and thermodynamic variables quite well in the
upper 50 meters but substantial observed variability is not
simulated in the lower layers.

The two-layer model appears to be quite useful for a
coastline like Peru where the shelf is very narrow and steep. We
would not suggest the use of this model for a broad, shallow shelf
regime. It is clear that a new level of complexity in model
formulation is required for further progress.

---

*Present affiliation, NORDA, Bay St. Louis, MS, USA

INTRODUCTION

The purpose of this study is to examine the Peruvian
upwelling system by means of a numerical model. The basis for the
study of upwelling lies in the pioneering work by Ekman (1905)
when he explained, by means of a balance between vertical friction
and earth rotation, the reason for the mass transport in the upper
ocean being directed to the right/left of the local winds in the
northern/southern hemisphere. Thorade (1909) was the first to
apply this reasoning to the upwelling problem, when he showed that
winds blowing equatorward parallel to the California coast would
cause an offshore transport in the surface layer and thus result
in upwelling near the coast.

Over the next six decades, the upwelling problem received
relatively little attention. A few of the notable contributions
during this period include Sverdrup (1938) and Sverdrup and
Fleming (1941). Using data gathered off Southern California, they
quantitatively applied Ekman's theory of upwelling to arrive at a
dynamical interpretation of coastal upwelling. Yoshida (1955a,b)
introduced first-order stratification into the upwelling problem
when he considered an ocean model with two homogeneous layers
which have a density difference between the layers of $\Delta \rho$. Yoshida
also considered the effects of time dependent winds.

In a comprehensive theoretical study of upwelling, Yoshida
(1967) used a two-layer quasi-steady model to examine the effects
of internal friction, the effect of the continental shelf, and
time variable upwelling. Smith (1968) presents an in-depth review
of these earlier studies.

Energetic low-frequency current fluctuations in the longshore
component of velocity, with periods on the order of several days,
are observed (Smith, 1974; Mittelstaedt et al., 1975; Brink et
al., 1978). The fluctuations off Oregon and Northwest Africa show
significant correlation with the local winds at a lag of less than
one day (Huyer et al., 1978; Badan-Dangon, 1978). Off Peru,
however, the longshore currents at depth are not significantly
correlated with the local winds (Brink et al., 1978). Smith
(1978) presents observations from the continental shelf and slope
off Peru which demonstrate that the fluctuations on the time scale
of days to weeks are coherent over several hundred kilometers
along the coast (between 12°S and 15°S) but are not correlated
with the local winds. He suggests that the fluctuations in the
currents are due to poleward propagating, coastal trapped waves
and that the phase speed and offshore length scale are consistent
with that for internal Kelvin waves.

The idea of coastal trapped waves propagating an upwelling

response into a region has previously been covered by Gill and
Clarke (1974). Adamec and O'Brien (1978) have demonstrated by
using a simple numerical model that the seasonal upwelling
observed in the Gulf of Guinea could result from the propagation
of internal Kelvin waves into the region. The Kelvin waves in
their model were excited by changes in the wind stress over the
western Atlantic.

The theoretical synthesis of the data collected in these
field experiments has primarily taken one of two approaches,
either numerical or analytical models. Both of these approaches
have their advantages and disadvantages. Analytical analysis is
primarily restricted to steady flow with highly idealized
geometries and boundary conditions, and to linear theory. Some of
the difficulties encountered with the analytical models are the
modelling of the singular coastal corner region, the consideration
of the physical processes introduced by time dependent forcing,
turbulent mixing, and the presence of thermoclines and
pycnoclines. Through careful numerical formulation of the
governing equations, application of efficient differencing
techniques, and utilization of plausible parameterization schemes
for various physical processes such as mixing, etc., numerical
models can be effective tools in simulating upwelling
circulations. The numerical approach allows us to consider
problems where the analytic solution to the problem is
untractable, but this is not without cost.

As the models become more sophisticated and realistic, the
cost of running the model escalates, and the interpretation of the
model results becomes more complex. The optimal approach to the
numerical formulation is to start with the simplest model,
preferrably one which can be verified analytically. The next step
would be to systematically examine the various physical factors
and forcing mechanisms which have a bearing on the problem. Then,
finally, building on the knowledge gained from these simpler
models, develop a model which includes the significant physical
parameters and can realistically simulate the observed upwelling
events. Ideally, a model which can adequately simulate upwelling
events can be used to provide flow field information to other
disciplines, such as biologists, when physical data are not
available. This approach has been the method used by the
Mesoscale Air-Sea Interaction Group at Florida State University in
developing numerical models to study the upwelling problem. The
first of these models was the two-layer, time-dependent, nonlinear
f-plane model by O'Brien and Hurlburt (1972). This model
successfully simulated the basic features of the upwelling
circulation and the equatorward coastal surface jet observed in
most upwelling regions, but it failed to reproduce the observed
poleward undercurrent.

Hurlburt and Thompson (1973), using a two-layer x-t β-plane model which neglected all longshore derivatives except for the longshore pressure gradients, successfully removed the barotropic mode and produced a poleward undercurrent.

Hurlburt (1974) used an x-y-t two-layer β-plane model to study the effects of various configurations of bottom topography and coastline geometry on wind driven eastern ocean circulations. He found that the topographic β-effect plays a fundamental role in the dynamics associated with mesoscale longshore topographic variations. All the foregoing numerical models were associated with upwelling along a meridional (N-S) coastline.

The first attempt to introduce realistic shoreline geometry and bottom topography was presented by Peffley and O'Brien (1976). They used a digitized version of the Oregon coast and offshore bottom topography in the Hurlburt (1974) model to simulate the coastal upwelling off Oregon. Peffley and O'Brien found that the topographic variations dominated over coastline irregularities in determining the longshore location of upwelling centers.

Preller and O'Brien (1980) used a version of the Hurlburt (1974) model with an idealized version of the Peruvian bottom topography and straight coastline to study the effects of various topographic features on the Peruvian upwelling system. In their model, Preller and O'Brien were unable to reproduce the poleward undercurrent with just wind stress and β-effect. To induce a steady poleward undercurrent, Preller and O'Brien introduced an external depth independent longshore pressure gradient in the form of an atmospheric pressure gradient. O'Brien et al. (1980) used this same model with a digitized version of the Peruvian coastline and bottom topography to simulate the major features of the Peruvian upwelling system.

A comparison of the output data from the O'Brien et al. model with objectively analyzed current meter data (O'Brien, Smith and Heburn, 1980) points out two significant shortcomings of this model. The first is an obvious lack of upper layer onshore flow (induced by horizontal density gradients) during the relaxation of upwelling favorable winds. The second is a rather anemic lower layer flow in the model as contrasted with observed energetic low-frequency current fluctuations. Brink et al. (1978) have shown that these fluctuations are not correlated with the local winds and Smith (1978) suggests that these are the result of remotely forced, coastal trapped waves propagating through the region.

To examine the effects of the interactions of a fluctuating undercurrent and the digitized Peruvian bottom topography, we will use the O'Brien et al. hydrodynamic model with a parameterization, based on coastal trapped Kelvin wave dynamics, for the longshore current. With the passages of one of these coastal trapped waves, we will see a direct relationship between the relative magnitude of the topographically induced upwelling center, as determined by the displacement of the model interface, and the strength of the undercurrent.

All of the foregoing models are strictly hydrodynamic and are unable to achieve a realistic steady-state. Furthermore, the interface is not allowed to intersect the surface, since numerical instabilities result when the layer thickness approaches zero. Therefore, they cannot simulate the observed upwelling frontal structure. O'Brien et al. (1977) have proposed one solution to this problem by allowing the uppermost layer to entrain mass from the lower layer just before the layer interface reaches the surface.

Another intrinsic assumption in these models is that the time scale for vertical advection is much shorter than that for turbulent vertical diffusion. Thompson (1974) suggests that this is valid for small departures of the interface from its initial state, but that the general validity is in question when one considers the studies of oceanic wind mixing by Kraus and Turner (1967), Denman (1973), Denman and Miyake (1973), and Pollard and Millard (1970). Thompson has shown, by using a x-t two-layer model which included thermodynamics and interfacial maxing, that vertical advection and vertical mixing may be comparable on a coastal upwelling event time scale. He defined an upwelling event time scale as "an interval longer than an inertial period, during which local winds initiate and maintain coastal upwelling." This particular parameterization of the interfacial mixing was restricted to the wind-dominated (forced convection) rather than the heat-dominated (free convection) regime, since the two-layer formulation of the model is not readily adaptable to thermocline reformation at shallower depths under conditions of strong solar heating.

In this study, we will first include a parameterization for the longshore current fluctuation into the x-y-t two-layer Peru model to examine the interaction of the currents with the bottom topography. We will also add thermodynamics and a parameterization for interfacial mixing to this model and use time-dependent forcing derived from JOINT II, March , April, and May (MAM) 1977 data to simulate the Peruvian upwelling system.

THE COASTAL UPWELLING PROBLEM

a. Bottom Topography Effects

     The inclusion of bottom topography into the study of ocean
circulations adds considerable difficulty to the analysis and
interpretation of the problem. Generally the approach to the
solution of ocean circulation problems is to consider a
flat-bottom, stratified ocean on a β-plane, and then decompose the
flow into barotropic and baroclinic modes. The vertical structure
of the horizontal velocities for these modes can then be
determined independently for each mode. Now with the addition of
a sloping bottom, the direct use of the barotropic and baroclinic
modes is no longer possible. The effect of the bottom topography
on this approach is to couple the modes such that the vertical
standing modes are no longer independent. This coupling makes the
problem extremely difficult to interpret.

     Since the observations off Peru tend to suggest that the
lower layer is isolated from the local surface forcing and wave
propagation is controlled by the topography, we will, for the
present, restrict our discussions of the effects of the bottom
topography to the lower layer while realizing that the layers are
still at least weakly coupled.

     We will now consider the lower layer as a homogeneous fluid
which conserves potential vorticity on a β-plane with the
interface as a free upper surface. Following Pedlosky (1979) we
find the conservation of potential vorticity can be expressed as

$$\frac{d}{dt}\left(\frac{f + \beta y - fH_B/D + \zeta + f\,\eta/D}{D}\right) = 0 \qquad (1)$$

where the layer thickness is

$$H = D - H_B + \eta \qquad\qquad (2)$$

and where D is the mean layer thickness, H_B is the height of the
bottom topography from some reference level, η is the displacement
of the free surface from its initial position (see Fig. 1), and ζ
is the relative vorticity.

For our study, we will consider a case where we have a poleward
longshore flow in the southern hemisphere (v<0, f<0) over a
mesoscale topographic feature where the magnitude of the
topographic beta $\left(\beta_T = \dfrac{\partial H_B}{\partial y}\right)$ is much greater than the planetary

beta. Then we can express the local time rate of change of
relative vorticity and free surface displacement as

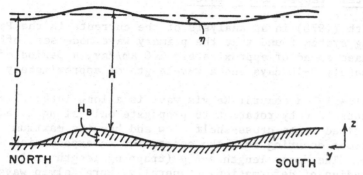

Figure 1. Schematic of lower layer showing the components which compose the thickness of the layer, i.e., $H = D - H_B + \eta$.

$$\frac{\partial}{\partial t}(\zeta + f\,\eta/D) = -v\frac{\partial}{\partial y}(\zeta + f\,\eta/D) + v\frac{\partial}{\partial y}\left(\frac{fH_B}{D}\right) \qquad (3)$$

where the first term on the right-hand side of equation (3) is the change due to wave propagation and the second term is the change due to the topographic beta effect. If we assume that the wavelength of the wave is much greater than the length scale of the topographic variation, we can, by appropriate scaling, separate the two effects. We will only consider the topographic effects for the present.

For a poleward flow (v<0), if we have $\beta_T = \frac{\partial H_B}{\partial y} > 0$ (i.e., downslope flow) we find that $v\frac{\partial}{\partial y}\left(\frac{fH_B}{D}\right) > 0$, which can be compensated for by either an increase in the relative vorticity, a negative free surface displacement or both. Now for v<0 and $\beta_T<0$ (upslope flow) we find $v\frac{\partial}{\partial y}\left(\frac{fH_B}{D}\right) < 0$, which implies a decrease in the relative vorticity and a positive displacement of the free surface. Similarly for an equatorward flow (v>0), we find that upslope flow ($\beta_T>0$) yields $v\frac{\partial}{\partial y}\left(\frac{fH_B}{D}\right)<0$ and downslope flow ($\beta_T<0$) yields $v\frac{\partial}{\partial y}\left(\frac{fH_B}{D}\right)>0$.

We can also see the relative magnitude of the local changes are directly proportional to the magnitude of the longshore velocity component.

b. Coastal Trapped Kelvin Waves

Smith (1978) in an analysis of the currents in the Peruvian upwelling system found that the primary wave mode seen off Peru has a phase speed of approximately 200 km/day, a period of approximately 8-10 days and a wavelength of approximately 2000 km.

Basically, a coastal Kelvin wave is a long internal gravity wave constrained by rotation to propagate only alongshore. It is characterized by no cross-shelf flow and has its maximum fluctuations occuring near the coast with an exponential decay offshore. The decay length scale (trapping length scale) is the Rossby radius of deformation. Generally, pure Kelvin waves occur only in the absence of bottom topography. Wang and Mooers (1976) examine the effects of bottom topography on the Kelvin waves and co-existence of Kelvin and topographic Rossby waves.

d. Vertical Mixing

With the exception of Yoshida (1967), Thompson (1974) and more recently Clancy et al. (1979), the role of turbulent vertical mixing and thermodynamics has been ignored in coastal upwelling models, based on the assumption that the time scale for vertical advection is much shorter than that for turbulent vertical diffusion. Thompson (1974), however, has demonstrated that the vertical advection and vertical mixing may be comparable on a coastal upwelling event time scale.

The major obstacle in including mixing into the upwelling problem is in the nature of mixing, in that mixing is basically turbulent. Within the present "state-of-the-art", turbulence is not well understood and the turbulence problem is unsolved. For that matter the methods to solve the problem are diversified and are the subject of much controversy within the research community. Niiler and Kraus (1977) examine the basic approaches to the problem of one-dimensional mixing in the upper ocean. They discuss the advantages and disadvantages of each of the most popular techniques and show that solutions to even the simplified one-dimensional problem are complex. Thompson (1978) presents a comprehensive review of the role of mixing in the upwelling systems and the problems involved in incorporating mixing into numerical models.

Most of the recent work relies on the theoretical models developed by Kraus and Turner (KT) (1967) or Pollard, Rhines and Thompson (PRT) (1973). The KT model considers the energy balance between the rate of turbulent kinetic energy produced by the

surface wind stress and the rate of increase in potential energy
due to mixing layer depth deepening, but does not consider the
wind effects on the mean velocities. The PRT model on the other
hand, considers the balance between the rate of work done by the
wind stress on the mean motions and the rate of increase of
kinetic and potential energies of the mean field. Niiler (1975)
has shown that the potential energy increase in the PRT model is
due to the shear production by the entrainment stress at the base
of the mixed layer. deSzoeke and Rhines (1976) demonstrate that
these two models are two asymptotic limits to the general balance
of the turbulent kinetic energy equations.

     Zilitinkevich, Chalikov and Resnyansky (1979) present sample
calculations (based on real data) for a number of the one
dimensional models, and compare the predicted depth and
temperatures of the homogeneous upper layer.

     Thompson (1974) introduced simple thermodynamics and
interfacial mixing into the upwelling problem in an x-t two-layer
time-dependent upwelling model. The interfacial mixing
parameterization for this model was prescribed as a function of
the Richardson number, Ri, based on the laboratory work of Kato
and Phillips (1969). Kato and Phillips, by examining a fluid with
a linear density gradient subjected to an applied surface stress,
found that a turbulent homogeneous layer bounded below by a sharp
interface quickly formed. Their observations of the rate
deepening of the homogeneous layer yield a relationship

$$Q = u_* f(Ri) \tag{4}$$

Where Q is the deepening rate, u_* is the friction velocity and is
related to the applied stress, τ, as

$$\tau = \rho_0 u_*^2 \tag{5}$$

and ρ_0 is the mean density. The Richardson number is given by

$$Ri = \frac{g \Delta \rho D}{\rho_0 u_*^2} \tag{6}$$

where $\Delta \rho$ is the density jump across the interface, D is the depth
of the homogeneous layer, and g is the gravitational acceleration.

     By dimensional analysis arguments, Turner (1973) found that

$$g \Delta \rho Q D \propto \rho_0 u_* \tag{7}$$

or

$$Q \propto u_* Ri^{-1}.$$ (8)

Which gives the same energy balance as the KT model.

Similar relationships between Q and Ri have been used by Pollard, Rhines and Thompson (1973), Denman (1973), and Niiler (1975).

The form used by Thompson (1974) for entrainment of lower layer fluid into the upper layer was

$$Q_1 = \phi u_* Ri^{-1} = \frac{\rho_0 \phi u_*^3}{g\Delta\rho h}.$$ (9)

where Q_1 is the entrainment rate of lower layer fluid to the upper layer and h_1 is the thickness of the upper layer. He used the value of the proportionality constant, $\phi = 2$, suggested by Denman (1973). Thompson also assumed that the entrainment of upper layer fluid into the lower layer by bottom generated turbulence could be expressed as

$$Q_2 = \frac{\rho_0 \phi u_*^{-3}}{g\Delta\rho h_2}$$ (10)

where h_2 is the thickness of the lower layer, u_*, given by

$$u_*^{-2} = \frac{|\vec{\tau}_B|}{\rho_0},$$ (11)

is the bottom friction velocity and $|\vec{\tau}_B|$ is the bottom stress. He found that $Q_2 \ll Q_1$ except when the interface is very near the bottom.

Based on the Melor-Yamada (1974) level 2 (MYL2) turbulence closure scheme, Martin and Thompson (1983) have developed a bulk model of the upper mixed layer which is compatible with a layered formulation for a numerical ocean model.

In this model the mixed layer depth was parameterized for each of two distinct regimes:

1) stratification limited - limited by the density stratification underlying the mixed layer.

2) heat flux limited – limited by the rate of surface heating due to a positive net surface heat flux.

In the stratification regime, as the mixed layer deepens through a stable density stratification, the stratification at the base of the layer will increase, while the velocity shear available to generate turbulent kinetic energy will decrease. Thus for a given stratification and wind stress, the mixed layer depth will stabilize when the turbulence at the base of the mixed layer becomes too weak to erode the stratified region below. To predict the layer depth for this regime Martin and Thompson used the PRT bulk mixed layer model to approximate the MYL2 model. For the PRT model, it is assumed that during mixed layer deepening into a stably stratified region, the bulk Richardson number, Ri_B, is equal to one:

$$Ri_B = \frac{\kappa g h \Delta T}{(\Delta u)^2 + (\Delta v)^2} = 1 \ , \tag{12}$$

where h is mixed layer depth, ΔT, Δu, and Δv are temperature and velocity differentials across the base of the mixed layer and κ is the coefficient of thermal expansion. Then letting $Ri_B = 1$ and rearranging, h becomes:

$$h = \frac{(\Delta u)^2 + (\Delta v)^2}{\kappa g \Delta T} \ . \tag{13}$$

In the heat flux limited regime, as the layer deepens, more work has to be done by turbulence to mix the surface heat flux throughout the layer. Thus for a given surface flux and wind stress, the mixed layer depth becomes established at a depth where a balance between production of kinetic and potential energy and dissipation is reached.

The layer depth predicted by the MYL2 model can be expressed as a function of dimensionless parameter Φ as

$$h = \frac{u_*}{f} h'(\Phi) \tag{14}$$

where Φ is the ratio of the Monin-Obukov and Ekman layer length scales and is defined as

$$\Phi = gH/fu_*^2 \ , \tag{15}$$

where H is the heat flux.

Briefly, when the mixed layer is limited by 1) underlying stratification, the depth is determined by the PRT bulk model with the bulk Richardson number equal to one, 2) surface heat flux, the depth is determined as a function of the dimensionless parameter Φ. When both wind deepening and surface heating are present, the mixed layer depth is taken to be the shallower of the two depths.

The next step is to couple the one-dimensional mixing models to a three-dimensional ocean circulation model which includes the effects of advection. One approach to this problem would be to compare the upper layer thickness (ULT) predicted by a 1-D mixed layer model (MLM) to that predicted by an advective 3-D model where the entrainment rate is predicted from a previous time step. If the ULT predicted by the 1-D MLM is greater than that predicted by the advection model then a new entrainment rate is calculated. If the ULT from the 1-D MLM is less than that for the advection model then we assume that the wind stress input is just mechanically stirring the upper layer and is not entraining mass across the interface (i.e., no interfacial mixing).

MODEL

For this study, we consider a two-layer vertically integrated, stably stratified, rotating, incompressible, hydrostatic, primitive equation model on a β-plane with a rotated coordinate system which is forced by local winds and externally excited coastal trapped waves. The latter forcing mechanism is new to coastal upwelling models. A thermodynamic version of the model, which assumes the fluid to be Boussinesq, is used to introduce the effects of thermodynamics and interfacial mixing.

Before discussing the specifics of these models, we will first describe the model geometry, basin orientation and location.

a. Model Geometry

As can be seen in Fig. 2, the Peru coastline in the JOINT II region is oriented northwest-southeast. In order to maximize the basin size and resolution within the eastern boundary region (upwelling zone), the model basin, (Fig. 3), is rotated 45° such that the righthand boundary approximates the coastline. The basin dimensions are 300 km in the y-direction (with y = 0 approximately along the C-line) and 1500 km in the x-direction.

A schematic of the model geometry, Fig. 4, shows the relative location of the vertically averaged variables u_i, v_i, ρ_i and the layer thicknesses h_i. The height of the bottom topography above the base of the model is represented by $D(x,y)$.

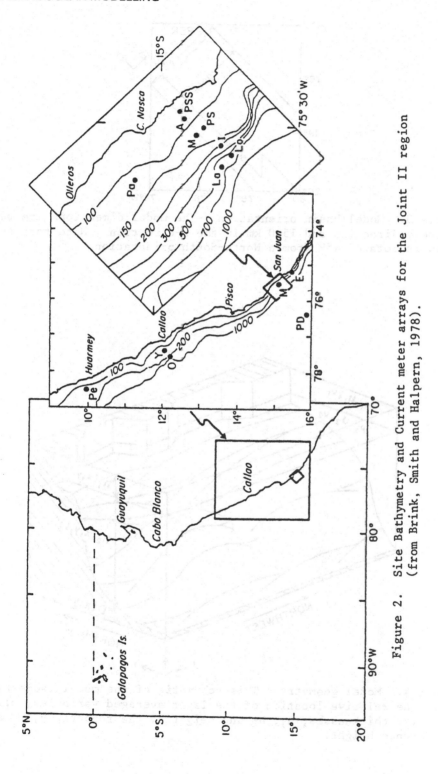

Figure 2. Site Bathymetry and Current meter arrays for the Joint II region
(from Brink, Smith and Halpern, 1978).

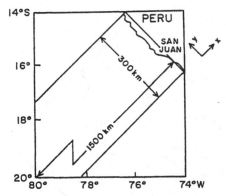

Figure 3. Model basin orientation. The model dimensions are 300 km
in the y-direction and 1500 km in the x-direction. Note that the
basin is rotated 45° from a North-South orientation.

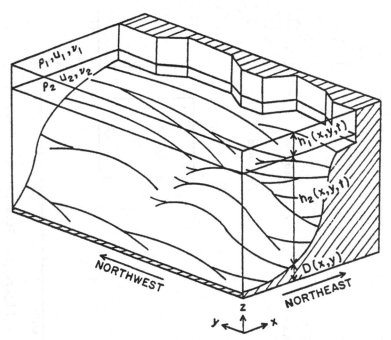

Figure 4. Model geometry. This schematic of the model geometry
shows the relative location of the layer averaged variables, u_i, v_i,
ρ_i, layer thicknesses, h_1, h_2 and height of the topography, D, above
a reference height.

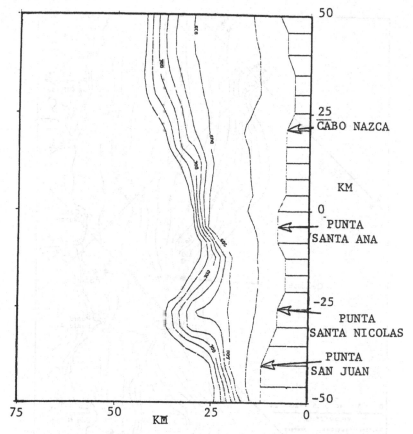

Figure 5. Digitized bottom topography. The bottom topography is
digitized from the Preller and O'Brien, 1977, Peruvian Near Shore
Bottom Topography and Coastline. The contour interval is 100 m.

     In order to resolve the solution in the upwelling zone
(generally on the order of 20 km in width), a fine resolution grid
(on the order of 2 km) is needed. Since a grid of this resolution
applied to a basin of this size is not economically feasible, and
the solution in the western part is of no interest for this
problem, the model uses a variable resolution grid which varies
discretely in the x-direction and is analytically stretched in the
y-direction. The grid spacing in the x-direction, Δx, varies from
2 km in the coastal region to 100 km in the western part of the
basin, and the grid spacing in the y-direction, Δy, varies from
approximately 4 km near y = 0 to approximately 13 km near the
northwestern and southeastern boundaries.

     The bottom topography (Fig. 5) used in this model, is a
rotated, digitized version of the Preller-O'Brien (1977) Peruvian

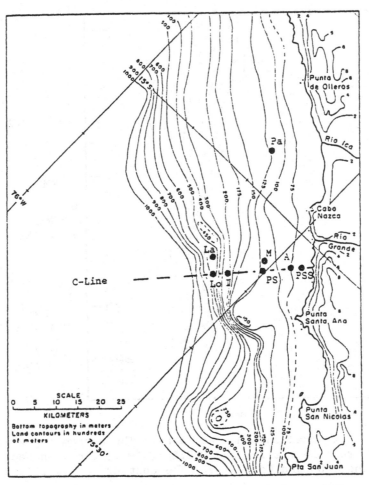

Figure 6. Peruvian Near Shore Bottom Topography and Coastline –
Preller and O'Brien, 1977. The location of the C-Line (y = 0 for
the model) is indicated by the dashed line. The location of
current meter arrays is plotted.

near shore bottom topography and coastline map, (Fig. 6). Some of
the salient features of this bottom topography are a broad shelf
(approximately 20 km) in the north, a narrow shelf (approximately
6 km) in the south, and a shelf slope which is variable in the
longshore as well as the offshore direction. Of particular note
in the longshore variability is the steep shelf slopes south of
Punta Santa Ana and Punta San Juan, and the presence of mesoscale

seamounts, which appear to play a significant role in the anchoring of the centers of enhanced upwelling, (Preller and O'Brien, 1980; O'Brien et al., 1983).

b. Governing Equations

To form the governing equations for the thermodynamic model, we need to formulate the mass continuity and momentum equations to consider horizontal variations in the layer averaged densities, ρ_i, and add the equations of state for each layer as well as the heat or energy equations. Furthermore, we also need to consider the parameterization of the interfacial mixing.

To start, we will assume that the fluid is Boussinesq, which allows us to neglect density variations in the momentum equations except when coupled with gravitational acceleration and to assume fluctuations in density are the result of thermal effects (Spiegel and Veronis, 1960).

Further, we will assume that in the equation of state, density is a linear function of temperature only. Brink et al. (1979) has shown, that for the region which we are modelling off Peru, the density variations are significantly correlated with the temperature variations and shows little correlation to the salinity variations. Therefore, for this region, density can be considered a function of temperature only.

With these additional assumptions, we now express the governing equations as:

$$\frac{\partial \vec{V}_1}{\partial t} + \vec{V}_1 \cdot \nabla \vec{V}_1 + \hat{k} x f \vec{V}_1 \; = \; -g\nabla(h_1+h_2+D) + \frac{gh_1}{2\rho_o}\nabla\rho_1 + \frac{\vec{\tau}_s - \vec{\tau}_I}{\rho_o h_1} + \vec{S}_1 + \vec{F}_1$$

$$+ \frac{(\nabla \cdot (\rho_1 h_1 A_H \nabla))\vec{V}_1}{\rho_o h_1} \tag{16}$$

$$\frac{\partial h_1}{\partial t} + \nabla \cdot (h_1 \vec{V}_1) = Q_1 - Q_2 + \frac{\gamma H}{\rho_1^2 C_p} \tag{17}$$

$$\frac{\partial \rho_1}{\partial t} + \vec{V}_1 \cdot \nabla \rho_1 \; = \; \frac{\Delta\rho\, Q_1}{h_1} + \frac{1}{h}\nabla \cdot (K_H h_1 \nabla \rho_1) - \frac{\gamma H}{\rho_1 C_p h_1} \tag{18}$$

$$\rho_1 = \rho_o - \gamma T_1 \tag{19}$$

$$\frac{\partial \vec{V}_2}{\partial t} + \vec{V}_2 \cdot \nabla \vec{V}_2 + \hat{k} x f \vec{V}_2 = -g\nabla(h_1 + h_2 + D) + g'\nabla h_1 - \frac{gh_1}{\rho_0}\nabla\rho_1 -$$

$$\frac{gh_2}{2\rho_0}\nabla\rho_2 + \vec{S}_2 + \vec{F}_2 + \frac{\vec{\tau}_I - \vec{\tau}_B}{\rho_0 h_2} + \frac{(\nabla \cdot (\rho_2 h_2 A_H \nabla))\vec{V}_2}{\rho_0 h_2} \qquad (20)$$

$$\frac{\partial h_2}{\partial t} + \nabla \cdot (h_2 \vec{V}_2) = Q_2 - Q_1 \qquad (21)$$

$$\frac{\partial \rho_2}{\partial t} + \vec{V}_2 \cdot \nabla\rho_2 = \frac{\Delta\rho\, Q_2}{h_2} + \frac{1}{h_2}\nabla \cdot (K_H h_2 \nabla\rho_2) \qquad (21)$$

$$\rho_2 = \rho_0 - \gamma T_2 \qquad (22)$$

where ϕ is a constant of proportionality and $\Delta\rho = \rho_* - \rho_1$ with ρ_* determined from a density function to be discussed in the mixing parameterization section.

The entrainment terms are given by:

$$\vec{S}_1 = \rho_* \frac{Q_1 (\vec{V}_2 - \vec{V}_1)}{\rho_1 h_1} \qquad (24)$$

$$\vec{S}_2 = -\rho_1 \frac{Q_2 (\vec{V}_2 - \vec{V}_1)}{\rho_2 h_2} \qquad (25)$$

$$Q_1 = \begin{cases} 0 & : \ h_1 > h_{MLD} \\[2em] \dfrac{\rho_0 \phi u_*^3}{g\Delta\rho h_1} & : \ h_{MIN} < h_1 < h_{MLD} \\[2em] \bar{u}_*\left(\dfrac{\partial D}{\partial x} + \dfrac{\partial D}{\partial y}\right) - h_2\left(\dfrac{\partial u_2}{\partial x} + \dfrac{\partial v_2}{\partial y}\right) & : \ h_1 = h_{MIN} \end{cases} \qquad (26)$$

$$Q_2 = \begin{cases} \dfrac{\rho_0 \phi \bar{u}_*^3}{g\Delta\rho h_2} & : \ h_2 < h_{BLD} \\[2em] 0 & : \ h_2 > h_{BLD} \end{cases} \qquad (27)$$

where

$$u_*^2 = \frac{|\vec{\tau}_s|}{\rho_0} \tag{28}$$

$$\bar{u}_*^2 = \frac{|\vec{\tau}_B|}{\rho_0} \tag{29}$$

and h_{MIN} is the minimum layer thickness for the model, h_{MLD} is the stabilized mixed layer depth, and h_{BLD} is the depth of the bottom mixed layer.

Equations (16) and (20) are the momentum equations where the additional terms involve the horizontal gradients of density and the entrainment ($\vec{S}_i$) of momentum from one layer to the other by some mixing process (Q_i). The density of the lower layer fluid which is entrained into the upper layer is represented by ρ_*. This is the density just below the interface and is determined from a density function which will be discussed in the mixing parameterization section. Equations (17), (18), (21) and (22) are derived from a combination of the mass continuity and energy equations for each layer. Equations (17) and (21) have the same form as the continuity equations in the hydrodynamic formulization except for a non-zero right-hand side. For ease of comparison between the two systems, we will refer to these as the continuity equations for this system and equations (18) and (22) as the thermodynamics equations. The term, H, represents a heating function, which we will discuss later.

Equations (19) and (23) are the equations of state for this system, where ρ_0 is a reference density and γ is the product of the reference density and the coefficient of thermal expansion for sea water.

Q_i represents an entrainment process due to mixing which we will discuss in detail in a later section.

c. Boundary Conditions

The boundary conditions applied at the coast are the no-slip and kinematic boundary conditions.

Hurlburt and Thompson (1973) have demonstrated that the solution in the upwelling zone is independent of that for the far-field boundary region providing the zonal extent of the model basin is at least 1000 km. Thus, the no-slip and kinematic boundary conditions can also be applied to the southwestern boundary without significantly affecting the solution within the upwelling zone since the zonal extent of this model is 1500 km.

The "quasi-symmetric" open boundary condition (Hurlburt, 1974) is used on the northwestern and southeastern boundaries. Basically this boundary condition is a modification of the Hurlburt and Thompson (1973) x–z model. Hurlburt and Thompson in their x–z model assumed all longshore variations to be negligible except for β and the longshore pressure gradient which can be derived from the Sverdrup balance,

$$\beta(v_1 h_1 + v_2 h_2) = \frac{1}{\rho} \operatorname{curl}_z \vec{\tau} \tag{30}$$

and shown to be given by, for the upper layer

$$g \frac{\partial}{\partial y}(h_1 + h_2 + D)\Big|_x = \int_{-L_x}^{x} \beta v_1 \, dx + g\frac{\partial}{\partial y}(h_1 + h_2 + D)\Big|_{-L_x} \tag{31}$$

and, for the lower layer,

$$g \frac{\partial}{\partial y}(h_1 + h_2 + D)\Big|_x - g'\frac{\partial h'}{\partial y}\Big|_x = \int_{L_x}^{x} \beta v_2 \, dx + g\frac{\partial}{\partial y}(h_1 + h_2 + D)\Big|_{-L_x} \tag{32}$$

$$- g'\frac{\partial h'}{\partial y}\Big|_{-L_x}$$

Hurlburt (1974) demonstrated that the application of the Hurlburt and Thompson x–z model as an open boundary condition allows the development of a Sverdrup interior flow even when the longshore portion of the basin is of mesoscale extent, provided the longshore flow is nearly geostrophic and that

$$\left| \int_{-L_x}^{x} f \frac{\partial v_i}{\partial y} dx \right| << \left| \int_{-L_x}^{x} \beta v_i dx \right| \; ; \quad i = 1, 2 \tag{33}$$

near the open boundary.

The modifications to the Hurlburt and Thompson model for the boundary conditions are that the $\dfrac{\partial h_1 v_i}{\partial y}$ terms are retained in the continuity equations and the longshore pressure gradients are given by:

$$g \frac{\partial}{\partial y}(h_1 + h_2 + D) = \int_{-L_x}^{x} \beta (v_1 - v_A) \, dx + g \frac{\partial}{\partial y}(h_1 + h_2 + D)\Big|_{-L_x} \tag{34}$$

$$g \frac{\partial}{\partial y}(h_1 + h_2 + D) - g' \frac{\partial h_1}{\partial y} = \int_{-L_x}^{x} \beta \left(v_2 + \frac{h_1}{h_2} v_A\right) dx + g \frac{\partial}{\partial y}(h_1 + h_2 + D)\Big|_{-L_x}$$

$$- g' \frac{\partial h_1}{\partial y}\Big|_{-L_x} \quad (35)$$

where

$$v_A = v_1 - v_1 h_1 + \left(v_2 + \frac{g'}{f}\frac{\partial h_1}{\partial y}\right)h_2/(h_1 + h_2) \quad (36)$$

The term v_A represents the ageostrophic part of the depth averaged flow. Hurlburt (1974) tested various techniques for removing the ageostrophic component and found this method to effectively remove the inertial oscillations and Ekman drift while suppressing the excitation of Kelvin waves.

Preller and O'Brien (1980) discuss in detail the development of the formulation of this method of removing the ageostrophic component by means of a modal analysis of the two-layer system.

The bottom boundary condition is kinematic and the upper surface is free.

d. Initial Conditions

The model is initially at rest and impulsively started at time zero by application of the local and remote forcing mechanisms to be discussed in a later section. The inertial oscillations, Rossby waves and gravity waves excited by the impulsive application of the forcing mechanisms quickly propagate out of the upwelling region. Since the basin width is chosen to be greater than 1000 km, they have no significant effect on the upwelling solution after approximately two and one-half days.

The upper layer, h_1, is initially 50 m thick and the maximum basin depth, H_{TOT}, is 800 m. The lower layer thickness, h_2, is

given by

$$h_2 = H_{TOT} - h_1 - D(x,y) \quad (37)$$

where $D(x,y)$ is the height of the bottom topography above the base of the model.

The upper layer density, in the thermodynamic model, is given by $\rho_1 = \rho_I(x)$ where $\rho_I(x)$ is constant over most of the basin and

has a slight gradient in the coastal region. The lower layer
density is estimated using

$$\rho_2 = \rho_I(x) + \alpha(\rho_B(x) - \rho_I(x)).$$ (38)

Unlike his atmospheric counterpart, the ocean modeller does
not have a dense synoptic data network available from which he can
derive an objectively adjusted initial state, and therefore must
use relatively simple, straightforward estimates for the initial
fields.

e. Forcing Mechanisms

Local Time dependent surface wind stresses, derived from
filtered meteorological buoy wind observations, are used as a
local forcing mechanism. The wind function used in this model is
designed to simulate the low frequency large scale and diurnal
land/sea-breeze wind fluctuations.

To simulate the time dependent amplitude of the large scale
wind forcing, the wind observations from Pacific Marine
Environmental Laboratory's PSS Mooring (15°03'S, 76°27'W, see Fig.
(2) and (6), are lowpass filtered such that periods greater than
36 hrs. are retained. Surface wind stresses calculated from the
filtered observations are then applied such that the curl of the
wind stress over the first 300 km is equal to the mean interior
longshore flow (i.e., Sverdrup balance) estimated from current
meter observations.

The offshore profile for this is given by

$$\frac{\partial \tau_y(x,t)}{\partial x} = \beta V_m(h_1(x,0) + h_2(x,0))$$ (39)

and

$$\tau_x(x,t) = F(t)$$ (40)

for x from 0 to 300 km offshore. The wind stress is then allowed
to slowly decay to zero over the remainder of the basin.

In Fig. (7), we have a sample of the cross-shelf profile of
the mean wind stress components where $|\tau_y| = 1$ dyne/cm^2 and $|\tau_x| =$
.5 dyne/cm^2 at the coast. In this figure, we see that the
offshore curl required to drive the mean longshore current (V_m) on
the shelf is relatively small (on the order 10^{-9} dyne/cm^2).

The diurnal land/sea-breeze regime is simulated by band-pass
filtering the PSS wind observations to allow periods between 12

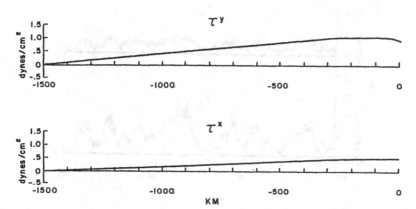

Figure 7. Wind Stress Offshore Profiles. Note that there is a
slight curl in the τ^y offshore profile within the first 300 km
from the coast.

and 36 hours to be retained. The wind stress amplitudes
calculated from these are then superimposed on the large scale
wind stresses such that they decay exponentially from a maximum at
the coast to the large scale at approximately 20 km off the coast.

The offshore profile for the total wind stress is given by

$$\vec{\tau}(x,t) = \vec{\tau}_L(x,t) + \vec{\tau}_D(t)e^{\delta x} \quad \text{for } x < 0 \tag{41}$$

where the e-folding distance is on the order of 20 km. The large
scale wind stress profile is given by $\vec{\tau}_L(x,t)$ and $\vec{\tau}_D(t)$ is the
diurnal wind stress amplitude.

Figure (8) presents time series plots of the large scale wind
stress amplitudes, the diurnal wind stress amplitudes, and
combined (large scale plus diurnal) wind stress amplitudes for the
longshore and cross-shelf components used to drive the model. Day
zero in the model integration corresponds to 0000Z on 22 March,
1977.

In this version of the model, no attempt is made to simulate
the longshore wind structure due to a lack of a detailed knowledge
of the horizontal longshore variability of the wind field and the
time evolution of the horizontal structure. The best available
source of information on the near shore horizontal structure are
the 500 ft. aircraft wind charts (Watson, 1978; and Moody, 1979).

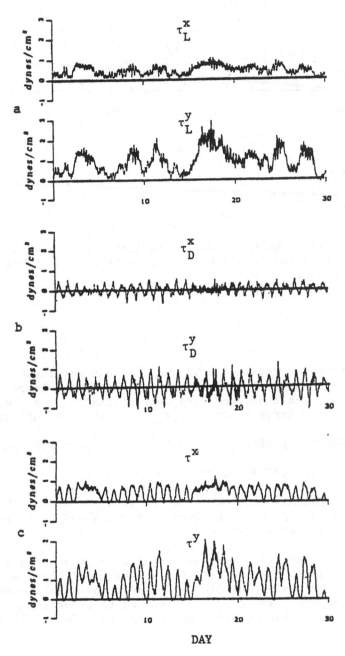

Figure 8. Wind Stress. This figure shows the time series of the wind stress amplitudes used to drive the model. a) The large scale forcing, b) the diurnal component and c) the total wind stress at the coast.

These charts, which are from flights centered on local noon for
selected days, give a good indication of the horizontal structure
during the sea-breeze portion of the diurnal land/sea-breeze
cycle. However, to parameterize adequately the effects of the
longshore variability, we need to know what the horizontal
structure is during the land-breeze portion of the diurnal cycle
and have some indication as to the time evolution of this
structure. These latter data are not available.

Remote A parameterization scheme based on a coastal trapped
wave dynamics is employed to simulate the longshore current
variability.

From the simplest set of equations which contain coastal
trapped Kelvin waves, (for any given mode);

$$fv = g \frac{\partial h}{\partial x} \qquad\qquad (42)$$

$$\frac{\partial v}{\partial t} = -g \frac{\partial h}{\partial y} \qquad\qquad (43)$$

$$\frac{\partial h}{\partial t} = -H_o \frac{\partial v}{\partial y} \qquad\qquad (44)$$

it can be seen that the local time rate of change of the longshore
velocity component is proportional to the local longshore pressure
gradient. Therefore, by using the appropriate equivalent depth,
phase speeds and wave lengths estimated from observations, it is
possible to parameterize the longshore pressure gradients induced
by the Kelvin waves using a time series of the longshore currents.
For this purpose, the time series of low-pass filtered, longshore
currents. For this purpose, the time series of low-pass filtered,
longshore velocity components from Oregon State University's
current meter mooring, Yucca-Too (12°05'X, 77°20'W, see Fig. (2),
located on the continental shelf equatorward (upstream) of the
model basin, are decomposed into their first and second empirical
orthogonal vertical modes.

Smith (1978) has shown that the primary Kelvin wave mode has
an equivalent depth of around 400 m, a phase speed of around 200
km/d and a trapping scale of about 70 km. This mode will appear
as depth independent motions over most of the shelf.

The amplitudes of the first empirical orthogonal mode are
time differenced, converted to local longshore pressure gradients
and then applied, with the appropriate phase shift (based on 200
km/d phase speed and assuming long wave lengths) as a depth

independent body force within the trapping length scale along the
northeastern boundary. Similarly, the second mode amplitudes
(which have an equivalent depth corresponding to the upper layer
thickness) are time differenced, converted to longshore pressure
gradients, and applied as a layer dependent force within the
appropriate trapping length scale (approximately 20-25 km).

f. Mixing Parameterization

The purpose for adding thermodynamics and interfacial mixing
to the model is twofold. The first reason is to include the
potentially important physics which have been neglected in most
upwelling models. The second is to provide a physical mechanism
to keep the interface from surfacing and thereby allow the
integrations to proceed further than the five to ten-day
limitation in the hydrodynamic models.

This second objective cannot be achieved by wind mixing alone
in a model which includes forcing mechanisms other than a constant
wind stress. Under the influence of variable winds and
propagating internal waves, there are times when the pycnocline
(which is represented in the model by the interface) will
intersect the surface. This condition cannot be modelled directly
with a layered model where the interface represents a Lagrangian
(material) surface. To model this situation, we must at some
point in the vertical replace the Lagrangian vertical coordinate
with an Eulerian or fixed level vertical coordinate.

We will for this model consider three regions in the
vertical. In the first region, where $h_1 > h_{MLD}$ and h_{MLD} is the
upper layer thickness determined from a mixed layer model, we
will assume that the wind stress is strong enough to only
mechanically stir the upper layer and that no interfacial mixing
is occurring.

For the second region, where $h_{MIN} < h_1 < h_{MLD}$, we will assume
that the wind stress is sufficient to erode the pycnocline and
that the entrainment rate is a function of the Richardson number.
Finally in the last region, where $h_1 < h_{MIN}$, we will set h_1 equal
to h_{MIN} and calculate the mass flux through the interface from the
vertically integrated three dimensional mass continuity equation.

Following the lead of Martin and Thompson (1983), we will use
the Pollard, Rhines, and Thompson (1973) (PRT) model to calculate
the h_{MLD} (which we will refer to as the stabilized mixed layer
depth) for a given wind stress and stratification.

In the PRT model, it is assumed that the bulk Richardson number,

$$Ri_B = \frac{g\Delta\rho\, h_{MLD}}{((\Delta u)^2 + (\Delta v)^2)\rho_0} \qquad (45)$$

(where $\Delta\rho$, Δu and Δv are the density and velocity differences, across the base of the mixed layer) is equal to one for complete wind mixing. Then by setting $Ri_B = 1$, we get for the stabilized mixed layer depth,

$$h_{MLD} = \frac{((\Delta u)^2 + (\Delta v)^2)\rho_0}{g\Delta\rho} \qquad (46)$$

To calculate the stratification, Martin and Thompson stored the perturbation temperatures at discrete grid points below the surface layer. In order to save storage space, instead of retaining the density at a series of grid points below the upper layer, we use an analytic density function which was derived by fitting CTD profiles, from the JOINT II area taken during MAM '77, to a hyperbolic tangent function. Fig. 9 is an example of the data fit using profiles picked at random. The profiles displayed were not used in the data fitting routine. The density function has the form

$$\rho(z) = \rho_B + C(\alpha z + \tanh\left(\frac{z - z_0}{s}\right) + b) \qquad (47)$$

$$b = \tanh\left(\frac{z_0}{s}\right) \qquad (48)$$

$$C = (\rho_T - \rho_B)/(\alpha + \tanh\left(\frac{1 - z_0}{s}\right) + b) \qquad (49)$$

where ρ_T and ρ_B are top and bottom densities respectively, z is the non-dimensional vertical coordinate, which is scaled by the total fluid depth, and z_0 is the scaled height of the pycnocline above the bottom. The parameters α and s determine the shape of the profile and are determined empirically from the CTD profiles.

After the mixed layer depth is determined in this manner, it is then compared with the thickness of the upper layer, h_1. If $h_1 > h_{MLD}$ no mixing occurs. If $h_1 < h_{MLD}$ and $h_1 > h_{MIN}$, where h_{MIN} is the minimum allowable upper layer thickness, then the entrainment rate is specified as

$$Q_1 = \phi\, u_* \, Ri^{-1} \qquad (50)$$

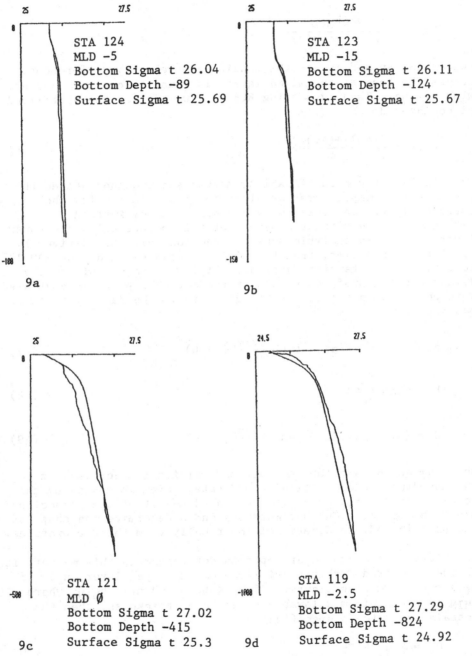

Figure 9. Examples of CTD data fit to the analytic density function.

where

$$Ri = \frac{g\Delta\rho h_1}{u_*^2} \qquad (51)$$

or

$$Q_1 = \frac{\phi \, u_*^3}{g\Delta\rho h_1} \qquad (52)$$

where ϕ is a constant of proportionality. Finally, if $h_1 < h_{MIN}$, then we let

$$Q_1 = \bar{u}_*\left(\frac{\partial D}{\partial x} + \frac{\partial D}{\partial y}\right) - h_2 \left(\frac{\partial u_2}{\partial x} + \frac{\partial v_2}{\partial y}\right) \qquad (53)$$

where $\bar{u}_*$ is the bottom friction velocity, D is the height of the bottom topography, h_2 is the thickness of the lower layer and u_2 and v_2 are the vertically averaged lower layer velocity components.

g. Heating

Thompson (1974) examined both heated, and non-heated versions of his x-z mixing model. He found that the primary effects of heating was to strengthen the zonal sea surface temperature (SST) contrast and enhance frontal intensification. His formulation of the heating function included diurnal variations, but he found that the effects of these diurnal variations were minimal. To keep an already complex model as simple as possible, we will use a simple approach to the heating, i.e., we will use a constant heating rate. In particular, we will use an estimate based on the observations from the JOINT II region.

h. Cross-shelf Physics

The primary premise for this model is that Ekman dynamics are important to the vertical structure of the flow field. Invoking this premise, and using a method developed by Thompson (1974), continuous analytical solutions can be obtained for the departures from the vertically integrated layer velocity fields, and thereby introduce boundary layers near the internal and external fluid interfaces. Defining $\vec{V}'_j$ as the departure from the vertically integrated velocity, $\vec{V}_j$, within each layer we can then determine the total velocity field,

$$\vec{V}_{Tj} = \vec{V}'_j + \vec{V}_j \qquad (54)$$

The departures can be calculated from the Ekman equations

$$k \times f\vec{V}_j = A_v \frac{\partial^2 \vec{V}_j}{\partial z^2} \; ; \quad j = 1,2 \tag{55}$$

Then by letting $z = 0$ be the sea surface, $z = b$ be the layer interface, $z = c$ be the bottom, and defining the complex velocities as

$$w_j = u'_j + iv'_j \; ; \quad i = \sqrt{-1} \tag{56}$$

the problem becomes

$$\frac{\partial^3 w_1}{\partial z^3} - s^2 \frac{\partial w_1}{\partial z} = G_1 \qquad b < z < 0 \tag{57}$$

$$\frac{\partial^3 w_2}{\partial z^3} - s^2 \frac{\partial w_2}{\partial z} = G_2 \qquad c < z < b \tag{58}$$

where

$$s = (1+i)(f/2 \, A_v)^{1/2} \tag{59}$$

$$G_j = -\frac{g}{\rho_0 A_v} \left[\frac{\partial p_j}{\partial x} + i \frac{\partial p_j}{\partial y} \right] ; \qquad j = 1,2 \tag{60}$$

with the integral constraint on the system that the vertically integrated perturbation velocities in each layer are zero, i.e.

$$\int_0^b w_1 \, dz = \int_0^c w_2 \, dz = 0 \, . \tag{61}$$

The boundary condition required to close the problem are (1) at the surface, $z = 0$, the perturbation velocity shear stress is set equal to the wind stress,

$$\frac{\partial w_1}{\partial z} = \frac{\tau_{sy}}{A_v} + i \frac{\tau_{sy}}{A_v} ; \tag{62}$$

(2) at the interface $z = b$, the total velocity is continuous and the perturbation shears are set equal,

$$w_1 = w_2 - [(u_1 - u_2) + i(v_1 - v_2)] \tag{63}$$

and

$$\frac{\partial w_1}{\partial z} = \frac{\partial w_2}{\partial z} ; \tag{64}$$

and (3) at the bottom, z = c, the total velocity is set equal to zero (no slip),

$$w_2 = -(u_2 + iv_2) \tag{65}$$

The total velocity for any point in the vertical therefore can be determined by adding the perturbation velocity to the vertically integrated layer velocity.

MODEL RESULTS

This thermodynamic model is started at 0000Z 22 March 1977 Each case covers a thirty day period. This period was selected since it allows the presentation of the model reaction to at least one cycle after the initiation of mixing. Also during this period we see a series of pulses in the upwelling favorable wind stress with one particular period (day 15 through 20) of strong sustained winds.

The longshore component of the flow is forced primarily by two factors, the wind stress curl and the externally forced Kelvin waves. The mean poleward flow is the result of the cross-shelf curl of the wind stress as specified in Eq. (40). The variability in the longshore component is induced by the Kelvin wave parameterization and the wave interaction with the bottom topography.

The cross-shelf component is primarily the result of the interaction of the Kelvin wave and the bottom topography, the required onshore flow to compensate for the offshore mass transport in the upper Ekman layer due to the upwelling favorable winds, and secondarily by the horizontal density gradients.

In the hydrodynamic model (Heburn and O'Brien, 1982) we could directly infer vertical motion from the movement of the interface. In the thermodynamic model, this is possible only when the upper layer thickness is greater than the stabilized mixed layer depth (i.e., when the interface represents a material (Lagrangian surface). When the interface ascends into the active mixing region, then the vertical motion is given by the mixing parameterization and cannot be directly inferred from the motions of the interface. Therefore, we must use a field other than the upper layer thickness field to identify the areas of enhanced upwelling. We will use the upper layer temperature field, which for a well mixed layer is the same as the sea surface temperature (SST), for this purpose. The SST is obtained from the upper layer density field by using the equation of state for the upper layer;

$$SST = (\rho_0 - \rho_1)/\gamma = T_1 \tag{68}$$

During the first 10 to 11 days, the thermodynamic model
behaves the same as the hydrodynamic model. Therefore, the upper
layer thickness can be used to identify the upwelling centers.
After day 11, the interface ascends into the active mixing region
and the sea surface temperature field begins to show more
horizontal structure. By day 17, (Fig. 10), we see the formation
of a "plume" like structure north of Punta San Ana, in the same
location as had been shown in the pycnocline anomaly field. Warm
pockets can be also seen in the bays between Punta San Nicolas and
Punta Santa Ana, Punta Santa Ana and Cabo Nazca, and Cabo Nazca
and Punta Olleros. These warm pockets indicate that there is
relatively less upwelling in the areas. The locations of these
relative minima in upwelling intensity are directly related to the
bottom topography.

One of the initial assumptions for this model was that
heating was not important. This was based on the comparison of
the results from the heated vs. unheated versions of Thompson's
(1974) model. Thompson found the primary result of heating in his
model was to strengthen the upwelling frontal zone with minimal
effect on the flow field. Results from an unheated version of

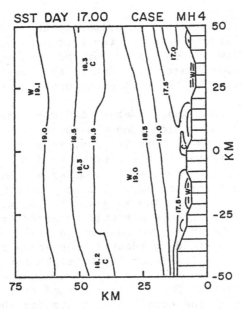

Figure 10. Sea Surface Temperature (SST) for model day 17.
Contour interval is .5°C.

this model show broad scale upwelling with little horizontal structure. In Fig. 11 we have x-t cross-sections of the SST field at y = 0, y = -20 km, and y = ±30 km. In this figure, we can see after the commencement of mixing that there is a continuous cooling of the upper layer and, generally, offshore advection of cold water. Furthermore, the horizontal SST plots for the unheated case display continuous offshore migration of the isotherms with only slight longshore variations.

If we add a positive heat flux (an observed feature in this region during this period) to the model, we then find the expected horizontal structure. That is, we recover that horizontal structure suggested by the pycnocline height field of a given situation and by the observations. Stevenson, Stuart, and Heburn (1980) report that this heat flux is fairly constant during March and April 1977.

Now in Fig. 12, we have cross-sections for y = 0, ±10, ±20, ±30, and 40 km. We can see that when a reasonable heating rate (on the order of .08°C day^{-1} m^{-1}) is added to the system we get a quasi-balance between the large scale wind driven upwelling and the heating. Then the fluctuations in the upwelling patterns due to internal wave action, wave-bottom topography interaction, and wind fluctuations becomes apparent.

Although the time periods are different, we see by comparing the y = 0 plot from Fig. 12 to Fig. 13 (from Hoover, 1980) a similar horizontal structure. Both show quasi-periodic fluctuations along the coast and at approximately 30 kms offshore. This quasi-periodic feature at 30 km offshore appears in all the cross-sections.

If we compare the time for the occurrence of the minima in the y = 0 plot with a time series for the long shore undercurrent as measured by the LAGARTA 92M current meter (Fig. 14) we see that the minima generally occur during the retardation of the poleward flow. Now, if we recall that with the passage of a crest of an internal wave we see an equatorward acceleration in the undercurrent and an uplifting of the pycnocline, then we might reasonably assume that fluctuations in the SST in this region are the result of the passage of internal waves. A comparison with the time series (Fig. 8) for the large scale wind stress also shows a relationship between the minima and the peaks in the wind stress thus indicating an interaction between internal waves and wind mixing.

Another feature which we can glean from Fig. 12 is the occurrence of a narrow upwelling zone (approximately 5-10 km) in the southern regions of the model where the shelf is narrow and a broadening of the upwelling zone (approximately 15-20 km) as we

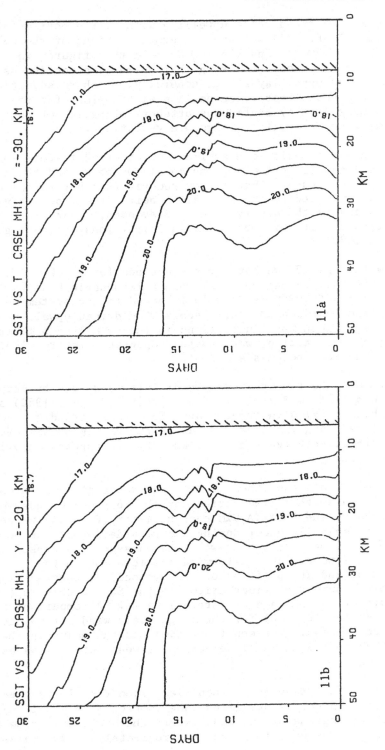

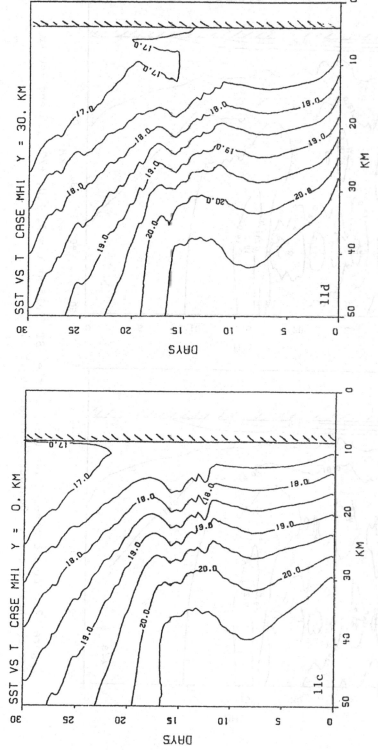

Figure 11. Time cross-section (x-t) of the SST field at y = 0, 20, ±30 km for the unheated case. Contour interval is .5°C.

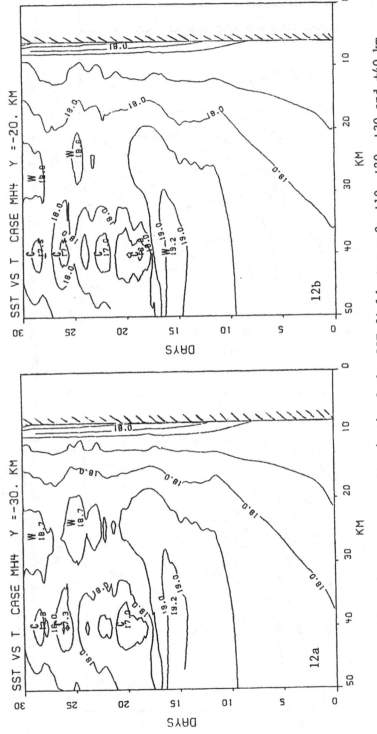

Figure 12.   Time cross-section (x-t) of the SST field at y = 0, ±10, ±20, ±30 and ±40 km for the heated case.   Contour interval is .5°C.

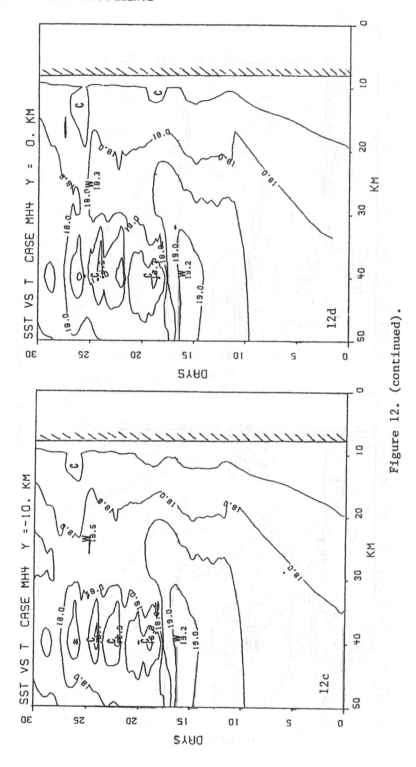

Figure 12. (continued).

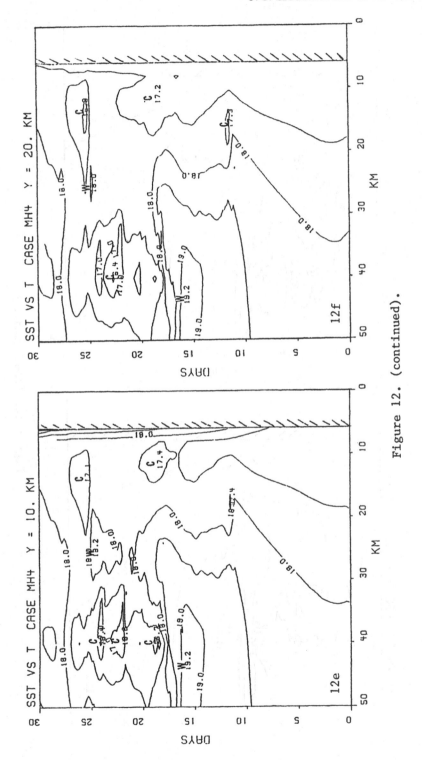

Figure 12. (continued).

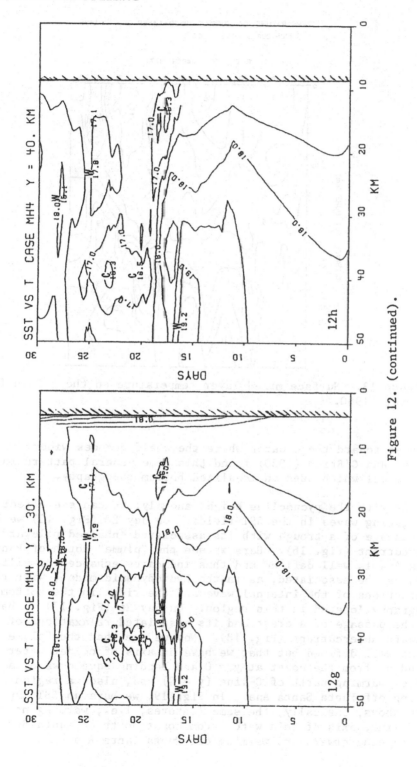

Figure 12. (continued).

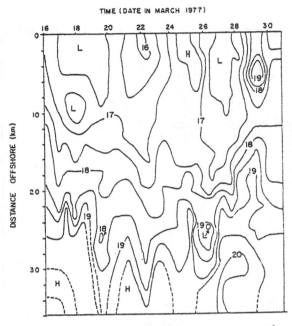

Figure 13. Surface mixed layer temperature on the C-line from Hoover 1980.

progress toward the equator where the shelf becomes wider. Preller and O'Brien (1980) found this same general pattern in their model which used an idealized bottom topography.

As with the pycnocline height anomaly, we can see effects of propagating waves in the SST fields. On day 20 (Fig. 15) we have the passage of a trough with its associated enhanced poleward undercurrent (Fig. 16). Here we see the "plume" north of Punta Santa Ana is well defined and thus indicates enhanced upwelling which can be associated, as in the hydrodynamic model, with the interactions of the internal wave and the ridge in the bottom topography located in this region. On day 26 (Fig. 17) we have now the passage of a crest and its associated relaxations of the poleward undercurrent (Fig. 18). Now we see that the "plume" is not as well defined but that we have an axis of colder water extending from the coast at y = 0 and bending equatorward, a large area of warming north of C-line (y = 0) and, also, a region of warming off Punta Santa Ana. In Fig. 19, we have an SST map which shows, basically, the same features, i.e., warming north of the C-line, axis of cold water from coast north of Punta Santa Ana bending equatorward and warming off Punta Santa Ana.

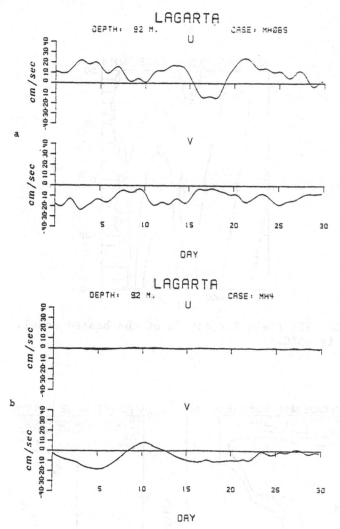

Figure 14.   Time series of the longshore (v) and cross-shelf (u) velocity.

        One final observation we wish to make concerns the so-called "two-cell" circulation pattern report by Mooers, Collins, and Smith (1976), and Johnson and Johnson (1979). On day 11 of the model run we see, in the cross-shelf flow field (Fig. 20), a pattern which resembles the "two-cell" pattern (Fig. 21, from Mooers, Collins, and Smith, 1976). This is a period when the interface is upwarped toward the surface due to the passage of an internal wave crest and when we have equatorward flow (Fig. 20) through the water column in the longshore component. The

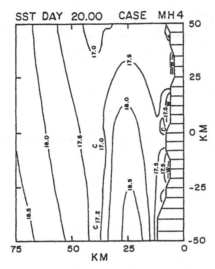

Figure 15.   SST field for day 20 of the heated model.   Contour
interval is .5°C.

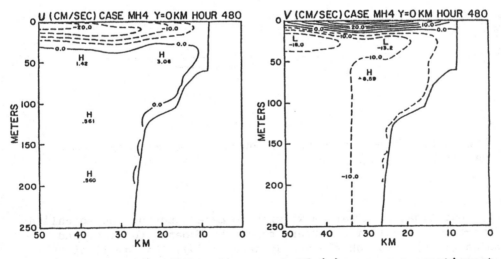

Figure 16.   Longshore (v) and cross-shelf (u) x-z cross-section at
y = 0 for model day 20.   Contour intervals are 5 cm/sec and dashed
contours represent negative values.

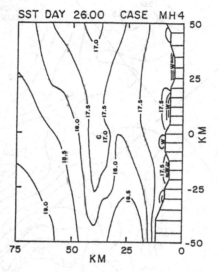

Figure 17.   SST for day 26.

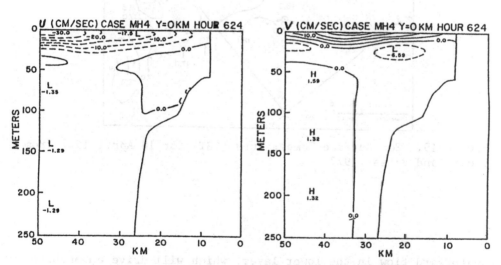

Figure 18.   Cross-shelf (x-z) sections of longshore (v) and cross-shelf (u) velocity components for model day 26.

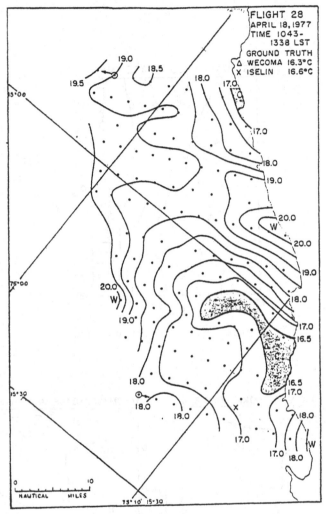

Figure 19.   Sea Surface Temperature (SST) for 18 April 1977 from
Stuart and Bates, 1977.

equatorward flow in the lower layer, which will drive an onshore
bottom Ekman layer (Fig. 20) appears to be a necessary requirement
for the observation of the "two-cell" pattern. Thus, as Smith
(1980) suggests, the "two-cell" circulation pattern is a transient
feature which could appear when some mechanism causes an
equatorward flow (including an onshore bottom Ekman layer) in the
lower layer.

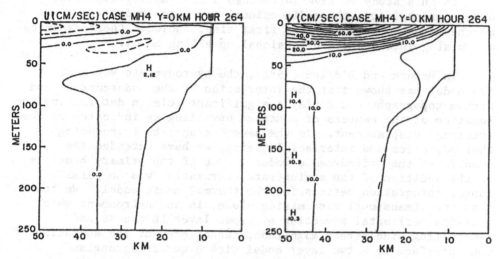

Figure 20. Cross-shelf (x-z) sections along the C-line (y = 0 in model) of the longshore (v) and cross-shelf (u) velocity components for model day eleven.

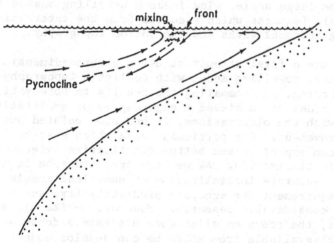

Figure 21. "Two-cell" upwelling circulation pattern from Mooers, Collins and Smith (1976).

SUMMARY

In this study we have introduced a fluctuating longshore
undercurrent into a three-dimensional upwelling model.
Furthermore, we have, for the first time, included thermodynamics
and mixing into a three-dimensional upwelling model.

In Heburn and O'Brien (1982), the hydrodynamic version, of
the model has shown that the interaction of the undercurrent and
bottom topography could play a significant role in determining the
location of the centers of enhanced upwelling as indicated by the
interface displacement. In the second stage, by introducing
thermodynamics and interfacial mixing, we have extended the
results of the hydrodynamic models. One of the primary benefits
of the addition of the mixing parameterization was to permit
longer integration periods for the thermodynamic model. We found
that one-dimensional wind mixing alone, in an environment which
includes horizontal advection of upper layer thickness and
fluctuating wind stress magnitude, cannot prevent the surfacing of
the interface in a two layer model with a quasi-Lagrangian
vertical coordinate. This was successfully overcome in the
present model by switching to an Eulerian vertical coordinate at a
predetermined minimum upper layer thickness. The results of this
model have shown that the features indicated by the pycnocline
height anomaly field in the hydrodynamic model can also be
observed in the sea surface temperature field in the thermodynamic
model. We have also seen that a positive heat flux was necessary
to recover the observed horizontal structure. Without heating we
saw that the large scale, wind induced upwelling masked the
smaller scale features which resulted from the interaction of the
propagating internal waves and the bottom topography.

This was a first attempt at using a three-dimensional
thermodynamics upwelling model with realistic topography and
external forcing, and comparing the results to observational data.
While this model has achieved a high degree of qualitative
agreement with the observations, it has also pointed out areas of
needed improvement. One particular area which must be improved
before we can expect to see better quantitative agreement is in
the model initialization. As we know from atmospheric modelling
experience, accurate initialization of numerical models is a
critical requirement for accurate predictability, and has been the
subject of considerable research. However, unlike his atmospheric
counterpart, the ocean modeller does not have a dense synoptic
data network available from which he can develop objectively
balanced initial states. In the area of oceanic data collection,
the JOINT II region is considered to be a heavily sampled area,
but one current meter array with ten moorings within the model
region and airborn SST measurements which cover less than one
tenth of the model area are not sufficient to derive an initial

state with much more than a guess. Due to the cost of obtaining
oceanic data, there will, in the foreseeable future, never be
enough data to initialize coastal oceanic models accurately.
Thus, alternative methods must be derived for initialization, if
we hope to achieve reasonable quantitative predictability in
future oceanic models.

APPENDIX A

Numerical Formulation

The numerical formulation for this model has evolved from
the previous upwelling models developed by the Mesoscale Air-Sea
Interaction Group. O'Brien and Hurlburt (1972) introduced a
highly efficient semi-implicit scheme into the upwelling model.
This scheme allows a larger maximum time step than the ordinary
Courant-Friedrich-Lewy (CFL) condition by treating the terms in
the equation which govern the external and internal gravity wave
modes (the fastest moving waves) with an implicit time
differencing scheme and the remaining terms with an explicit
scheme. From this treatment a set of coupled Helmholtz equations
for the vertically integrated cross-shelf flow, u_1 and u_2, can be
obtained. This set of equations can then be solved iteratively
using the tridiagonal variant of Gaussian Elimination.

Thompson (1974) introduced and thoroughly tested the
discretely telescoping grid used in the x-direction in the model.
He found considerable saving in computer time by using this
technique versus a fine mesh grid over the entire basin. Also he
found less than one percent difference between the two solutions
in the upwelling region.
Hurlburt (1974) extended the model to three dimensions and
included a variable coastline geometry. The irregular basin
geometry limited the effectiveness of using the semi-implicit
technique in both horizontal directions in that the advantage
gained by using a large time step was negated by the necessity of
using relaxation methods to solve the coupled two-dimensional
Helmholtz equations. Thus Hurlburt used the semi-implicit
technique in the x-direction and an explicit method in the
y-direction.

To reduce the number of grid points, he used the discretely
varying grid in the x-direction and an analytical stretched
variable grid in the y-direction. The stretching function used is
given by

$$S(a) = c(\alpha a + \sum_{k=1}^{N} \tanh(\frac{a-a_k}{\gamma_k}) + b \qquad (67)$$

where $S(0) = 0$ and $S(1) = 1$ and

$$b = \sum_{k=1}^{N} \tanh\left(\frac{a_k}{\gamma_k}\right) \tag{68}$$

$$c = \left[\alpha + b + \sum_{k=1}^{N} \tanh\left(\frac{1-a_k}{\gamma_k}\right)\right]^{-1} \tag{69}$$

The variables α and γ_k are stretching parameters which can be varied to give the desired variable resolution. Now if we let $q = q(S(a(y)))$ be any dependent variable and normalize a by the basin width, $a = y/L_y$, then we find the first and second derivatives in y are given by

$$\frac{dq}{dy} = \frac{1}{L_y} \frac{ds}{da} \frac{dq}{ds} \tag{70}$$

and

$$\frac{d^2q}{dy^2} = \frac{1}{(L_y)^2} \left(\frac{dS}{da}\right)^2 \frac{d^2q}{dS^2} + \frac{1}{(L_y)^2} \frac{d^2S}{da^2} \left(\frac{dq}{dS}\right) . \tag{71}$$

Because of the explicit treatment in the y-direction, the most severe constraint on the maximum allowable time step is given by the linear CFL stability condition;

$$\Delta t < MIN \Delta y / [g \cdot MAX \ (H_1 + H_2)]^{1/2} . \tag{72}$$

The diffusive terms are treated implicitly in the x-direction using the Crank-Nicholson (1947) scheme, while the other frictional terms are lagged in time. Leapfrog time differencing is used for the Coriolis and nonlinear terms. The advection terms are approximated using Scheme F from Grammeltvedt (1969).

ACKNOWLEDGMENTS

This work was a part of George Heburn's Ph.D. dissertation at the Florida State University. This work was supported by the Coastal Upwelling Ecosystems Analysis (CUEA) project through the International Decade of Ocean Exploration (IDOE) program of the National Science Foundation (NSF Grant OCE 78000611). Partial funding was provided by the Equatorial Ocean-Climate Air-Sea Interactions program of the National Science Foundation (NSF Grant ATM-7920485). The computations for the model were performed on the CRAY 1/CDC 7600 at the National Center for Atmospheric Research (NCAR) Boulder, Colorado, and the CDC Cyber 170-730 at Florida State University, Tallahassee, Florida. NCAR is sponsored by the National Science Foundation. The final preparation and travel to Os, Norway, was provided by the Office of Naval Research. ONR has been extremely generous in their support of the MASIG's since 1970.

Sincere appreciation is extended to Mrs. Ruth Pryor and Mrs. Pat Teaf for typing manuscripts for this study, and Dewey Rudd for drafting many of the figures.

REFERENCES

Adamec, D., and O'Brien, J. J., 1978, The seasonal upwelling in the Gulf of Guinea due to remote forcing, J. Phys. Oceanogr., 8:1050.

Badan-Dangon, A., 1978, Principal components of the velocity field off Northwest Africa, paper presented at the symposium on the Canary Current: Upwelling and Living Resources, International Council for the Exploration of the Sea, Las Palmas, Canary Islands.

Brink, K. H., Allen, J. S., and Smith, R. L., 1978, A study of low frequency fluctuations near the Peru coast, J. Phys. Oceanogr., 8:1025.

Brink, K. H., Gilbert, W. E., and Huyer, A., 1979, Temperature sections along the C-line over the shelf off Cabo Nazca, Peru, from moored current meters, 18-March - 10 May, 1977, and CTD observations, 5 March - 15 May, 1977, CUEA Tech Rept. 49, 78 pp., School of Oceanography, Oregon State University, Corvallis, Oregon.

Brink, K. H., Smith, R. L., and Halpern, D., 1978, A compendium of time series measurements from moored instrumentation during the MAM '77 phase of JOINT - II, CUEA Technical Report 45, School of Oceanography, Oregon State University, 72 pp.

Clancy, R. M., Thompson, J. D., Hurlburt, H. E. and Lee, J. D., 1979, A model of mesoscale air-sea interaction in a sea breeze-coastal upwelling regime, Mon. Wea. Rev., 107:1476.

Crank, J., and Nicholson, P., 1947, A practical method for numerical evaluation of solutions of partial differential equations of heat-conduction type, Proc. Camb. Philos. Soc., 43:50.

Denman, K. L., and Miyake, M., 1973, Upper layer modification at ocean station PAPA: Observation and simulation, J. Phys. Oceanogr., 3:185.

Denman, K. L., 1973, A time-dependent model of the upper ocean, J. Phys. Oceanogr., 3:173.

deSzoeke, R. A., and Rhines, P. B., 1976, Asymptotic regimes in mixed layer deepening, J. Mar. Res., 34:111.

Ekman, V. W., 1905, On the influence of the earth's rotation on ocean currents, Arkiv. Mat. Astron. Fysik, 12:1 (Reprinted in Royal Swedish Acad. of Sci., 1963).

Gill A. E., and Clarke, A. J., 1974, Wind-induced upwelling,
    coastal currents and sea-level changes, Deep Sea Res.,
    21:325.
Grammeltvedt, A., 1969, A survey of finite-difference schemes for
    the primitive equations for a barotropic fluid, Mon. Wea.
    Rev., 97:384.
Heburn, G. W., and O'Brien, J. J., 1982, Numerical study of the
    influence of longshore current fluctuations on coastal
    upwelling off Peru, J. Phys. Oceanogr., (in press).
Hoover, S. T., 1980, Mixed layer variability in the upwelling
    region off Peru - March 1977, M.S. Thesis, University of
    Delaware.
Hurlburt, H. E., 1974, The influence of coastline geometry and
    bottom topography on the eastern ocean circulation, Ph.D.
    Dissertation, Florida State University, Tallahassee.
Hurlburt, H. E., and Thompson, J. D., 1973, Coastal upwelling on a
    β-plane, J. Phys. Oceanogr., 3:16.
Huyer, A., Smith, R. L., and Sobey, E. J. C., 1978, Seasonal
    differences in low-frequency current fluctuations over the
    Oregon continental shelf, J. Geophys. Res., 83:5071.
Johnson, D. R., and Johnson, W. R., 1979, Vertical and cross-shelf
    flow in the coastal upwelling region off Oregon, Deep Sea
    Res., 26:399.
Kato, H., and Phillips, O. M., 1969, On the penetration of a
    turbulent layer into a stratified fluid, J. Fluid Mech.,
    37:643.
Kraus, E., and Turner, J. S., 1967, A one-dimensional model of the
    seasonal thermocline II. The general theory and its
    consequences, Tellus, 19:98.
Martin, P. M., and Thompson, J. D., 1983, Formulation and testing
    of a layer-compatible upper ocean mixed-layer model,
    submitted to J. Phys. Oceanogr.,.
Mellor, G. L., and Yamada, T., 1974, A hierarchy of turbulence
    closure models for planetary boundary layers, J. Atmos. Sci.,
    31:1791
Mittelstaedt, E., Pillsbury R. D., and Smith, R. L., 1975, Flow
    patterns in the Northwest African upwelling area, Deutsches
    Hydrographisches Zeitschrift, 28:145.
Moody, G. L., 1979, Aircraft derived low level winds and upwelling
    off the Peruvian coast during March, April and May, 1977,
    CUEA Technical Report 56, Dept. of Meteorology, Florida State
    University, Tallahassee.
Mooers, C. N. K., Collins, C. A., and Smith, R. L., 1976, The
    dynamic structure of the frontal zone in the coastal
    upwelling region off Oregon, J. Phys. Oceanogr., 6:3.
Niiler, P. P., and Kraus, E. B., 1977, One-dimensional models of
    the upper ocean, Modeling and Prediction of the Upper layers
    of the Ocean, Ed. by E. B. Kraus, Pergamon Press, New York,
    143-172.

Niiler, P. P., 1975, Deepening of the wind-mixed layer, J. Mar. Res., 33:405.

O'Brien, J. J., Clancy, R. M., Clarke, A. J., Crepon, M., Elsberry, R., Gammelsrod, T., MacVean, M., Roed, L. P., and Thompson, J. D., 1977, Upwelling in the ocean: Two and three-dimensional models of upper ocean dynamics and variability, in: "Modelling and prediction of the upper layers of the ocean", E. Kraus, ed., Pergamon Press, New York.

O'Brien, J. J., and Hurlburt, H. E., 1972, A numerical model of coastal upwelling, J. Phys. Oceanogr., 2:14.

O'Brien, J. J., Smith, R. L., and Heburn, G. W., 1983: Determination of vertical velocity on the continental shelf, (submitted to J. Phys. Oceanogr.).

Pedlosky, J., 1979, "Geophysical Fluid Dynamics", Springer-Verlag, New York.

Peffley, M., and O'Brien, J. J., 1976: A three-dimensional simulation of coastal upwelling off Oregon, J. Phys. Oceanogr., 6:164.

Pollard, R. T., and Millard, R. C., 1970, Comparison between observed and simulated wind generated inertial oscillations, Deep Sea Res., 17:813.

Pollard, R. T., Rhines, P. B., and Thompson, R. O. R. Y., 1973, The deepening of the wind-mixed layer, J. Geophys. Fluid Mech., 4:381.

Preller, R., and O'Brien, J. J., 1977, Peruvian bottom topography and coastline map, available from Mesoscale Air-Sea Interaction Group, Florida State University, Tallahassee.

Preller, R., and O'Brien, J. J., 1980, The influence of bottom topography on upwelling off Peru, J. Phys. Oceanogr., 10:1377.

Smith, R. L., 1968, Upwelling, Oceanogr. Mar. Biol. Ann. Rev., 6:11.

Smith, R. L., 1974, A description of current, wind, and sea level variations during coastal upwelling off the Oregon coast, July - August, 1972, J. Geophys. Res., 79:435.

Smith, R. L., 1978, Poleward propagating perturbations in currents and sea level along the Peru coast, J. Geophys. Res., 83:6083.

Smith, R. L., 1980, A comparison of the structure and variability of the flow filed in the three coastal upwelling regions: Oregon, Northwest Africa and Peru, submitted to IDOE International Symposium on Coastal Upwelling.

Spiegel, E. A., and G. Veronis, 1960, On the Boussinesq approximation for a compressible fluid, J. Astrophys., 131:442.

Stevenson, M. R., Stuart, D. W., and Heburn, G. W., 1980, Short term variations observed in the circulation, heat content and surface mixed layer of an upwelling plume off Cabo Nazca, Peru, submitted to IDOE International Symposium on Coastal Upwelling.

Stuart, D. W., and Bates, J. J., 1977, Aircraft sea surface temperature data - JOINT II 1977, CUEA Data Report 42, Dept. of Meteorology, Florida State University, Tallahassee, 39 pp.

Sverdrup, H. U., 1938, On the processes of upwelling, J. Mar. Res., 1(2):155.

Sverdrup, H. U., and Fleming, R. H., 1941, The water off the coast of Southern California, Bull. Scripps Inst. Oceanogr., 4:261.

Thompson, J. D., 1974, The coastal upwelling cycle on a Beta-plane: Hydrodynamic and thermodynamics. Tech. Report, Mesoscale Air-Sea Interaction Group, Florida State University, Tallahassee.

Thompson, J. D., 1978, Role of mixing in the dynamics of upwelling systems, 203-221, in: "Upwelling Ecosystems", E. M. Tomczak and F. Boje, eds., Springer-Verlag, New York.

Thorade, H., 1909, Die Kalifornischen Meeresstromungen, Ann. Hydrogr. Bul., 37:17&63.

Turner, J. S., 1973, Buoyancy effects in fluids, Cambridge monographs on mechanics and applied mathematics, Cambridge Univ. Press.

Wang, D.-P., and Mooers, C. N. K., 1976, Coastal-trapped waves in a continuously stratified ocean, J. Phys. Oceanogr., 6:853.

Watson, A. I., 1978, A study of the low-level mesoscale winds observed off the Peruvian coast during March and April, 1976, CUEA Technical Report 41, Department of Meteorology, Florida State University, Tallahassee.

Yoshida, K., 1955a, Coastal upwelling off the California Coast, Rec. Oceanogr. Works Japan, 2(2):8.

Yoshida, K., 1955b, An example of variations in oceanic circulation in response to the variations in wind field, J. Oceanogr. Soc. Jap., 11(3):103.

Yoshida, K., 1967, Circulation in the eastern tropical oceans withspecial references to upwelling and undercurrents, Japan J. Geophys., 4:1.

Zilitinkevich, S. S., Chalikov, D. V., Resnyansky, Yu. D., 1979, Modelling the oceanic upper layer, Oceanol. Acta., 2:219.

THE CURRENTS IN A SHALLOW COASTAL CORNER REGION –

THE GERMAN BIGHT – MODEL, MEASUREMENTS AND FORECAST

Ekkehard Mittelstaedt

Deutsches Hydrographisches Institut
Hamburg, West Germany

A model and measurements are used to describe some features of the residual currents in the German Bight. The tidal currents are excluded because this part of the currents is already presented by Brockmann et al. (1981), who base their investigations on the same model and the same observations.

THE MODEL

The numerical model is explicit, three dimensional (grid size, 11 km) and has 4 layers. As long as the water depth is great enough, the upper three layers have each a thickness of 8 meters. The lowest layer extends from the bottom of the third layer (at 24 m depth) down to the bottom. In order to take approximately into account the influence of the whole North Sea, the three dimensional model of the German Bight is coupled with a two dimensional (one layer) model of the North Sea (grid size, 22 km). The external forces, which drive the coupled models, are tides, wind stress and atmospheric pressure.

The tidal excitation takes place along the open boundaries of the North Sea model, in the English channel at 4° W and in the North at $59^\circ 30'$ N. At present, there are no density gradients incorporated in the models.

This brief report presents some results of a simplified diagnostic version of the model, in which only the dominant semi-diurnal M_2 tide is used to excite the model, a uniform wind field of constant direction and speed (7.5 ms^{-1}) is assumed, and the averaged currents of the sixth tidal cycle are taken to represent the residual currents.

Figures 1 and 2 show the simulated currents of the surface
layer (0 to 8 m), the second layer (8 to 16 m) and the local
"bottom" layer. At water depths less than 8 m the bottom layer
coincides with the surface layer of the model.

THE MEASUREMENTS

Current measurements throughout 20 years (1959 - 1979) are
used to compile a unique data set of the currents in the German
Bight. The measurements have been carried out by means of different current meters. Most of the data stem from near surface depths
between 8 and 15 m and from depths of 1 to 5 m above the bottom.

To estimate the residual currents, the following procedures
are used:

- current records within vicinity of 10 km are considered to represent the water movements at one and the same position.

- independent of the exact observation depth, the records are associated either with near surface or near bottom measurements.

- the daily vector mean (average over 24.84 hours) is considered to
  be the residual current within either layer (near surface or near
  bottom).

- all daily means of the currents are associated with the corresponding daily means of the 3-hourly wind observations from 1959 to
  1979 aboard the lightvessel ELBE 1 (near the mouth of the river
  Elbe).

The statistical treatment of the data, independent of time,
yields typical tendencies of the residual current directions for a
given wind direction. Yet, from the data available, the residual
currents are not always locally unidirectional but are in many
cases considerably scattered. Among other reasons, this occurs
because we did not take account of:

- the largescale wind and atmospheric pressure fields over the entire North Sea, as well as their temporal development.

- the exact depth of observation within the near surface and near
  bottom layers.

- remote forcing, such as external surges..

The relationship between residual currents and wind speed is
not investigated here. The current velocities used are mean values
independent of wind speed.

In Figs. 3 and 4 velocities and the estimated directions (black
lines) of the observed residual currents for south-easterly or north-
westerly winds are shown. The circles indicate the positions of

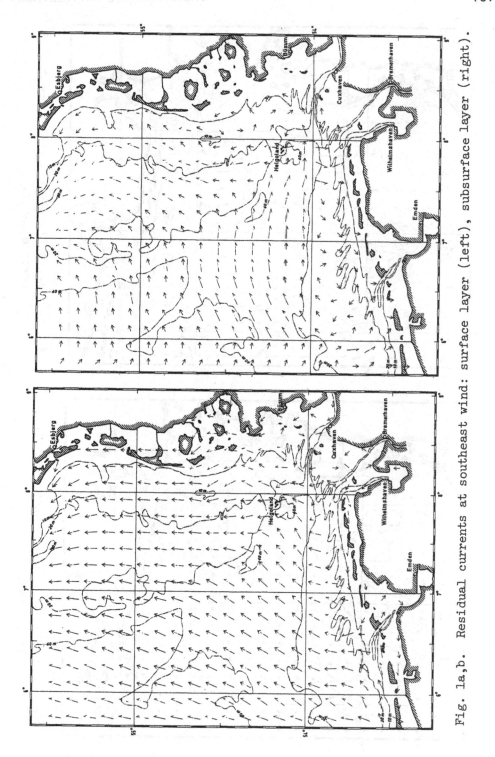

Fig. 1a,b. Residual currents at southeast wind: surface layer (left), subsurface layer (right).

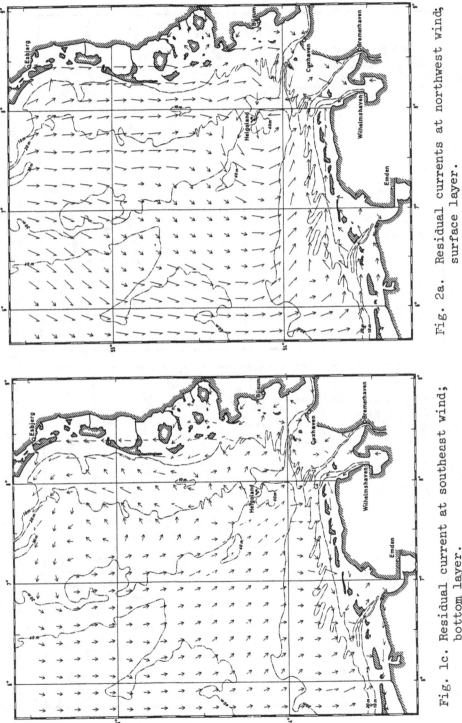

Fig. 2a. Residual currents at northwest wind;
surface layer.

Fig. 1c. Residual current at southeast wind;
bottom layer.

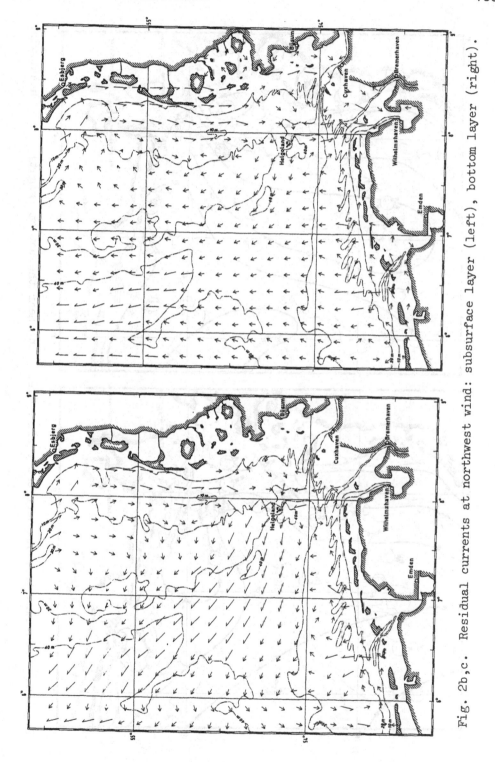

Fig. 2b,c. Residual currents at northwest wind: subsurface layer (left), bottom layer (right).

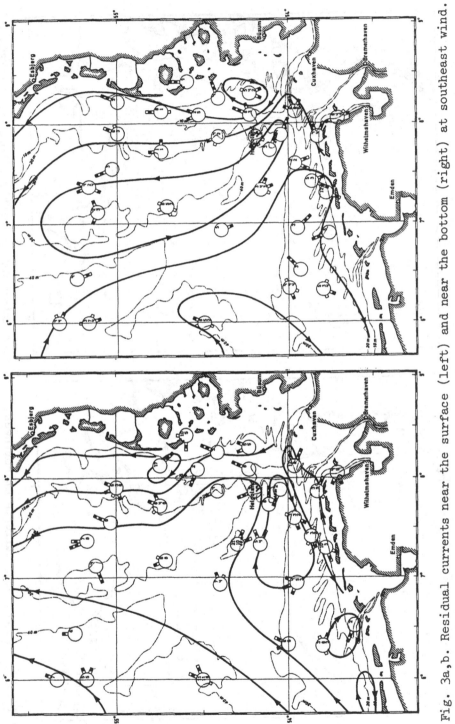

Fig. 3a,b. Residual currents near the surface (left) and near the bottom (right) at southeast wind.

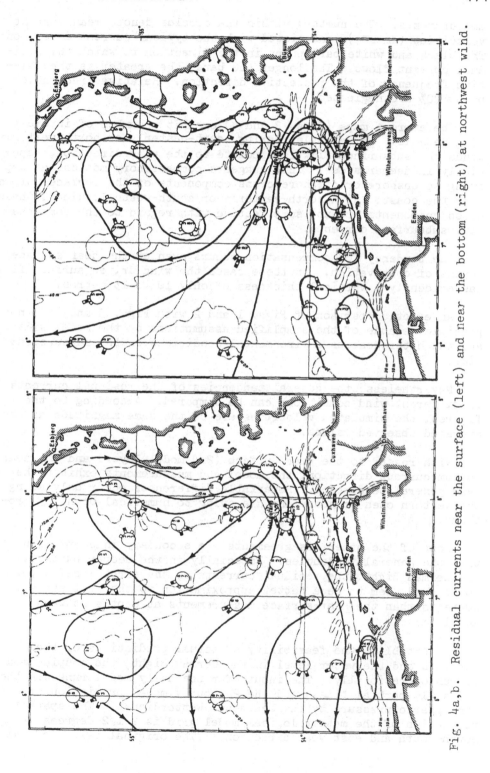

Fig. 4a,b. Residual currents near the surface (left) and near the bottom (right) at northwest wind.

measurements. The numbers within the circles denote mean current
velocities (cm s^{-1}), associated with the typical current directions.
The black and white bars point in the direction to which the residual current flows. The longer the bars, the greater is the relative frequency of the respective direction. Relative frequencies
below 20% are neglected.

The current maps (Figs. 3 and 4) exhibit a number of interesting details. For example, during south-easterly or north-westerly
winds the subsurface flow is opposite to the wind direction, especially in deeper areas. This suggests compensatory motions at depth
owing to onshore or offshore cross components of the surface currents
near the coast. With north-easterly or south-westerly winds, these
cross components are not so effective with regard to the compensatory subsurface currents.

The tendency for compensation occurs even at the near surface
depths of observation. In these cases the wind driven surface flow
has evidently a vertical thickness of only 10 to 15 metres.

An exact comparison of Figs. 1 and 2 with Figs. 3 and 4 is not
possible because of the simplified assumptions on the model side
and the heterogeneous environmental conditions on the measurement
side.

Nevertheless, the general tendencies of the residual currents
in different wind conditions can be compared. According to the
figures, the simulated velocities are of the same magnitude as the
averaged observed values.

With regard to the near surface (2nd model layer) and the near
bottom current direction, the model and measurements exhibit the
same general tendencies. Remarkable differences occur only along
the western open boundary and are owing to numerical boundary problems.

Most of the other disagreements are secondary in comparison
with the general tendencies. Especially in the deeper parts of
the German Bight, the residual currents of the first model layer
(0 to 8 m) seem to be a better approximation to the real surface
currents than the near-surface measurements at depths between 8
and 15 m.

To establish the feasibility of actual predictions of the
currents and the water level in the German Bight, the coupled model
is run operationally. The inputs for the daily model runs are the
tides (10 partial tides) and the 24-hourly model prediction of
atmospheric pressure by the Deutscher Wetterdienst. The spatial
resolution of the meteorological model grid is 2 1/2 degrees in
north-south and east-west directions. The original values of the

predicted atmospheric pressure are interpolated onto the 22 km grid of the North Sea model in order to estimate the surface winds via the geostrophic winds over the entire North Sea. The stress of the surface winds then generates the wind induced currents of the model.

In connection with the observational data the model appears to be a useful tool for the actual forecasts. If in real cases the local basis of observational data is insufficient or uncertain (e.g. at the sea surface) the operational model may provide the required information on the currents (tidal and residual currents). In regions where the model is wrong (e.g. along the western open boundary) the measurements will be the only basis for prognostic estimates. In practice, local forecasts of the currents will be in every case a subjective combination of estimates from the measurements and the results of the model.

ACKNOWLEDGEMENT

The numerical simulations were carried out by K. Ch. Soetje.

REFERENCE

Brockmann, Ch., Lange, W., Mittelstaedt, E., and Soetje, K.Ch., 1981, The Tidal Stream in the German Bight. A comparison of measurements and numerical model results, Dt. hydrogr. Z., 34, 2:56-60.

UPWELLINGS AND KELVIN WAVES GENERATED BY TRANSIENT ATMOSPHERIC FRONTS

Michel Crépon and Claude Richez

Laboratoire d'Océanographie Physique
Muséum National d'Histoire Naturelle
Paris, France

ABSTRACT

Motions generated by transient atmospheric fronts are investigated. Using asymptotic expansions in time, it is shown that the variation of the longshore component of the wind stress generates an interface elevation which propagates along the coast as an internal Kelvin wave front. Strong horizontal baroclinic motions are linked to this wave front. The velocity within the deep layer is in the opposite direction to that of the wind on the surface. Far from the coast, the wind stress curl, associated with the atmospheric front, produces strong interface elevations which increase with time and combine with coastal elevation to give peculiar behaviour of the interface associated with strong currents at the discontinuities.

INTRODUCTION

In the present paper, we develop a linearized model of two-dimensional transient upwelling generated by variability in the wind stress. Following R.R. Thompson (1970) who first showed that meteorological forcing could generate Kelvin waves, we use an integral transform technique. Asymptotic expansions of oceanic parameters for large time are calculated for a wide range of atmospheric fronts. Full details of the calculations are given in Crépon and Richez (1982).

EQUATIONS OF MOTION

The variation in ocean density is idealized in terms of a homogeneous two layer fluid. This is a crude but realistic assumption

175

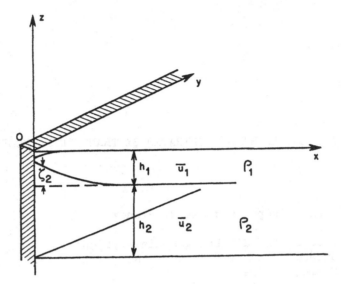

Fig. 1. Definition of parameters

for studying the first internal mode which is the most important for motions of small amplitude (Lighthill, 1969). Hydrostatic and Boussinesq assumptions are employed, and we deal with vertically-averaged equations of motion in each layer. We consider small perturbations from rest, hence the equations can be linearized. We assume that coupling between the upper and lower layers is only due to pressure forces. The wind stress is represented by a body force acting through the surface layer. The total depth $(h_1 + h_2)$ is held constant. The Coriolis parameter is considered constant since we are dealing with phenomena whose spatial extent is of the order of the internal radius of deformation.

The problem can be reduced by the standard separation into normal modes. Using a technique similar to that described by Csanady (1975), the elevation $\overline{\zeta}_n$ and the velocity $\overline{u}_n$ of the upper (n=1) and the lower (n=2) layers are given in terms of the barotropic (n=1) and baroclinic (n=2) normal mode elevation ζ_n and velocity $\vec{u}_n$ by:

$$\overline{\zeta}_1 = \zeta_1 + \varepsilon \left(h_2 / (h_1 + h_2) \right)^2 \zeta_2$$

$$\overline{\zeta}_2 = h_2 / (h_1 + h_2) (\zeta_1 - \zeta_2) \qquad (2.1)$$

$$\vec{\overline{u}}_1 = \vec{u}_1 + h_2 / (h_1 + h_2) \vec{u}_2$$

$$\vec{\overline{u}}_2 = \vec{u}_1 - h_1 / (h_1 + h_2) \vec{u}_2 \qquad (2.2)$$

where h_1 and h_2 are the thickness of the upper and lower layers respectively, ρ_1 and ρ_2 are the corresponding densities and ε is the density ratio $(\rho_2 - \rho_1)/\rho_2$.

The elevation ζ_n is a solution of the equation:

$$\frac{\partial}{\partial t}\left[c_n^2 \, \nabla^2 \zeta_n - \left(\frac{\partial^2}{\partial t^2} + f^2 \right)\zeta_n \right] = \frac{1}{\rho}\left[\frac{\partial}{\partial t}\, \nabla . \vec{\tau} + f\, \vec{z}\, \nabla x\, \vec{\tau} \right] \quad (2.3)$$

where $c_1^2 = g(h_1 + h_2)$, $c_2^2 = g\varepsilon h_1 h_2/(h_1 + h_2)$, and $\vec{\tau}$ is the wind stress.

The corresponding velocity $\vec{u}_n$ is a solution of:

$$\left(\frac{\partial^2}{\partial t^2} + f^2 \right) \vec{u}_n =$$

$$\frac{1}{H}\left[- c_n^2 \left(\frac{\partial}{\partial t}\, \nabla \zeta_n + \nabla \zeta_n \, x \, f\, \vec{z} \right) + \frac{1}{\rho}\left(\frac{\partial}{\partial t}\, \vec{\tau} + \vec{\tau}\, x\, f\, \vec{z} \right) \right] \quad (2.4)$$

where $H_1 = h_1 + h_2$, $H_2 = h_1$ (depths relative to the barotropic and baroclinic modes). $\vec{z}$ is the unit vector along the z axis.

In the following, normal mode solutions are calculated for different wind stresses. For simplicity, the indices are omitted in the equations.

In order to solve (2.3), a Fourier transformation (FT hereafter) is applied on y:

$$Z\,(t,x,k) = \int_{-\infty}^{+\infty} \exp\,(- \, i2\pi ky)\, \zeta(t,x,y)\, dy$$

and a Laplace transformation on t:

$$\tilde{Z}\,(p,x,k) = \int_{0}^{+\infty} \exp(- \, pt)\, Z(t,x,k)\, dt$$

The solutions are obtained in terms of Fourier-Laplace transforms (FLT). Laplace-Fourier inversion (LFT) is then applied to come back to physical space.

SOLUTION FOR IRROTATIONAL WIND STRESS (CLASS A)

Analytical Solution

Consider a wind stress $\vec{\tau}(t,x,y)$ perpendicular to and constant along the straight line given by the equation $y - x\, tg\, \alpha = 0$ (Fig.2). The wind starts from rest at $t = 0$ and is represented by

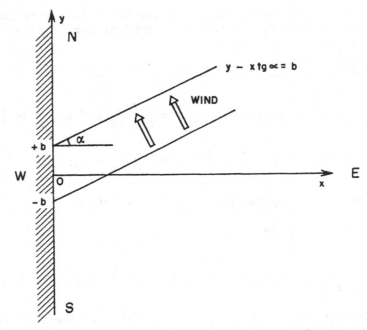

Fig. 2. Wind pattern corresponding to the A case (irrotational wind).

$$\vec{\tau} \ (\ - \sin \alpha \ \vec{x} + \cos \alpha \ \vec{y} \ ) \ \hat{T}(t) \ \tau(y - x \ \text{tg} \ \alpha) \tag{3.1}$$

where $\hat{T}(t)$ is the step function, $\vec{x}$ and $\vec{y}$ are unit vectors along the x and y axis.

The solution of the Fourier-Laplace Transform of (2.3) yields:

$$\tilde{Z} = - \frac{1}{\rho \ c} \ \frac{T(k)}{p(p^2 + f^2 + (2\pi kc')^2)} \Bigg[i \ 2\pi kc' \ \exp(-i2\pi kx \text{tg} \alpha)$$

$$- \frac{(p^2 + f^2)(p \ \sin\alpha - f \ \cos\alpha \ ) \ \exp(\ - \gamma \ x)}{(\ p \ \gamma - 2\pi ikf \ ) \ c} \Bigg] \tag{3.2}$$

where $T(k)$ is the Fourier Transform of $\tau(x,y)$.

A Laplace inversion of $\tilde{Z}$ is difficult. Nevertheless, we can obtain an asymptotic expansion as (ft) tends to infinity by using Sutton's algorithm (1934), based on algebraic expansions around Laplace Transform singularities which have the largest real values. Let us denote $r = c/f$ and $r' = c'/f = r \ / \cos \alpha$. The radius of deformation r is implicitly indexed (r_1 is the baro-tropic radius of deformation, r_2 is the baroclinic one,

As $t \to \infty$ the leading term of the asymptotic expansion of $\tilde{Z}$ is given by:

$$
Z(t,x,k) \underset{t \to \infty}{=} -\frac{T(k)}{\rho c f} \left[\overbrace{i\, 2\pi k r' \; \frac{\exp(-i2\pi kx\, tg\alpha)}{1+(2\pi k r')^2}}^{(1)} \right.
$$

$$
+ \overbrace{\frac{i}{2\pi k r'} \cdot \frac{\exp(-1+(2\pi kr)^2)^{1/2}\, x/r)}{1+(2\pi k r')^2} - \exp(-x/r)}^{(2)}
$$

$$
\left. + \overbrace{\cos\alpha \, \exp(-x/r) \left\{ \frac{i}{2\pi kr} - \left(\frac{i}{2\pi kr}+tg\alpha\right) \frac{\exp(i2\pi krft)}{1+(2\pi krt\, tg\alpha)^2} \right\}}^{(3)} \right]
$$

$$(3.3)$$

The first term of (3.3) represents the FT of the response of the ocean to the same forcing in the absence of a coast. The second term represents a set of Poincaré waves of second kind of zero frequency generated at the coast in order to satisfy the boundary condition $(u(0,t) = 0$. At $x = 0$ (at the coast), the first two elevations cancel each other exactly. The third term is the FT of a Kelvin wave.

By substituting (3.2) into the FLT of (2.4) one obtains the FLT of the velocity. Sutton's algorithm is then applied to obtain the asymptotic expansion of the velocity at large (ft) (Crépon and Richez, 1982).

Physical Interpretation

Wind blowing over a strip. Let us consider a wind constant over the strip bounded by the lines $y - x\, tg\alpha = \pm b$ (Fig. 2). The wind is perpendicular to these two lines and zero outside the strip. The FT of the wind modulus is

$$T(k) = \tau_0 \sin(2\pi kb) / \pi k \qquad (3.4)$$

The $\overline{FT}$ (Fourier Transform inversion) of the first two terms of (3.3) yields elevations $\zeta(1)$ and $\zeta(2)$ which cancel each other

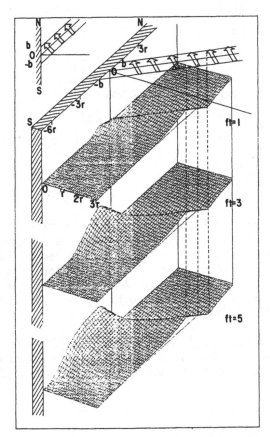

Fig. 3. Wind blowing over a strip with α = π/4. Three-dimen-
         sional view of the interface elevation at three
         different times (the width of the strip is equal to
         $2r_2$ = 2b).

at the coast. The $\overline{\text{FT}}$ of the third term is:

$$\zeta(3) = - \frac{\tau_0}{\rho c f} \cos\alpha \, \exp(- x/r) \left[(|y| - |ct + y|)/r \right.$$

$$- \, tg\alpha\{(sgn \; \alpha + sgn(ct+y)) \; \exp(-|ct + y|/(r|tg\alpha|))$$

$$\left. - \, sgn(ct + y)\} \right]$$

$$\overset{*}{\underset{y}{}} \; 0.5 \; (\; \delta(y - b) - \delta(y + b)) \qquad\qquad (3.5)$$

where $\overset{*}{\underset{y}{}}$ represents the convolution product with respect to y.

The elevation $\zeta(3)$ is the sum of a Kelvin wave front (first term of (3.5) which is proportional to the component of the wind stress parallel to the coast, and of an evanescent wave (second term of (3.5)) which is proportional to the component of the wind stress perpendicular to the coast.

For $t \gg (b - y)/c$, $\zeta(3)$ becomes:

$$\zeta(3) = - \frac{\tau_0}{\rho c f} \cos\alpha \, \frac{2 \; b}{r} \; \exp(- x/r) \qquad\qquad (3.6)$$

In a two layer ocean and close to the coast, the interface elevation is mainly driven by the opposite of the baroclinic mode, $\zeta_2(3)$, the amplitude of which is much larger than the barotropic one. In the above case, an upwelling is generated at $y = \pm b$, and propagates towards $y < 0$ as an internal Kelvin wave front. Its amplitude is proportional to the longshore integral of the wind stress. It is fully developed and reaches its maximum intensity in the region $y < - b$, outside the strip over which the wind blows (Fig.3).

By substituting (3.2) into the FLT of (2.4), one obtains the FLT of the velocity. Sutton's algorithm is then applied to obtain the asymptotic expansion of the FT of the velocity at large (ft), Crépon and Richez (1982). The most important term of the longshore component of the velocity is linked to the Kelvin wave front. For $y < - b$ (outside the strip) and for $t > (b - y)/c$ (for large time),

$$v = \frac{\tau_0}{\rho H f} \cos\alpha \, \frac{2 \; b}{r} \; \exp(- x/r) \qquad\qquad (3.7)$$

In a two layer ocean, it is found that the baroclinic velocity is much larger than the barotropic one. Substituting (3.7) into (2.2), we find in the region $y < - b$ and for large time (ft > 10)

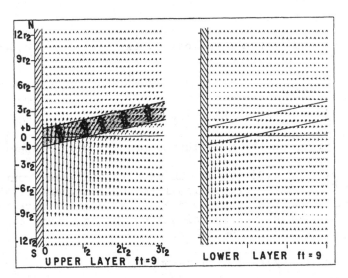

Fig. 4. Velocity fields in the two layers after ft = 9
(h_1 = 50 m, h_2 = 150 m, ε = 0.00266)

that the velocity components parallel to the coast in the upper
$(\overline{v}_1)$ and lower $(\overline{v}_2)$ layers are of the form

$$\overline{v}_1 = \frac{\tau_0}{\rho H_1 f} \; \frac{h_2}{h_1} \frac{2b}{r_2} \; \exp(-x/r_2) \qquad\qquad (3.8)$$

$$\overline{v}_2 = -\frac{\tau_0}{\rho H_1 f} \; \frac{2b}{r_2} \; \exp(-x/r_2) \qquad\qquad (3.9)$$

In the region u < -b, the velocity parallel to the coast
forms a baroclinic coastal jet. In the surface layer, the velocity is in the same direction as the wind. In the lower layer, it
is in the opposite direction (Fig. 4). The vertical structure of
the current is different from the case in which the longshore
component of the wind is constant in space. In this case, the
current in both layers flows in the direction of the wind.

From (3.3) it is possible to obtain the elevation for a wide
variety of forcings.

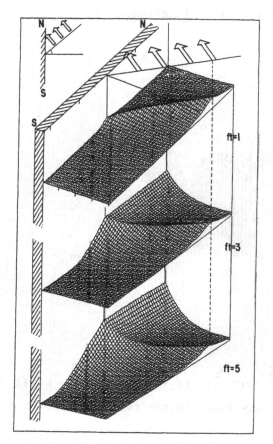

Fig. 5. Wind blowing over a northern half plane with $\alpha = \pi/4$.
Three dimensional view of the interface elevation at
three different times.

<u>Wind blowing over a northern half plane</u>. We have:

$$\vec{\tau} = \tau_0 \ (-\sin\alpha \ \vec{x} + \cos\alpha \ \vec{y}) \ \hat{T}(y - x \ tg \ \alpha) \ \hat{T}(t)$$

and

$$T(k) = \frac{\tau_0}{2} \ (\ \delta(k) - \frac{i}{\pi k} \ )$$

where $\delta(k)$ is the Dirac function.

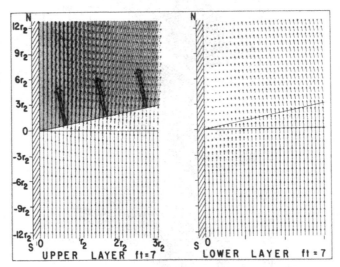

Fig. 6. Velocity field in the two layers at ft = 7 with $\alpha=\tau/4$.

By a FFT of (3.3), using (2.1), we obtain the interface elevation. In the northern half plane, an upwelling which increases
with time is generated at the coast by the longshore component of
the wind (Fig. 5). At the wind discontinuity, the mass continuity
of the fluid generates a Kelvin wave front which propagates southwards along the coast.

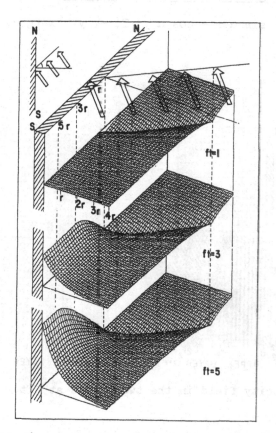

Fig. 7. Wind blowing over a southern half plane with $\alpha = \pi/4$.
Three dimensional view of the interface elevation at
three different times.

In the region $y - xtg\alpha > 0$, the velocity field (Fig. 6)
shows a boundary coastal jet in the upper layer and a very weak
current in the lower layer near the coast. Far from the coast, in
the upper layer, the current corresponds to the Ekman transport.
In the region $y - xtg\alpha < 0$, the streamlines are parallel to the
coast.

Wind blowing over a southern half plane. We have:

$$\vec{\tau} = \tau_0 \ (- \sin \alpha \ \vec{x} + \cos \alpha \ \vec{y}) \ \hat{T}(-(y - x \ tg \ \alpha) \ \hat{T}(t)$$

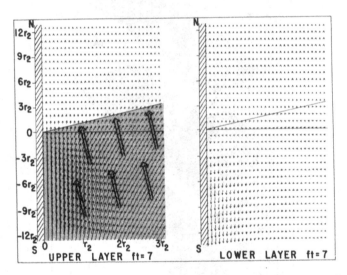

Fig. 8. Velocity field in the two layers at $ft = 7$ with $\alpha=\tau/4$.

and
$$T(k) = \frac{\tau_0}{2} \ (\ \delta(k) + \frac{i}{\pi k} \ )$$

Fig. 7 shows a three dimensional view of the elevation at three
different times and Fig. 8 shows the velocity field in each layer
at $ft = 7$. In the region where the wind is blowing, the Ekman

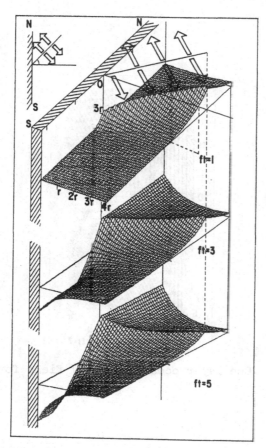

Fig. 9. Divergent wind front with α = π/4. Three dimensional
    view of the interface elevation at three different times.

flux is deflected by the coast and generates a boundary coastal jet
in the upper layer. At the discontinuity of the wind, a Kelvin wave
front is generated which propagates southward along the coast and
stops the increase of the coastal jet. In the lower layer, the
current flows in the opposite direction to the wind (Crépon and
Richez, 1982).

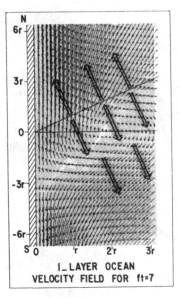

Fig. 10. One layer ocean: Velocity field for ft = 7.

    Divergent Wind Front. Let us consider the case of a wind
blowing perpendicular to the line y - xtgα = 0, northwards in the
northern region and southwards in the southern region.

$$\vec{\tau} = \tau_0 \quad sgn(\ y - xtg\alpha)\ (-\sin\alpha\ \vec{x} + \cos\alpha\ \vec{y})\ \hat{T}(t)$$

$$T(k) = -i\frac{\tau_0}{\pi k}$$

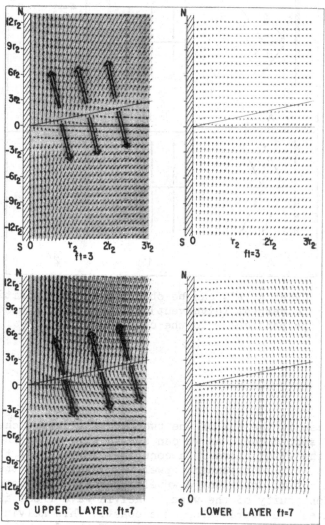

Fig. 11. Velocity field in the two layers at ft = 3 and ft = 7.

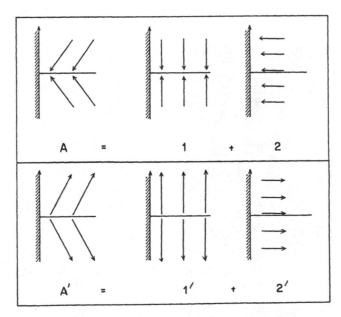

Fig.   12.   Cases A (A') can be obtained by superposition of a
             convergent (divergent) (1 or 1') and a wind blowing
             perpendicular to the coast (2 or 2').

This case is the sum of the two previous cases, changing the
sign of the second one. As we can see in Fig. 9, at the discontinuity of the wind, there is a constant elevation of the interface.
At the coast, a time dependent upwelling develops in the northern
region, and a time dependent downwelling in the southern region.
At the discontinuity of the wind, a Kelvin wave front is generated
and propagates southwards along the coast. This Kelvin wave front
stops the increase with respect to time of the downwelling. Far
offshore, the atmospheric front acts as a coast.

In a one layer ocean, this wind produces a complicated velocity field (Fig. 10) with a loop on the southern side of the atmospheric front, a northward boundary coastal jet in the northern
region and a southward coastal jet in the region downwelling. Far
from the coast, we observe the Ekman flux.

In a two layer ocean, the current flows in the opposite direction to the wind in the lower layer along the coast (Fig. 11). With respect to time, the line of divergence of the current, linked to the Kelvin wave, propagates southwards.

We remark as we deal with linearized equations, it would be easy to solve more realistic atmospheric forcings by linear combinations of the solutions for the previous ones. For instance, solutions for the cases presented in Fig. 12 can be easily obtained by adding solutions of cases (1) and (2) or (1') and (2').

SOLUTION FOR A ROTATIONAL WIND STRESS

Analytical Solution

The calculations are similar to those of the previous section. Let us suppose a wind stress $\vec{\tau}$ parallel to a straight line of the form $y - x \, tg\alpha = 0$ and constant as we move along it. The wind starts from rest at $t = 0$:

$$\vec{\tau} = (\cos\alpha\, \vec{x} + \sin\alpha\, \vec{y})\, \tau(y - x\, tg\alpha)\, \hat{T}(t) \qquad (4.1)$$

Solution of the Fourier-Laplace Transform of (2.3) yields:

$$\tilde{Z} = \frac{1}{\rho c}\, \frac{T(k)}{p^2(p^2+f^2+(2\pi kc')^2)}\left[i2\pi kfc'\, \exp(-\, i2\pi kxt g\alpha) \right.$$

$$\left. -\, \frac{(p^2+f^2)(pf\, \sin\alpha + (p^2+(2\pi kc')^2)\, \cos\alpha)\, \exp(-\,\gamma x)}{c\,(p\gamma - i2\pi kf)} \right] \quad (4.2)$$

Using Sutton's algorithm, it is found that the leading term of the asymptotic expansion of (4.2) as t tends to infinity is:

$$Z(t,x,k) \underset{t \to \infty}{=} \frac{T(k)}{\rho cf}\left[\overbrace{i2\pi kr'ft\, \frac{\exp(-\, i2\pi kxt g\alpha)}{1+(2\pi kr')^2}}^{(1)} \right.$$

$$\overbrace{-\, i2kr'ft\, \frac{\exp(-1+(2\pi kr)^2)^{1/2}\, x/r)}{1 + (2\pi kr')^2}}^{(2)}$$

$$\left. \overbrace{-\, \frac{1}{\cos\alpha}\, \frac{(1+(2\pi kr)^2)^{1/2}}{1+(2\pi kr')^2}\, \exp(-(2\pi kr)^2)^{1/2}x/r)}^{(3)} \right]$$

$$
\begin{aligned}
& \overbrace{- \frac{i\sin\alpha}{2\pi kr} \left\{ \frac{\exp(-(1+(2\pi kr)^2)^{1/2} x/r)}{1 + (2\pi kr')^2} - \exp(- x/r) \right\}}^{(4)} \\
& \overbrace{- \sin\alpha \, \exp(- x/r) \left\{ \frac{i}{2\pi kr} -(\frac{i}{2\pi kr} + tg\alpha) \, \frac{\exp(i2\pi krft)}{1+(2\pi krtg\alpha)^2} \right\}}^{(5)} \Bigg]
\end{aligned}
$$

$$(4.3)$$

The first term of (4.3) represents the FT of the response of the ocean to the wind given by (4.1) in the absence of a coast. The second and the fourth terms represent Poincaré waves of second kind of zero frequency generated at the coast in order to satisfy the boundary condition $u(0,t) = 0$. At $x = 0$, terms (1) and (2) of (4.3) cancel each other exactly. The third term of (4.3) is the FT of the elevation due to the component of the wind perpendicular to the coast. The fifth term of (4.3) is the FT of a Kelvin wave front. It is equal to the third term of (3.3) multiplied by $tg\alpha$. When the wind is perpendicular to the coast, the Kelvin wave front cannot be generated.

We obtain the solutions for the velocities by substituting (4.2) in the FLT of (2.4) and by applying Sutton's algorithm (Crépon and Richez, 1982).

Physical Interpretation

Wind blowing over a northern half plane. Let us consider a wind blowing offshore and parallel to the line $y - x \, tg\alpha = 0$, in the region $y - x \, tg\alpha > 0$.

$$
\vec{\tau} = \tau_0 \, (\cos\alpha \, \vec{x} + \sin\alpha \, \vec{y}) \, \hat{T}(y - x \, tg\, \alpha) \, \hat{T}(t) \qquad (4.4)
$$

and

$$
T(k) = \frac{\tau_0}{2} \left(\delta(k) - \frac{i}{\pi k} \right) \qquad (4.5)
$$

The curl of the wind stress is equal to:

$$
- \tau_0 \, \sin\alpha \, \delta(y - x \, tg\, \alpha) \qquad (4.6)
$$

The first term of (4.3) yields for the interface:

$$
\zeta(1) = - \frac{\tau_0}{2\rho c f} \, ft \, \exp\left(- \frac{|y - x \, tg\alpha|}{r} \right) \qquad (4.7)
$$

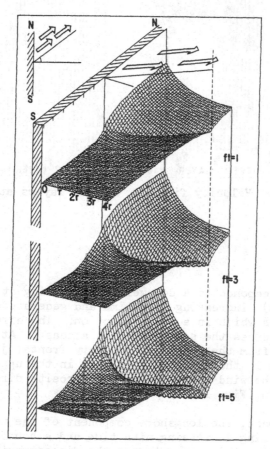

Fig. 13. Wind blowing offshore over a northern half plane with
$\alpha = \pi/4$. Three dimensional view of the interface
elevation at three different times.

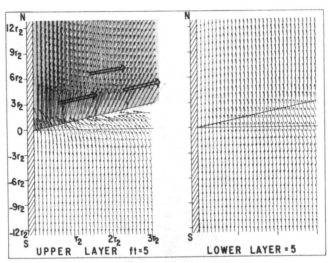

Fig. 14. Velocity field in the two layers at ft = 5.

This corresponds to a downwelling occurring at the atmospheric front (Fig. 13), increasing with time and generated by the curl of the wind stress which is a Dirac function. The effect of the wind stress curl induces the same motion as a coast. At the atmospheric front and far from the coast, a "boundary frontal jet" is generated. The current is in the offshore direction in the upper layer in the region where the wind blows and in the opposite direction in the southern region (Fig. 14).

At the coast, the longshore component of the wind stress generates an upwelling increasing with time and a boundary coastal jet. A Kelvin wave front is generated at the discontinuity and propagates southwards. It is interesting to remark that, near the coast, in the region where the wind blows, the current flows on the left of the wind.

In the lower layer, the current is in the same direction as in the upper layer but weaker. There is a deflection of the current at the discontinuity line.

Let us suppose that the wind blows in the opposite direction, that is towards the coast. A downwelling is generated at the coast in the northern region, and an important upwelling, increasing with

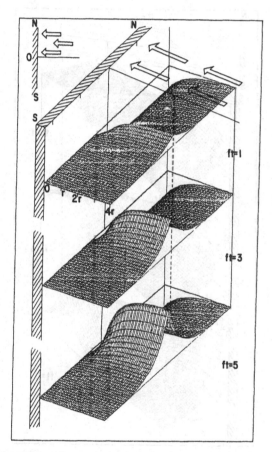

Fig. 15. Wind blowing onshore over a northern half plane.
Three dimensional view of the interface elevation
at three different times.

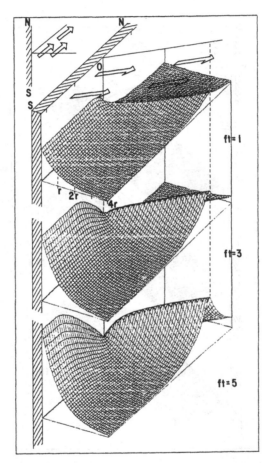

Fig. 16. Wind blowing offshore over a southern half plane with
α = π/4. Three dimensional view of the interface
elevation at three different times.

time, appears at the atmospheric front at an offshore distance of
the order of the internal radius of deformation r_2 . When α is
equal to zero, the downwelling at the coast is steady and there is
no Kelvin wave front (Fig. 15).

Wind blowing over a southern half plane.

$$\vec{\tau} = \tau_0 \, (\cos\alpha \, \vec{x} + \sin\alpha \, \vec{y}) \, \hat{T}(-(y - x \, tg\alpha))\hat{T}(t) \qquad (4.8)$$

$$T(k) = \frac{\tau_0}{2} \, (\, \delta(k) + \frac{i}{\pi k} \,) \qquad (4.9)$$

At the coast, in the southern region, the longshore component
of the wind raises a time dependent upwelling. A Kelvin wave front
is generated at the discontinuity, propagates southwards and stops
the increase with time of the upwelling (Fig. 16). At the atmospheric front, at some distance from the coast ($x > r_2$), the baroclinic mode generates a large elevation of the interface, increasing

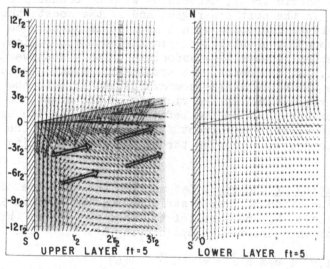

Fig. 17. Velocity field in each layer at $ft = 5$.

with time and exponentially decreasing with the distance from the front.

As in the previous case, a "boundary frontal jet" appears along the discontinuity line (Fig. 17), in opposing directions in the two regions $(y - x \, tg \, \alpha) \gtrless 0$. Near the coast, in the southern region, the current flows as a boundary coastal jet, but is on the left of the wind. In the lower layer, the current flows in the opposite direction to the boundary coastal jet and is deflected when it flows through the discontinuity line of the wind.

When α is equal to zero, the motion generated by the atmospheric front is the same as in the previous case (Fig. 15) but at the coast we would observe a steady upwelling in the southern region. In particular, the Kelvin wave front no longer exists in either case. This phenomenon could be a mechanism for the formation of offshore "plumes" which are often encountered in upwelling regions.

Wind Shear. Consider now a shear of the wind stress along the line $y - x \, tg \, \alpha = 0$

$$\vec{\tau} = \tau_0 \, \sin(y - x \, tg \, \alpha) \, (\cos\alpha \vec{x} + \sin\alpha \vec{y}) \, \hat{T}/t \tag{4.10}$$

$$T(k) = - i \frac{\tau_0}{\pi k} \tag{4.11}$$

This case can be solved by a linear combination of the two previous ones. A time dependent downwelling is observed (Fig. 18) at the atmospheric front, associated with a "boundary frontal jet" (Fig. 19). In the northern region and at the coast, the elevation of the interface increases with time. It is associated with a northward boundary coastal jet. In the southern region, there is a time dependent downwelling associated with a southward boundary coastal jet. At the coast, the atmospheric front generates a Kelvin wave which propagates southwards and arrests the downwelling. In the two regions and in the vicinity of the coast, we observe that the current flows on the left of the wind in the upper layer. In the lower layer, the current flows northwards in the two regions and is deflected when passing through the discontinuity line of the wind.

Relaxation of the wind. Let us consider the case of a class B wind, blowing offshore over a strip defined by the two lines $y - x \, tg \, \alpha = \pm \, b$, and parallel to these two lines. We now study what happens when the wind stops blowing at $ft = 5$.

$$\vec{\tau} = \tau(y) \, (\cos \alpha \, \vec{x} + \sin \alpha \, \vec{y}) \, \hat{T}(t) \tag{4.12}$$

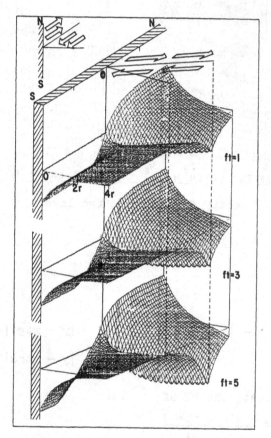

Fig. 18. Wind shear front with α = π/4. Three dimensional
          view of the interface elevation at three different times.

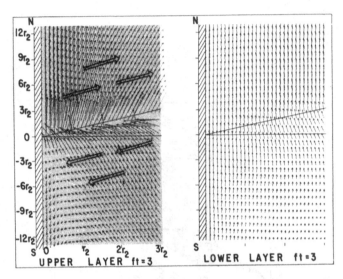

Fig. 19. Velocity field in the two layers at ft = 3.

with

$$\tau(y) = \tau_0 (1 - | y - x \, tg\alpha | / b) \qquad |y - x tg\alpha| < b$$

$$\tau(y) = 0 \qquad\qquad\qquad\qquad\qquad |y - x tg\alpha| > b \qquad (4.13)$$

Hence, at the coast, the FT of $\vec{\tau}$ is:

$$T(k) = \frac{\tau_0}{b} \; \frac{\sin^2(\pi b k)}{(\pi k)^2} \qquad\qquad\qquad (4.14)$$

With this T(k), the FT of the analytical solution is given by (4.3). Far from the coast, the curl of the wind stress generates a jet, the amplitude of which increases with time. The elevation is of the form:

$$\zeta(1) = \frac{\tau_0}{\rho c f} \; ft \; \exp(- |y - x \, tg\alpha|/r' \underset{y}{*} (- \frac{r'}{\tau_0} \frac{\partial}{\partial y} \tau(y))$$

$$(4.15)$$

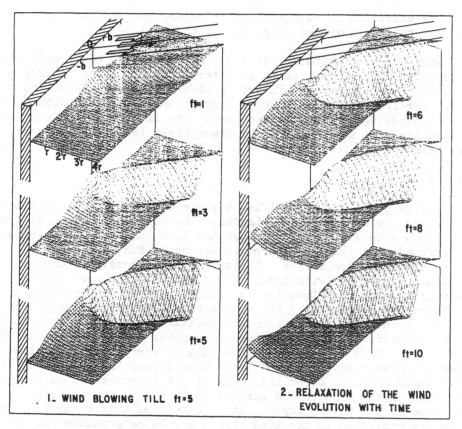

Fig. 20. Wind blowing offshore over a strip (width $2b = 2r_2$).
At ft = 5, the wind stops blowing.

The larger the wind stress curl, the larger $\zeta(1)$ is. In a
two layer ocean, the baroclinic mode $\zeta_2(1)$ generates an offshore
upwelling of a width of the order of r_2 (Fig. 20 a) which increases with time. This phenomenon could also be a mechanism for the
formation of offshore "plumes".

At the coast, $\zeta(1)$ is cancelled by $\zeta(2)$. The dominant motion
becomes the Kelvin wave front $\zeta(5)$, generated at $y = b$, $y = 0$,
$y = -b$, and propagating towards $y < 0$.

For $y < -b$ and $t > (b - y)/c$, $\zeta(5)$ may be written:

$$\zeta(5) = -\frac{\tau_0}{\rho c f}\ \frac{b}{r}\ \sin \alpha\ \exp(-x/r) \qquad\qquad (4.16)$$

The elevation is constant with time and proportional to the integral of the longshore component of the wind.

Now, the wind stops at ft = 5. At the coast, the steady part of the upwelling propagates as an internal Kelvin front towards y < 0 (Fig. 20 b), and the interface elevation becomes equal to zero. Far from the coast, the offshore plume associated with $\zeta_2(1)$ stops increasing and remains constant. The offshore plume may depend on the past structure of the atmospheric forcing and not on the present one.

CONCLUSION

The response of a stratified f-plane ocean to transient atmospheric fronts has been investigated. The asymptotic state at large time for a wide variety of forcings has been calculated analytically. It is found that only longshore variations in the wind stress generate Kelvin wave fronts. Horizontal two dimensional circulations are associated with them. For a wind blowing over a strip, the amplitude of the elevation is proportional to the integral of the longshore component of the wind stress and does not depend on time as in the case of the boundary coastal jet. The elevation is maximum outside the wind strip and propagates as an internal Kelvin wave front. This could explain the lack of correlation sometimes observed between upwelling and local wind stress (Smith, 1978). In a two layer ocean, a baroclinic coastal jet is generated in the direction of the wind in the surface layer and in the opposite direction in the bottom layer. It is concluded that a large scale longshore variation in the wind stress may be responsible for the poleward current which is observed in the upwelling areas located on western coasts of continents (Hart and Currie, 1960; Smith, 1968; Huyer, 1976).

The wind stress curl can generate offshore plumes, with width of the internal radius of deformation and increasing with time. In the upper layer, a "boundary frontal jet" is generated along the atmospheric front and the currents are in opposite direction on each side of the front.

REFERENCES

Crépon, M., and Richez, C., 1982, Transient upwelling genera-
    ted by two-dimensional atmospheric forcing and variabi-
    lity in the coastline, J. Phys. Oceanogr., 12,(in press).
Csanady, G.T., 1975, The coastal jet conceptual model in the
    dynamics of shallows seas, in:"The Sea, Ideas and Ob-
    servations on Progress in the Study of the Sea", 6,
    Marine Modeling, 117-144, Goldberg, Mc Cave, O'Brien
    and Steele Ed., Wiley Interscience Pub., J. Wiley and
    Sons, New York, London, Sydney, Toronto.

Hart, T.S., and Currie, R.I., 1960, The Benguela current,
    Discovery Reports, 31: 123-298.

Huyer, A., 1976, A comparison of upwelling events in two loca-
    tions: Oregon and Northwest Africa, J. Mar. Res., 34,
    4:531-546.

Lighthill, M.J., 1969, Dynamic response of the Indian Ocean to
    onset of the Southwest Monsoon, Phil. Trans. Roy. Soc.
    London (A), 265, 1159, 45-92.

Smith, R.L., 1968, Upwelling, in: "Oceanogr. Mar. Biol.", Ann.
    Rev., 6: 11-46.

Smith, R.L., 1978, Poleward propagating pertubations in cur-
    rents and sea-levels along the Peru Coast, J. Geophys.
    Res., 83, C 12:6083-6092.

Sutton, W.G., 1934, The asymptotic expansion of a function
    whose operational equivalent is known, J. London Math.
    Soc., 9:131-137.

Thomson, R.E., 1970, On the generation of Kelvin-type waves
    by atmospheric disturbances, J. Fluid Mech., 42. 4:
    657-670.

WIND-FORCED SHELF BREAK UPWELLING

Brian D. Petrie

Atlantic Oceanographic Laboratory, Bedford Institute
of Oceanography
P.O. Box 1006, Dartmouth, N.S., Canada, B2Y 4A2

ABSTRACT

An analysis of current meter and hydrographic data collected
from an array of moorings situated at the shelf break off Nova
Scotia is presented. In particular, the focus is on the ocean
response to moderate (10 to 20 m s^{-1}), transient winds of two or
more days duration blowing parallel to the local bathymetry. Under
these conditions, upwelling occurs at the shelf break from depths
to be confined to within 10 km of the slope. Acceleration, Coriolis
and pressure gradient terms are important for the along-and cross
slope momentum balances. Local topographic variations cause anoma-
lously large bottom currents on the shoreward side of the shelf
break.   Both coastal sea level and the shelf break circulation
respond coherently to synoptic wind forcing but quantitative agreement with 2 layer shelf models is poor.

INTRODUCTION

The continental shelf (Fig. 1) off Nova Scotia is broad, about
200 km wide, and after an initial sharp dropoff near the coast,
doesn't exhibit any continuous increase in depth until the shelf
break is reached. However, the topography of the shelf is complicated by the presence of numerous shallow banks and deep basins.
In December 1975 Bedford Institute of Oceanography began an observation program consisting of a number of current meter moorings
and hydrographic cruises (Fig. 1). The main mooring line runs
down the axis of the deepest channel connecting the inner shelf to
the deep ocean. Another line of two sites lies to the west, offshore of LaHave Bank. Note that in this region the shelf break
runs approximately east-west.

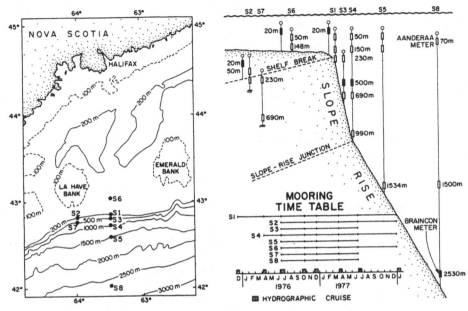

Fig. 1. Schematic diagram of shelf break array.

This paper will sketch the response of shelf water to transient
wind events with particular emphasis on processes occurring at the
shelf break. A recent review by Huthnance (1981) has outlined the
numerous physical processes which can occur in such a region. With
reference to shelf break upwelling, he notes that observations of
the phenomenon are considerably outnumbered by the models. From
the data collected on the Scotian Shelf the discussion will deal
with the response of sea level, shelf and shelf break currents to
atmospheric forcing during fall-winter periods when winds are
strongest, the number of upwelling events over a season, upwelling
velocities, the depth on the continental slope to which upwelling
effects are felt and the momentum balances. Finally, a comparison
of the current observations from one event particularly amenable
to modelling will be made to the response predicted by a two layer
numerical model similar to that of O'Brien and Hurlburt (1972).

A more complete discussion of these data is given by Petrie
(1982). Complete data summaries including filtering techniques,
in situ calibrations, instrument accuracies, time series plots and
overall statistics have been given by Lively (1979 a,b). Meteorological data are primarily from an Atmospheric Environment Service
station at Sable Island (43°55'N, 60°W), about 300 km east-northeast
of the mooring array.

RESPONSE ON THE SHELF

    Winds on the Scotian Shelf during fall and winter are more
energetic than those during spring and summer with enhanced variance in the band 2-10 d (Smith and Petrie, 1982). The period of
October-December 1976 was characterized by a mean wind of about
5.4 m/s directed towards 72°T and by a number of storms with winds
directed basically eastwards along the shelf break. Strong pulses
of wind toward the east should result in setdown of coastal sea
level, enhanced current variance in the 2-10 d band, increased offshore Ekman surface drift at the mooring sites, strengthened alongslope currents and upwelling at the shelf break. Each of these
aspects of the flow will be examined in turn.

(a) Coastal Sea Level

    Adjusted sea level at Halifax (Fig. 2) for the period October-
December 1976 shows that in general eastward winds do result in
setdown at the shore in keeping with the results of Sandstrom
(1980). In Fig. 2 eastward winds have been plotted positive downwards and the scale adjusted to conform to the average regression
slope of 0.025 m setdown of sea level per meter per second eastward wind as observed by Sandstrom. The comparison shown in Fig. 2
is not as good as that of Sandstrom (1980). The discrepancies are
discussed by Petrie (1982) but will not be pursued here.

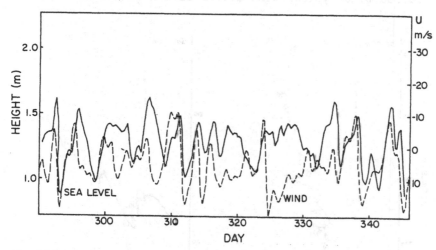

Fig. 2. Adjusted sea level measured at Halifax and eastward wind
        speed observed at Sable Island for the period October-
        December 1976. The broken line indicates mean sea level
        over the period shown. Wind speed is plotted positive
        downward with the scale adjusted to conform to
        Sandstrom's (1980) observation of about 0.025 m setdown
        of sea level per meter per second eastward wind.

(b)   Current Variance

The total (eastward plus northward) current spectrum for the
October–December period at the shelf site S6 (148 m) has variance
($1.9 \times 10^{-2} m^2/s^2$) in the 2-10 d band ten times larger than at the
same site during the following summer. Similarly, the variance
($2.5 \times 10^{-2} m^2/s^2$) of the 20 m current data was about five times
greater than for summer periods.

(c)   Outer Shelf Currents

Wind-induced circulation is marked most dramatically on the
outer shelf by large pulses of onshore bottom currents which
occasionally peak (unfiltered data) to 1.0 m/s or greater.   The
progressive vector diagrams (Fig. 3) for the 20 m and 148 m sensors
at S6 show four distinct episodes of exceptionally strong onshore
flow commencing on days 289, 312, 325 and 338.   These events result
from strong, persistent winds (Fig. 2) which blow essentially to
the east-northeast or east with mean speeds of 10 to 15 m/s.   The
intense flows at the deeper instrument were directed along 32°T to
within 6°.   On the other hand, the shallow instrument, which had a
decided tendency toward southward (0.10 m/s) flow for much of the
time, exhibited eastward to north-eastward currents during the
major events, especially the last three.   Both the inordinate
strength and direction of the bottom currents (a 2 layer shelf
model forced by a typical wind stress of 0.3 Pa for the events would

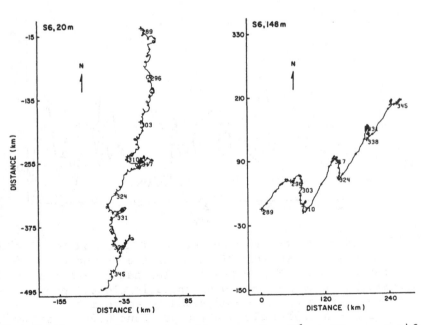

Fig. 3. Progressive vector diagrams for S6 at 20 m and 148 m,
        October–December 1976.

give a current of about 0.3 m/s in the direction of the forcing
after two days at the S6 site) and the direction of the near-
surface flow (an onshore component of about 0.10 m/s instead of
offshore at a similar rate) suggest that topographic effects play
a significant role in determining the circulation in this channel
between two shallow banks.

    Figure 4 shows the temperature and currents resolved along
32 and 122°T for the 148 m sensor at S6. Note that the event which
began on day 259 is lost due to the filtering procedure used to
remove high frequency motions. It is apparent that the pulses of
strong current are associated with lowered temperatures. Salinity
(not shown) remains roughly constant at approximately 35 ppm implying that the cooler water has upwelled from greater depths on the
slope and is not from the fresher surface layer. It is evident
also from Fig. 4 that the decrease in temperature lags the increase
in current. This lag (roughly 18 hr) corresponds to the time from
the beginning of an event it takes the upwelled water at the shelf

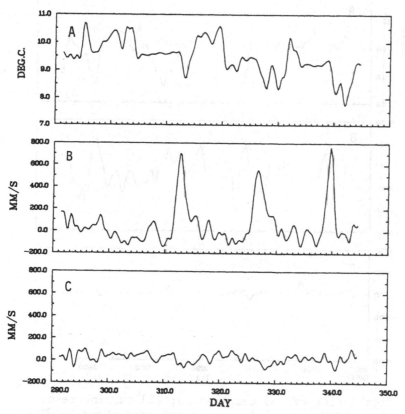

Fig. 4. Time series of (a) temperature; (b) current resolved
        along 32°T; and, (c) current resolved along 122°T at
        S6 (148 m) for October–December 1976.

break to reach the S6 site. The delay of 18 hr corresponds to an onshore velocity of 0.34 m/s and implies that strong flows must prevail over at least the distance from the break to the S6 mooring.

The onshore flow of slope water could conceivably reach the deep inner basins of the shelf and partially flush them. Such occurrences have been documented (Petrie and Smith, 1977).

RESPONSE AT THE SHELF BREAK

During the major wind events in the October–December period vertical shear developed for both components of current at the shelf break moorings S1 and S2. In general, eastward and northward flows decreased between 50 and 220 m at S2 and between 50 and 150 m at S1. Note that the direction vane at S1, 220 m became tangled with the ground line rendering the data unreliable. The response at these

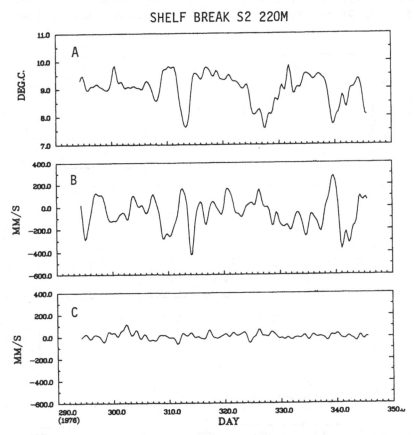

Fig. 5. Time series of (a) temperature; (b) current resolved along 72°T; and, (c) current resolved along 162°T at S2 (220 m) for October–December 1976.

locations is typified by the data from S2, 220 m shown in Fig. 5
which feature strong alongslope flows and upwelling of cooler water.
The response of the alongslope current decreased offshore but onshore flow was evident at the 230 m depth on the 700 m isobath (S3
and S7) and at 150 m at the 1000 m contour S4.

(a)   Frequency of Upwelling Events

     During the two fall and one winter period when moorings were
in place, at least 20 upwelling events were indicated by coincident
extrema of the temperature and velocity fields. These could be
matched to strong (> 10 m/s), persistent (> 2 d), eastward winds
with about 85% accuracy.

(b)   Upwelling Velocities

     Based on hydrographic data taken during cruises and the time
series of temperature and salinity recorded by the current meters,
water from as deep as 400 m was observed to reach the 220 m shelf
break sensors. Mean and peak vertical velocities could be estimated
by examining the rate of change of the filtered data (low passed to
remove variability with period of less than 28 hr) at 0.5 mm/s and
2 mm/s respectively.

(c)   Depth of Upwelling on the Slope

     Upwelling was apparent at least to 230 m and upwelled water
came from as deep as 400 m. There were occasions during eastward
winds when the flow at S4 (1000 m isobath), 500 and 690 m, turned
from a very persistent one nearly due west to a northerly current.
At S3 (700 m isobath) during these events the onshore flow at 670 m
was weaker and was accompanied by a drop in temperature possibly
indicative of upwelling. Upwelled water came from about 50 m deeper
than the S3 sensor based on the vertical temperature distribution.
Upwelling appeared to be confined to the shallow site.

MOMENTUM BALANCES

     Momentum balances were examined in three ways:

  i) Comparison of the vertical shear in the alongslope current
     calculated from a north-south line of hydrographic data
     which by chance were collected during an upwelling event.

 ii) Comparison of geostrophic velocity shear based upon tempera-
     ture and salinity data with the alongslope current shear,
     both from the moored sensor observations.

iii) Scaling all terms in the momentum equations relative to the
     Coriolis term using data from different moorings, depths and

upwelling events. The cross-slope momentum balances may be
evaluated by all three of these means but the alongslope
balance can only be addressed by iii).

At the shelf break, acceleration, Coriolis force and pressure
gradient terms appear to be important. At the shelf site S6, it
appears that in addition to these terms at least one nonlinear term
(VV_y) is important. This is probably the result of the complex
topography giving rise to significant cross-slope gradients in an
already intense flow.

DISCUSSION

The analysis sketched here has led to the following
conclusions:

(1) The coastal sea level and the shelf break circulation respond
     coherently to synoptic wind forcing.

(2) Strong alongslope currents with large vertical shears are
     confined to the shelf and shelf break area.

(3) At the shelf break intense upwelling is observed which
     originates from at least 400 m with peak vertical velocities
     of 2 mm/s.

(4) During 1977 when currents were measured at 50, 100 and 150 m
     on the 1000 m isobath, there was suggestion of uniform onshore
     flow during upwelling events.

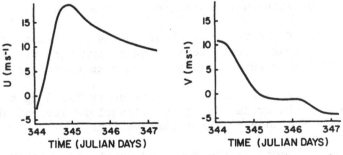

Fig. 6. Sable Island wind components for days 344 to 347, 1977.

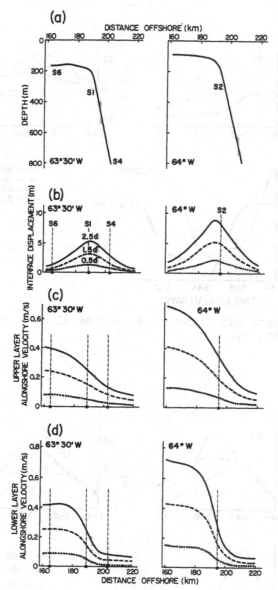

Fig. 7. (a) Topography along 63°30'W and 64°W. Mooring positions
are indicated. (b) Interface displacement at times 0.5,
1.5 and 2.5 d after the onset of forcing. The light vertical lines indicate mooring positions. (c) Alongslope
current for the upper layer at the same time periods.
Inertial period motions have been removed by harmonic
analysis. (d) Alongslope current for the lower layer at the
same time periods. Positive currents are in the direction
of wind forcing.

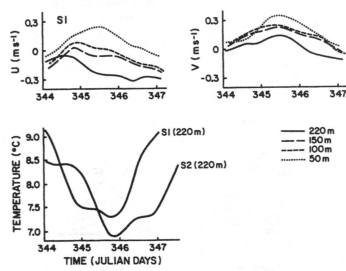

Fig. 8. Currents and temperatures at the S1 mooring for days 344
         to 347, 1977.

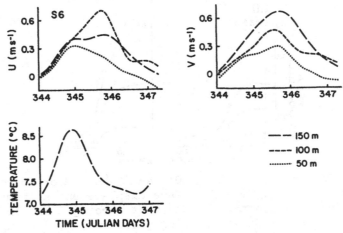

Fig. 9. Currents and temperature at the S6 mooring for days 344
         to 347, 1977.

(5) The leading terms in the alongslope and cross-slope momentum
     balances at the shelf break are acceleration, Coriolis force
     and pressure gradients. At the shelf site (S6), non-linear
     terms are particularly important.

     O'Brien and Hurlburt (1972) have considered two layer numerical
models of the response of continental shelf water to steady wind
forcing. A similar model was applied to the Scotian Shelf and compared to observations. The wind event of Fig. 6 was modelled by
a steady alongslope wind of 13.5 m/s beginning on day 344.5 and
lasting for 2.5 d. The primitive equations were solved numerically
for the topography along 63°30' (the main mooring line) and 64°W
(secondary mooring line) with a horizontal resolution of 1 km, an
upper layer depth of 50 m with ρ = 1024 kgm/m^3, and a lower layer
depth of the total depth minus 50 m with ρ = 1027 kgm/m^3. The
model results are shown in Fig. 7 (inertial period waves have been
removed by harmonic analysis) and may be compared to the data from
S1 and S6 in Figs. 8 and 9 respectively. The observations, unlike
the model, show the alongslope current decreasing starting about
day 346 at S6 and S1, differences at S1, 50 and 150 m of 0.30 m/s
(maximum model difference at this site is about 0.02 m/s after 60
hr with the alongslope current in the lower layer exceeding that
in the upper layer), and cross-slope currents at S1 differing by
0.20 m/s in the vertical (model predicts 0.05 m/s between the two
layers). The two layer model predicts (Fig. 7) more intense upwelling at S2 than at S1 and this may be reflected in the colder
water observed (Fig. 8) at S2. The model, however, is inadequate
to assess the depth that upwelled water would come from on the
slope. There is the expectation, though, from relevant models of
coastal upwelling and from the data that the onshore flow in the
deeper water would compensate the offshore near surface Ekman flux
and be approximately uniform with depth. The upwelled water
detected at the shelf break would then come from a wedge of fluid
defined by the thermocline and the physical bottom of the continental slope. After 2d, under the influence of the wind event modelled
above, water from about 340 m should be detected at S1 (220 m) in
good qualitative agreement with upwelling events.

     The topography of the outer Scotian Shelf which is inade-
quately described by 2 dimensional models has led to poor quantitative and qualitative agreement between predicted and observed
flow. Investigation of a continuously stratified (or multiple
layered) 3 dimensional model of shelf break upwelling would appear
timely.

REFERENCES

Huthnance, J.M., 1981, Waves and currents near the continental
     shelf edge, Prog. Oceanog., 10:193-226.

Lively, R.R., 1979a, Current meter and meteorological obser-
      vations on the Scotian Shelf, December 1975 to January
      1978. Volume 1: December 1975 to December 1976. Bedford
      Institute of Oceanography Data Series, BI-D-79-1, January
      1979, 280 pp.

Lively, R.R., 1979b, Current meter and meteorological observa-
      tions of the Scotian Shelf: December 1976 to January
      1978. Volume 2: December 1976 to January 1978. Bedford
      Institute of Oceanography Data Series, BI-D-79-1,
      January 1979, 368 pp.

O'Brien, J.J., and Hurlburt, H.E., 1972, A numerical model of
      coastal upwelling, J.Phys.Oceanogr., 2:14-26.

Petrie, B., 1982, Current response at the shelf break to
      transient wind forcing, J.Phys.Oceanogr., submitted.

Petrie, B., and Smith, P.C., 1977, Low-frequency motions on
      the Scotian Shelf and Slope, Atmosphere, 15:117-140.

Sandstrom, H., 1980, On the wind-induced sea level changes on
      the Scotian Shelf, J.Geophys.Res., 85:461-468.

Smith, P.C., and Petrie, B., 1982, Low-frequency circulation
      at the edge of the Scotian Shelf, J.Phys.Oceanogr.,
      12:28-46.

A SEASONAL UPWELLING EVENT OBSERVED OFF THE WEST COAST OF

BRITISH COLUMBIA, CANADA

Howard J. Freeland

Institute of Ocean Sciences
P.O. Box 6000
Sidney, B.C. V8L 4B2

INTRODUCTION

This paper discusses observations of an upwelling centre observed on several occasions over the last 3 years. The paper represents only a summary of the work completed and a discussion in greater depth is to appear elsewhere, Freeland and Denman (1982). The latter paper, even though more detailed than the present one, is based on a body of data considerably greater than that actually presented. Hence I will attempt to illustrate the processes observed with observations not explicitly used in Freeland and Denman (1982)

The region to be discussed, see Fig. 1, lies off the west coast of southern Vancouver Island. Besides being exposed to the eastern Pacific Ocean this area is also exposed to forcing from the Strait of Juan de Fuca. The outflow from that Strait is predominately in the upper 100m of the water column and is the upper layer of a classical estuarine circulation pattern, driven ultimately by the fresh water of the Fraser River. This outflowing water, which has a salinity contrast with the deep layers of only about 1½⁰/oo by the time it reaches the shelf, turns northwards along the coast to form a Vancouver Island coastal current. The coastal current is, presumably, fed by buoyancy sources (fresh water outflow from rivers and inlets) to form a direct analogue to the Alaska Coastal Current and the Norwegian Coastal Current. (See papers by Royer and by Mork in this volume.)* This outflow

_____

*It would be interesting to attempt a detailed comparison of the different buoyancy driven coastal currents, augmented if possible by laboratory models.

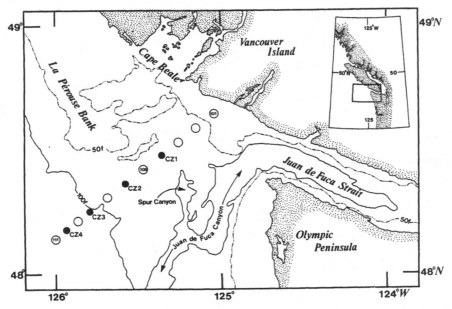

Fig. 1. The study region off southern Vancouver Island

driven current appears to be a permanent feature of the local
oceanography but has, presumably, a large seasonal variation.

The local continental shelf is heavily dissected by deep
submarine canyons. The main channel is the Juan de Fuca Canyon.
However, a second canyon which we nicknamed the "Spur Canyon"
appears to play a very substantial role in determining the local
physical and biological oceanography. The changes in the density
structure which we have attributed to an interaction with this
canyon in turn have great impact on the deep water masses in a
nearby fjord, Alberni Inlet, Stucchi (1982).

DISSOLVED OXYGEN AND DENSITY

Freeland and Denman (1982) show time series of dissolved
oxygen concentration and density at one place on the continental
shelf (station 105). The data show unambiguous evidence of an
abrupt seasonal event each spring which delivers dense, low-oxygen
water onto the continental shelf in June. The annual cycles of
density and oxygen are very large and in some general sense
represent a single annual upwelling event that typically lasts for
4 months.

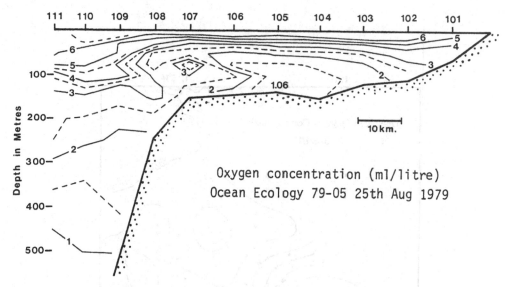

Fig. 2.   A typical cross section of dissolved oxygen concen-
          tration across the continental shelf. The line of
          stations 101 to 111 is shown on Fig. 1.

        Fig. 2 shows a vertical section of dissolved oxygen
concentration through the anomaly and distinctly shows the pool
of low oxygen water on the continental shelf. A confused area
over the shelf edge is evident where the inversion in dissolved
oxygen appears to be real and not a result of mixing up sample
bottles. The section shown is fairly typical of summer conditions
on the southern Vancouver Island shelf; see Freeland and Denman
(1982) for other examples.

        Figure 3 shows horizontal sections of water properties
contoured to exhibit the horizontal extent of the low-oxygen
anomaly. On the diagrams the 180m depth contour has been marked
lightly; note that the property contours always close around
the northern terminus of the Spur Canyon. Only two examples are
shown here from a single cruise; others are shown by Freeland
and Denman (1982). Other properties have been mapped and the same
conclusion is reached regularly. We conclude that the head of
Spur Canyon is a major local source of dense low oxygen water
that is also particularly rich in nutrients. The nutrient source
in turn triggers high phytoplankton productivity so that the feature
can be seen in blue-green ratio satellite photography, representative of chlorophyll concentration, even though there is no
surface manifestation in temperature or other physical variables.

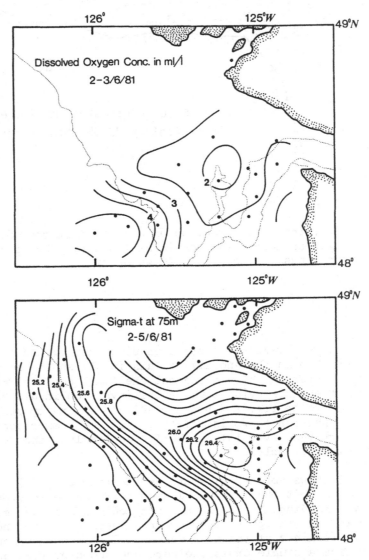

Fig. 3.   Typical distributions of oxygen (upper panel) and sigma-t
(lower panel) in the horizontal plane. Note the relationship of the innermost contours to the northern terminus
of the Spur Canyon.

Biologically the area influenced by the presence of this feature
is highly productive. Mackas and Sefton (1983) have analysed zooplankton species variation and find the same patterns emerge as
appear from the physical variables. In particular the statistical
structures of physical and biological fields appear to be related.

VELOCITY STRUCURE

     As mentioned in the introduction, there is an outflow from the
Strait of Juan de Fuca which appears to drive a coastal current.
Current meters located inshore show a weak seasonal cycle, as
might be expected from the nature of the current. At the shelf
edge however, we observe a strong northward flow at all depths
in the winter. In the spring of each year the shelf edge currents
reverse producing what Huyer et al (1979) call the "spring
transition". After the spring transition a barotropic eddy appears
on the southern Vancouver Island shelf. At first sight it might be
that the lateral shear (northward flow inshore, southward at the
shelf edge) would be sufficient to spin up such an eddy. However,
suppose that at time t=0 we initiate narrow currents V_o and $-V_o$ at
opposite sides of a continental shelf of width 2L. By scaling,
the time taken to spin up the central region is $T=L^2/A$, where A is
the horizontal eddy diffusion coefficient. Taking L=30km, and
$A \sim 10^5 cm^2/sec$ (Okubo, 1971) we find T=3 years. Even if A
approaches oceanic values of, say, $5.10^6 cm^2/sec$, T is still much
too long. The flow out of the Juan de Fuca Strait is however
strongly sheared, see Hugett et al (1976) for example, and must
be advecting cyclonic vorticity onto the shelf. It also seems
likely that the shelf edge current will advect cyclonic vorticity
from the north. Then there seems little difficulty in getting
vorticity to the centre of the shelf so that the spin-up time
is governed by an advective time scale.

     By whatever means, it does appear that an eddy with weak
vertical shear is initiated at the time of the spring transition.
The centre seems to locate itself over the head of the Spur Canyon
which then serves as a source of upwelled water. Presumably the
geometry is accidental and the eddy is there because of vorticity
source constraints that have nothing to do with the canyon.
However, let us now consider the interaction between the (approximately) barotropic eddy and a narrow canyon.

     The internal Rossby radius on the shelf is about 15 km, the
canyon is therefore narrow, having a width of typically 5 km.
We suggest that to first order the eddy is not influenced by the
canyon because of this scale disparity. A parcel of water in mid
depth is in approximate geostrophic balance in the eddy; so a
radial inward pressure gradient balances the Coriolis force
produced by the azimuthal component of velocity. Deeper, the
pressure gradient remains continuous, but as we pass into the

canyon the azimuthal velocity must drop to zero. Hence, water inside
the canyon is unbalanced geostrophically but sees an inward pressure
gradient. This, we believe, drives the upwelling. Can this mechanism
supply water from a sufficiently great depth?

Let us suppose that the eddy does work on water parcels as they
move along the canyon system from the shelf edge, distance L from
the centre of the eddy. That work is W_1, where

$$W_1 = \int_0^L \frac{\partial p}{\partial x}\, dx$$

and $\frac{\partial p}{\partial x}$ is the geostrophic pressure gradient associated with the
eddy. Let us balance this against the work done by raising a
water parcel from an arbitrary depth H to the depth of the
continental shelf, 150m, i.e. W_2, where

$$W_2 = \int_H^{150} g\frac{\partial \rho}{\partial z}\, zdz$$

and $\rho(z)$ is the background density profile at the mouth of the
canyon. If $W_1 = W_2$ then H is the depth from which water can be
raised to the shelf level, arriving with zero kinetic energy.
Hydrographic data suggest a value of $H \sim 450\,m$, so it is energeti-
cally possible to raise water by this means from such large depths.

DISCUSSION

The energetic arguments above are interesting but not com-
pelling. The process is strongly non-linear because the water
advected from large depths up the canyon will tend to produce an
outward baroclinic pressure gradient that will limit the flow.
However, this baroclinic gradient will make a steady state possible
and in that sense will tend to simplify the problem. A more
detailed theoretical analysis is carried out in Freeland and Denman
(1982).

REFERENCES

Freeland, H.J. and Denman, K.L. 1983, A topographically controlled
        upwelling centre off southern Vancouver Island, J. Mar. Res.,
        (in press)
Huggett, W.S., Bath, J.F. and Douglas, A., 1976, Data record of
        current observations, Volume XV, Juan de Fuca Strait 1973,
        Institute of Ocean Sciences, Sidney, B.C., Canada.

Huyer, A., Sobey, E.J.C. and Smith, R.L., 1979, The spring
    transition in currents over the continental shelf.
    J. Geophys. Res., 84 (C11), 6995-7011.
Mackas, D.M. and Sefton, H.A., 1983, Species assemblages of zoo-
    plankton and phytoplankton off the southern British
    Columbia coast. I. Geographic patterns, J. Mar. Res.,
    (sub judice).
Okubo, A., 1971, Oceanic diffusion diagrams, Deep Sea Res.,
    18 (8), 789-802.
Stucchi, D.J., 1982, Shelf-fjord exchange on the west coast of
    Vancouver Island. Proceedings of the Coastal Oceanography
    Workshop, Os, Norway, June 6-11, 1982.

SHELF WAVES OF DIURNAL PERIOD ALONG VANCOUVER ISLAND

William R. Crawford, Richard E. Thomson and
W. Stanford Huggett

Institute of Ocean Sciences
Sidney, B.C., Canada V8L 4B2

ABSTRACT

We observe strong diurnal period tidal currents over the continental shelf in current meter records from the west coast of Vancouver Island. The clockwise rotation of these currents, their
rapid decay seaward of the shelf break and northward phase propagation suggest they are due to low mode number continental shelf
waves driven by tidal currents. We compare the current ellipses
given by harmonic analysis of current meter records off Estevan
Point at the center of the Island with theoretical waves of two
forms: a Kelvin wave described by the model of Henry et al. (in
preparation) and the baroclinic continental shelf wave described
by the model of Brink (1982). Excellent agreement between observed
currents and the sum of these two waves is obtained for each of four
mooring periods for the last deployment period, when 95% of the
variance was explained by these waves.

INTRODUCTION

Continental shelf waves are oscillations of subinertial
frequency trapped along continental margins where topographic
gradients provide a restoring mechanism for horizontal fluid displacements. First detected by Hamon (1962) in sea level records
along the east coast of Australia, these motions continue to be the
focus of numerous theoretical and observational investigations.
Recent knowledge of the generation, propagation and dissipation of
these waves has been summarized in review articles by Allen (1980),
Mysak (1980) and Huthnance (1981).

For most part, research has been confined to quasi-nondisper-
sive wind-generated motions. These have wavelengths of order 100 km
and periods of 2 to 10 days, corresponding to the time intervals
between travelling extra-tropical cyclones. Considerably less attention has been devoted to tidally-forced continental shelf waves that
can exist at diurnal periods poleward of 30° latitude. Such waves
were first observed by Cartwright (1969) in current records on
the St. Kilda Shelf off Scotland and have been subsequently detected
in current meter records along the northeast coast of United States
(Daifuku, 1981) and on the west coast of Vancouver Island (Thomson
andCrawford, 1982; Crawford and Thomson, 1982, hereafter called
CT82).

In the latter paper we discuss observations of these currents
along Vancouver Island and compare measured currents with predictions
of a simple two-dimensional numerical model of barotropic currents.
Here we extend the theory to show that a baroclinic shelf wave
offers a superior fit to the observed current ellipses over the
shelf and to the vertical structure of the currents in deep water.

MEASUREMENTS

We present here a brief description of the measurement program,
described as well in CT82 and by Freeland (1983) in this volume.
Current meter moorings were deployed along the west coast of Vancouver Island between May 1979 and September 1980, as illustrated
in Fig. 1. The southern line of moorings off Carmanah Point was maintained by H. Freeland of I.O.S. while we placed meters along the
Estevan and Brooks lines. Current meters on the continental shelf
were at the nominal depths of 50 and 100 m, suspended from sub-surface floats. We set shallow meters at 15-20 m below the surface at
E0, E1, E2 and B1 in the summer periods (May-September). The meters
were serviced at 4 month intervals (May, September, January) and
water properties were obtained along the coast to a distance of 100
km offshore during each servicing cruise.

The continental shelf broadens from north to south along
Vancouver Island, from a width of 10 km off Brooks Peninsula to
70 km off Carmanah Pt. The continental slope is of uniform width
(50 km) and outer depth (2500 m) along the Island, with a break at
200 m depth separating the shelf and slope regions. The rather flat
Cascadia Basin extends 100 to 200 km offshore from the outer slope,
and for continental shelf waves can be considered as the abyssal
plain. Several canyons cut into the shelf at the northern and
southern portions, and the deepest is Juan de Fuca Canyon which
cuts through the shelf to join the 200-300 m deep Juan de Fuca
Strait with the Northeast Pacific Ocean. Canyons are absent in the
central region within 50 km of either side of Estevan Pt., where
depth contours are relatively regular.

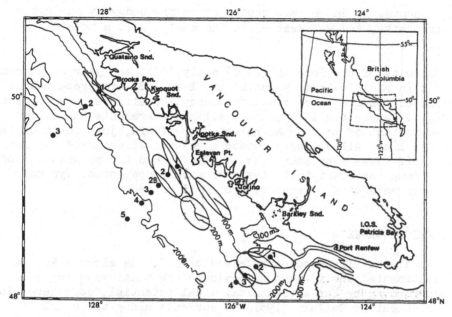

Fig. 1. Positions of current meter moorings along the continental
        shelf and slope of Vancouver Island, and observed K_1, KK_K
        current ellipses at shelf moorings. The line in each ellipse
        represents the current at the time of high tide at shore.
        Meters were at 15 and 35 m depth on the line near Tofino,
        near 50 m depth at other positions.

OBSERVATIONS AND ANALYSIS

     In summer the motion of shelf waters is dominated by 10-20
cm/sec diurnal period currents; in winter strong SE winds
accelerate the mean flow to comparable speeds. Accordingly the
summer currents reverse twice daily while those in winter pulse in
strength but are generally uniform in direction. Because the main
tidal constituents K_1 and O_1 clearly emerge above the background
spectral values, we conclude that these strong currents are of tidal
origin. Observations in Queen Charlotte Sound (Thomson and Huggett,
1981) and to the south of Juan de Fuca Strait (Lagerloef, personal
communication) show weaker diurnal currents. This suggests that the
shelf waves either begin or amplify off the southern end of Vancouver Island and vanish or attentuate to the north of the Island.

     As in CT82 we have used the harmonic analysis scheme of
Foreman (1978) to obtain the observed current ellipses for the
diurnal frequencies. Only the K_1, P_1 and O_1 constituents in the
current meter records have a signal to noise ratio greater than 1,
and the following discussion is restricted to the K_1 and O_1

constituents. P_1 and K_1 are separated in frequency by one cycle
in six months, and we treat P_1 in the analysis as a modulation
of K_1.

Results of the Foreman (1978) analysis of one year of current
records in shelf waters (9 months at C1 and C2) are presented as
current ellipses at each tidal frequency; K_1 ellipses are plotted
in Fig. 1, where the line in each ellipse represents the current
at the instant of maximum K_1 height at Bamfield, just north of the
Carmanah line. All ellipses rotate clockwise as expected for 1st
mode shelf waves near shore (Hsieh, 1982), and the progression of
phase along the coast indicates northward propagation. Typical
shelf currents at K_1 frequency are 10-15 cm/sec.

THEORETICAL WAVES

At diurnal frequencies poleward of 30°, one expects Kelvin
and continental shelf waves to dominate the tidal wave regime.
Direct forcing by the astronomical tidal potential, considered in
the treatment by Daifuku (1981) of currents along the coast of
North America, is of smaller relative magnitude here. Along the
west coast of North America, within several hundred kilometers
from shore, the diurnal period tidal heights can be fitted well to
Kelvin and Poincaré-like waves, with the Kelvin wave alone giving
a good fit to the phase speed (Munk et al., 1970; Crawford et al.,
1981). Kelvin waves are surface waves modified by the rotation of
the earth, of $O(10^4)$ km in wavelength and in offshore decay scale.
Their height to current speed ratio is $O(40/1)$ sec and they are
clearly observed in tidal height records, less well observed in
records of tidal currents.

We employ a numerical scheme of Henry et al. (in preparation)
to determine the height and current structure for such waves.
This scheme solves the shallow water wave equations (LeBlond and
Mysak, 1978), Eqns 25.1,2) subject to the requirements of no alongshore variation in topography and no stratification to give the
structure of such a wave for input values of frequency and cross-
shelf topography. At K_1 frequency, (=0.04178 c/hr) we find solutions
of wavelength 1.7×10^4 km with counter-clockwise rotating currents
over the continental shelf and slope. (The classical Kelvin wave,
defined for a wave in a rotating flat bottom basin, has no flow
normal to the coast at any position. This feature is not found for
such waves over a sloping bottom, but the phase speed and alongshore currents are modified little, and we follow the convention
of Munk et al. (1970) in calling these disturbances Kelvin waves.)

Baroclinic shelf waves are waves of the second class, or
gyroscopic waves, whose surface displacement is small and can be
neglected in analysis. Such waves have an offshore extent of $O(L)$

where L is the shelf width, and wavelength of $O(10^2$ to 10^3 km).
Their height to current ratio is $O(1/3)$ sec; hence their dominance
in current meter records. To examine the theoretical nature of
these waves we employ a model kindly given to us by K. Brink who
has examined the shelf wave-like nature of the currents off Peru
(Brink, 1982). He solves the shallow water baroclinic wave equations
(Brink, 1982) for a shelf wave, again subject to the requirement
of no variations in bottom topography in the along shore direction.
For computational stability, we impose a rigid lid in the solution,
a restriction which alters the dispersion curves by a few percent.

Dispersion curves for Kelvin wave, and first and second mode
shelf waves are presented in Fig. 2. At K_1 frequency, all three may
exist, although the wavelength of the second mode shelf wave of 70
km is close to the width of the shelf and comparable to the scale
of irregularities in the bottom topography of the continental shelf.
Solutions of the numerical scheme of Brink for second mode shelf
waves are sensitive to the structure of the bottom near shore and
near the shelf edge, features which change significantly along the
entire island and are difficult to represent accurately in this
numerical scheme, which permits no such variations.

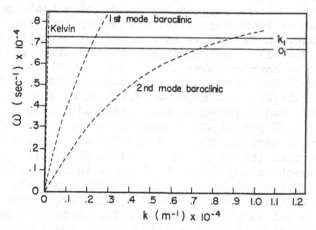

Fig. 2. Dispersion curves for Kelvin wave and first and second mode
   baroclinic shelf waves for the bottom topography along the
   Estevan line of current meters.

RESULTS

In Fig. 3 we show the resulting current ellipses for the
Estevan line of current meters. For the K_1 currents of each of
the 18 current meter records we plot ellipses of the Kelvin wave,
stratified (baroclinic) shelf wave, their sum and the observed
current ellipse. Kelvin and shelf wave ellipse shape and orientation are given by the numerical schemes described above. We fit
each by a single least squares method to all 18 observed ellipses
with only two degrees of freedom for each wave: an amplitude and a
phase. In this way the orientation, shape and relative size of
each ellipse is that of the theoretical wave.

Within each ellipse is a line which represents the magnitude
and direction of the K_1 current at the time of maximum tidal
potential. A small arrow above each ellipse gives the sense of
rotation, either clockwise or counterclockwise.

Near shore, at E1 and E2, the shelf determines the K_1 currents,
while at E5, 100 km from shore, the relative importance of the
Kelvin wave increases. The largest discrepancy lies at E1 where the
shelf wave undergoes a rapid shift in phase with depth. At other
points the fit is excellent. At E4 and E5 it is obvious that both
Kelvin and baroclinic shelf waves are required to explain the observed ellipses. Here, neither of the two waves alone can account
for the shift in ellipse orientation with depth. The shift in the
current at maximum tidal potential also requires the presence of
both waves.

These two waves alone account for 95% of the variance in the
current meter records. Including the second mode wave improves the
fit by less than 1% and supports our contention that continental
shelf waves along Vancouver Island are mainly confined to the first
baroclinic mode. In Table 1 we have given characteristics of
fitted waves for all 4 deployment periods. Clearly, best results
are obtained for May-September 1980 when spatial coverage was
greatest. The fitted waves in Table 1 suggest that the K_1 shelf wave
has a surface displacement amplitude of 5 cm at shore, in agreement with CT82, and that Kelvin wave amplitudes were 26 cm for the
May-September 1980 period and between 23 to 31 cm for the remaining
three deployment periods. These values for the diurnal surface
tide are less than the 43 cm value used in CT82 based on tide
gauge measurements along Vancouver Island. In their study of bottom
pressure and current meter records from waters off the southwestern
coast of the United States (south of 30°N, where diurnal period
shelf waves do not exist) Munk et al (1970) attribute much of the
diurnal tidal energy to Poincaré-like waves, and such disturbances
may account for the lack of agreement between observed and model
Kelvin waves here.

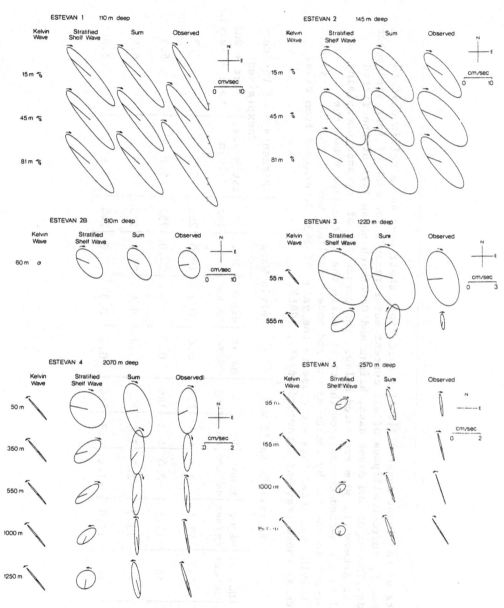

Fig. 3. A comparison of model K_1 current ellipses of Kelvin waves and baroclinic shelf waves with observed K_1 current ellipses along the Estevan line for the period May–September 1980. Model ellipses fitted to the observed by a least squares fit, and their sum is plotted in column three. Shelf wave ellipses derive from a model of Brink (1982). Kelvin wave ellipses are given by the model of Henry et al. (in prep.).

W. R. CRAWFORD ET AL.

Table 1. Results of the least squares fit of a baroclinic shelf wave and a Kelvin wave to observed K_1 current ellipses of all current meter records for each of four deployment periods. The major and minor axes, height and Greenwich phase lags of the fitted waves at Estevan 1, 50 m are noted below. The percentage of variance unexplained is listed in the second column from the right, while the last column shows the improvement to the fit by including the 2nd mode baroclinic wave in the least squares fit. Kelvin wave phase lags are close to the value of 243° determined from tidal heights at shore stations (Crawford et al., 1981).

Estevan 1 50 meters	No. of meters	Baroclinic Shelf Wave				Kelvin Wave				Percentage of Variance Unexplained	
		Major cm/sec	Minor cm/sec	Ht. cm	G deg	Major cm/sec	Minor cm/sec	Ht. cm	G deg.	1st Mode+ Kelvin	1st + 2nd Mode + Kelvin
May–Sept. 1979	13	11.8	-2.0	4.4	175	0.6	0.3	19.0	260	7.2	7.2
Sept.79–Jan.80	16	11.5	-2.0	4.3	146	0.7	0.4	24.0	257	5.8	5.5
Jan.–May 1980	17	13.2	-2.3	5.0	144	0.8	0.4	25.0	253	7.9	6.9
May–Sept. 1980	18	12.9	-2.2	4.9	164	0.6	0.3	21.0	244	5.6	5.4

The shelf wave phases undergo a distinct summer-winter phase shift, noted in Table 1. We have experienced some success in attributing this shift to a change in the mean alongshore currents over the shelf, and discuss the topic in a paper presently in preparation.

ACKNOWLEDGEMENTS

Dr. H. Freeland has kindly permitted us to use the data from his current meter survey of the Carmanah line. M. Woodward, J. Love, F. Hermiston and F. Stephenson assisted with the current meter survey of the Estevan and Brooks lines. Data reduction and analysis was handled by K. Lee, K. Booth, A. Douglas, D. Francis and D. Ramsden. We are indebted to Dr. K. Brink who graciously provided his model of baroclinic shelf waves.

REFERENCES

Allen, J. S., 1980, Models of wind-driven currents on the continental shelf. Ann. Rev. Fluid Mech., 12:389-433.

Brink, K. H., 1982, A comparison of long coastal trapped wave theory with observations off Peru, J. Phys. Oceanogr. 12:897-913.

Cartwright, D. E., 1969, Extraordinary tidal currents near St. Kilda, Nature, London, 223:928-932.

Crawford, W. R., Rapatz, W. J. and Huggett, W. S., 1981, Pressure and temperature measurements on seamounts in the North Pacific, Mar. Geod. 5:43-54.

Crawford, W. R. and Thomson, R. E., 1982, Continental shelf waves of diurnal period along Vancouver Island, J. Geophys. Res. (in press).

Daifuku, P. R., 1981, The diurnal tides on the northeast continental shelf off North America, M.Sc. Thesis, Dept. of Earth and Planetary Sciences, Mass. Inst. of Tech., Cambridge, Mass. 97 pp.

Foreman, M. G. G., 1978, Manual for tidal current analysis and prediction, Pac. Mar. Sci. Rep. 78-6, Inst. Ocean Sci., Sidney, B.C., Canada.

Freeland, H. J., 1983, A seasonal upwelling event observed off the west coast of British Columbia, Canada, this volume.

Hamon, B. V., 1962, The spectrums of mean sea level at Sydney, Coff's Harbour, and Lord Howe Island, J. Geophys. Res., 67:5147-5155.

Henry, R. F., Crawford, W. R. and Thomson, R. E., Calculation of dispersion curves and modal shapes of continental shelf waves, Pacific Marine Science Report (in preparation), Sidney, B.C.

Hsieh, W. W., 1982, On the detection of continental shelf waves, J. Phys. Oceanogr., 12:414-427.

Hsieh, W. W., 1982, On the detection of continental shelf waves,
    J. Phys. Oceanogr. 12:414-427.
Huthnance, J. M., 1981, Waves and currents near the continental
    shelf edge, Prog. Oceanogr., 10:193-226.
LeBlond, P. A. and Mysak, L. A., 1978, Waves in the Ocean,
    Elsevier Oceanography Ser., Vol. 20, Amsterdam.
Munk, W., Snodgrass, F. and Wimbush, M., 1970, Tides offshore:
    Transition fram California coastal to deep-sea waters,
    Geophys. Fluid Dyn., 1, 161-235.
Mysak, L. A., 1980, Recent advances in shelf wave dynamics, Rev.
    Geophys. and Space Physics, 18:211-241.
Thomson, R. E. and Crawford, W R., 1982, The generation of
    diurnal period shelf waves by tidal currents, J. Phys.
    Oceanogr., 12: 635-643.
Thomson, R. E. and Huggett, W. S., 1981, Wind-driven inertial
    oscillations of large spatial coherence, Atmosphere-
    Ocean, 19:281-306.

TOPOGRAPHIC INFLUENCES ON COASTAL CIRCULATION: A REVIEW

G.A. Cannon and G.S.E. Lagerloef

Pacific Marine Environmental Laboratory, NOAA
Seattle, Washington 98105, USA

INTRODUCTION

Submarine canyons are major topographic features of the continental shelf at its seaward edge. Canyons on narrow shelves have
their heads very close to the beach, whereas others may be 100 km
or more offshore. In addition, some canyons connect with glacial
troughs or shelf channels (from earlier subaerial erosion) which
extend entirely across the shelf and intersect major estuaries.
Observations of flow in canyons were first made in the late 1930's
(Stetson, 1937; Shephard et al., 1939). Subsequent observations
have focused on the role of submarine canyons in sediment transport.
Excellent reviews of this work have described processes in canyons
on the southern California coast (Inman et al., 1976), the northeast U.S. coast (Keller and Shephard, 1978) and a large variety of
shelves worldwide (Shephard et al., 1979). Thus, our discussion of
this work is limited to noting briefly that the dominant processes
causing sediment movement in those studies appeared to be gravity
waves and winds in shallower canyons and tides and internal waves
in deeper canyons. Also, the east coast canyons primarily transported fine-grain sediments, whereas the California canyons were very
active in transporting coarse-grained sediment. Most of the observations supporting these conclusions, however, were limited to a
few days duration.

During the past 10-12 years, observations of flow for longer
durations both in the canyons and on the surrounding continental
shelves have shown that canyons, shelves and, in some cases, nearby
estuaries can interact on larger scales (Cannon, 1972). Tidal flow
dominates many of these observations. However, significant time-
averaged flows exist in many of the canyons and vary with changing

235

flow conditions on the surrounding shelf. Variations in wind stress over the shelf appear to be a major factor causing many of the flow variations both in the canyons and in the coupling with estuaries.

This paper is a summary of studies of variations of coastal flow caused by major across-shelf topographic features and subsequent variability related to changing wind patterns. These studies can be grouped into three general categories: 1) west coast upwelling-downwelling; 2) east coast flow over a shelf valley; and 3) flow modifications by banks and troughs. In addition, we have attempted to show some details utilizing examples from our own work.

UPWELLING-DOWNWELLING

Upwelling studies off the African coast in 1972 occurred over very irregular topography at the shelf break consisting of a quasi-continuum of short, narrow and deep canyons (Shaffer, 1976). These canyons were found greatly to influence local upwelling by guiding the onshore flow along the canyon axes in a zone about 20 km wide. Fixed centers of upwelling were located in a confined region of order 3 km at the heads of the canyons, and the upwelled water diverged from the canyon heads and was redistributed southward (equatorward) in a coastal jet. Vertical velocities occurred within but not between the canyons and this flow became more intense near the head of the canyon. Horizontally divergent flow was found on the banks between canyons.

The main results of this work showed that the time response of up-canyon upwelling was well-correlated with the local trade wind and that a wind-driven regime with Ekman transport was the essential driving mechanism. One major variation was observed during trade wind relaxation, which was not a complete reversal. Flow in the canyon reversed and a burst flow occurred down and out the canyon and then along the continental slope. Slopes of all pertinent isopleths also reversed during the event which occurred over about a two-day interval. With increase in the trades, the downwelling subsided and the upwelling recurred.

O'Brien and his colleagues have developed numerical models to help determine the effects of topography on upwelling (Hurlburt, 1974; Peffly and O'Brien, 1976; Preller and O'Brien, 1980). Hurlburt (1974) modelled a symmetric canyon and found: 1) decreased upwelling north (poleward) of the canyon axis; 2) increased upwelling south of the axis; and 3) enhanced upwelling over the axis nearshore. For a symmetrical ridge, the effects were opposite.

A two-layer model of the Oregon coast upwelling was the first utilizing real topography (Peffly and O'Brien, 1976) and was an adaptation of the idealized study of Hurlburt (1974). Their main result of the application to Oregon was that the distribution of upwelling alongshore was more a product of bottom topography than of the shoreline configuration. Also, upwelling tended to favour the equatorward side of canyons but the source appeared to be equatorward flowing water being deflected upwards. The equatorward side of a ridge also tended to accent upwelling with onshore transport in the lower layer.

Preller and O'Brien (1980) used the same model applied to real topography with a ridge in an upwelling area of the Peru coast. They also found enhanced upwelling on the equatorward side of the ridge, the source being the poleward undercurrent which was deflected upwards.

Freeland and Denman (1983) have shown that canyons can augment upwelling other than as a local response to local wind. Their studies in the northern spur of the Juan de Fuca submarine canyon (Fig. 1) and on the neighbouring continental shelf indicate that the pressure gradients along this canyon spur can raise water from about 450 m onto the shelf whereas local winds can only raise water from more moderate depths of about 250 m. Their calculations indicate water could be raised from 400 m onto the shelf in one to three months depending whether flow was barotropic or baroclinic respectively. At two different spring transitions, they observed 78 and 97 days. Because this canyon connects with an estuary through a glacial trough, it is possible that estuarine effects (inflow in the lower layer) could speed up the process (Cannon, 1972). See their paper in this volume for additional details.

The only studies involving downwelling have been in the Juan de Fuca canyon system by Cannon and his colleagues (Cannon, 1972; Cannon et al., 1972; and Cannon and Holbrook, 1981). This canyon intersects a glacial trough which has depths exceeding 200 m extending completely across the continental shelf into the Strait of Juan de Fuca (Fig. 1). The trough-canyon intersection forms a sill-like feature of about 230 m near the outer edge of the shelf. During winter 1970 currents measured on this sill implied excursions which exceeded the shelf width both in and out canyon at the bottom and at 50 m above the bottom (Cannon, 1972). The out-canyon flow was correlated to southerly winds during storms as a compensating current to onshore Ekman surface flow. It was speculated that the normal flow was in canyon as part of the Strait of Juan de Fuca estuarine flow except during major storms. One wind reversal was not accompanied by a flow reversal indicating that along-canyon density gradients were also important. Finally, it was speculated that the large out-canyon excursions which has speeds sufficient to transport sediment could be a mechanism and route for transportation of sediment found in the Nitinat fan at the base of the continental slope.

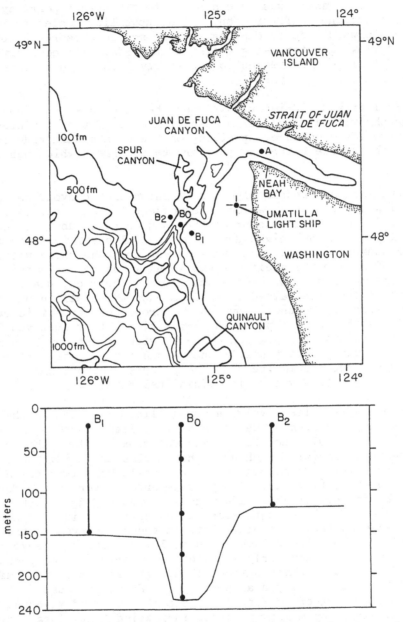

Fig. 1. Juan de Fuca submarine canyon showing mooring locations
(upper) and cross section of self moorings showing meter
locations (lower).

During October-November 1971 a second experiment was carried out to examine spatial variations in the canyon, on the nearby shelf and in the mouth of the estuary (Fig. 1; Cannon et al., 1972). Four wind-related events were observed, two in-canyon and two out-canyon (Figs. 2 and 3). The implied out-canyon excursions below shelf level were equal to the shelf width of 80-85 km and occurred during southerly winds and onshore flow in the surface layer (Cannon and Holbrook, 1981). The flow regime at the mouth of the estuary was also shown to reverse during the out-canyon excursions with inflow at surface and outflow at the bottom. Water properties within the canyon also indicated the changing excursions (Fig. 4). Colder water was brought into the basins within the trough during inflow and was flushed out during the out-canyon excursions. These events have been shown to be coupled directly to the changing coastal wind utilizing a two-layered upwelling-fjord interaction model (Klinck et al., 1981).

Bottom currents on the surrounding shelf changed markedly during the two flow regimes (Fig. 3). Flow tended to be toward the canyon from the north side during normal canyon inflow and toward the canyon from the south during the reversed out-canyon events. Both cases can be interpreted as cyclonic vortices over the canyon at the shelf bottom similar to that shown by Lagerloef (1983) in a study on the Alaskan shelf (summarized below). Scaling of stratification applied here implies vertical decoupling and limitation to the lower layer. The observations here also show inertial currents confined to the upper layer following the storm events.

During normal in-canyon flow, the currents in the canyon at 175 m were greater than at 125 m which could imply that they are pressure gradient driven as shown by Nelson et al. (1978) in a study of the Hudson shelf valley. However, the currents 5 m off the bottom were less than at 175 m, but this may be a result of bottom friction. Currents at the bottom (225 m) and 50 m off the bottom (175 m) were approximately equal during the reversed out-canyon events. Nelson's study showed that a decrease with depth would be expected if the flow were caused by wind stress alone. Thus, a combination of effects is probably occurring in this case.

Winter measurements in the shelf-break Quinalt canyon on the Washington coast have recently shown the existence of two flow regimes (Hickey and Baker, 1981). In the deep, narrow part of the canyon, the flow was up-down canyon. However, over the broader upper canyon the flow was approximately geostrophic following isobaths. Concurrently, their sediment observations showed a lack of suspended particles near bottom relative to 220-300 m above the bottom.

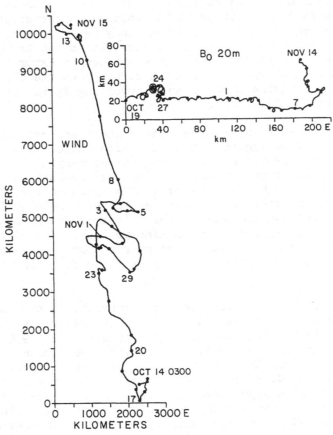

Fig. 2. Progressive vector diagrams for winds at Umatila light
ship (left) and near surface, upper layer currents at
20 m a B_o (right).

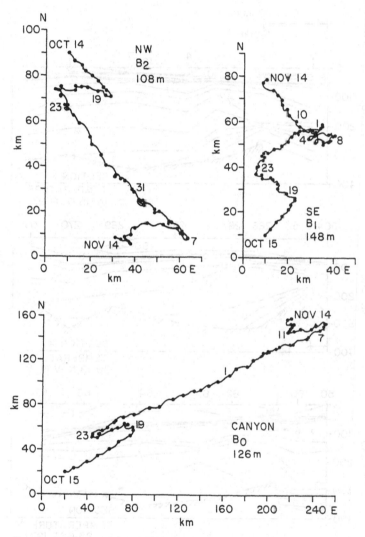

Fig. 3. Progressive vector diagrams of currents at nearly the
same depths at moorings B_2, B_0, B_1 across the canyon.
The shelf mooring diagrams are 5 m above bottom but the
canyon record is about 100 m up.

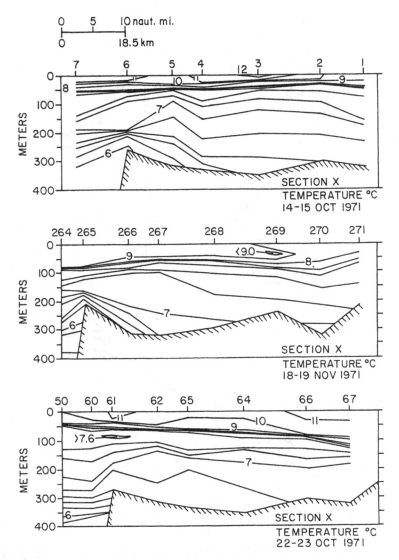

Fig. 4. Temperature sections along the Juan de Fuca canyon glacial trough. The right end is near mooring A in Figure 1, and the large change in bottom depth at the left is near B_0.

SHELF VALLEY

The Hudson Shelf Valley off the U.S. east coast is a major depression transecting the continental shelf and intersecting a submarine canyon. It extends from near New York City at the apex of the New York Bight to the Hudson Canyon at the shelf break. The entire Middle Atlantic shelf with emphasis on the New York Bight has been the subject of considerable attention for a number of years and several studies have shown the effects of the depression (Nelson et al., 1978; Lavelle et al., 1975; Hsueh, 1980; Han et al., 1980; Hopkins and Dieterle, 1981; Mayer et al., 1982).

Nelson et al. (1978) made observations in the upper part of the valley very near the coast where the axis tends more nearly north-south. Their observations consisted of near-bottom currents in the valley and on the shelf partway to shore for about 50 days in autumn 1973. Winds measured at Kennedy Airport were used to develop an empirical correlation to determine the onset of main-tenance of up or downchannel flow. Then, utilizing historic monthly average wind data, they showed that upchannel flow should dominate in October-April with minor downchannel flow in May-September, except for July.

The dominant physical process for upchannel flow was westerly winds causing an offshore Ekman transport accompanied by a near-shore sea surface setdown. Increased speeds with depth implied the flow was pressure driven. Northeasterly winds occur primarily during strong storms which result in increased sea surface height at coast and downchannel flow. Earlier, work with more limited data had also indicated the importance of piling up water on the Long Island coast in causing downchannel flow (Lavelle et al., 1975). During downchannel flow, however, currents decrease with depth which is more indicative of being wind stress caused. Finally, monthly average winds rarely are northeasterly. Thus, the normal flow is probably shoreward with downchannel flow occurring only during storms. This is similar to the flow in the Juan de Fuca canyon-trough system described above.

Hsueh (1980) investigated the flow over a larger area out to the shelf break for an isolated 3-day interval with eastward wind and upchannel flow. He developed a barotropic model balancing vorticity induced by flow over the topography with bottom Ekman suction. The model was driven first by wind stress induced divergence at the coast and second by flow through a cross-shelf open boundary to the north. His results showed flow following bathymetry around the canyon. The wind stress case had a stronger effect inshore and the shelf flow had stronger effects offshore. Shoreward turning of flow as the channel was encountered was sharper in the wind-driven case and appeared to be in better agreement with data. By testing with other wind directions, he

concluded that northeastward winds would be most effective in generating upchannel flow. Hopkins and Dieterle (1981) have recently
used a similar model to demonstrate downchannel flow with winds
from a northerly direction.

Han et al. (1980) extended the work of Hsueh to include
stratification utilizing a vertically averaged density field. They
studied four cases and the results indicated that upchannel bottom
flow resulted from strong eastward winds and from weak and moderate
northeastward winds. The strongest currents occurred with the
strongest winds. Southwestward winds, however, forced downchannel
flow in deep water with an opposite surface flow. The model compared best with observations for the two cases with strongest winds
(both had eastward components) and worst for weaker winds.

Mayer et al. (1982) have synthesized the Hudson Shelf Valley
studies which include a combination of effects. They describe a
regime which is in part quasi-geostrophic-topographically controlled
and in part estuarine. The mean cross-shelf pressure-gradient
force is seaward over the Middle Atlantic Bight in conjunction with
the mean southwestward geostrophic flow. However, the vertically
averaged density gradient is seaward which produces a shoreward
pressure-gradient force below about 45 m depth. Therefore, in the
valley below the shelf horizon, pressure forces deep flow upchannel
balanced by friction with a mean velocity of a few centimeters per
second. Storm forcing causes large pertubations. Sea level set-up
enhances the barotropic gradient while set-down reinforces the baroclinic gradient. Upchannel flow is associated with reducing or
reversing the southwestward flow on the shelf and vice-versa.
Topographic effects appear to be consistent with the linear vorticity balance used by Hsueh (1980) and by Han et al. (1980), where
cross-isobath flow was balanced by bottom Ekman suction. Upchannel
(downchannel) turning results from northeastward (southwestward)
variations in flow on the shelf. Northeastward wind stress can
induce upchannel flow without being strong enough to reverse the
mean southwestward flow on the shelf.

BANKS AND TROUGHS

In several of the previous examples, the circulation and its
interactions with topography were studied in the context of forcing
or adjustment by local winds. Although intense storms usually have
an effect on any coastal regime, there are cases where long-term
mean circulation patterns are not apparently wind forced. For
example, the influence of the larger scale ocean circulation applied
at the shelf break appears to dominate flow in the Middle Atlantic
Bight (Beardsley and Boicourt, 1981).

A variety of studies of such regimes has addressed the
interaction of the mean flow with topography. There is evidence

that inertial (potential vorticity) effects can be important on
shelves with large bathymetric features of short length scales.
Eide (1979) described an anticyclonic vortex over the Halten Bank,
Norway, which was well described by a stratified potential vorticity
model. Butman et al. (1982) observed an anticyclonic circulation
around the shoaler Georges Bank on the east shelf of North America.
In both, the anticyclonic flow indicated a tendency toward potential
vorticity conservation with relative vorticity being more anticylonic
toward shallower water. Lagerloef (1983) has shown for the Kiluida
Trough off Alaska that the observations are suggestive of a Taylor-
Proudman column with maximum cyclonic relative vorticity associated
with greatest depth.

Two other studies of topographic effects on the residual
circulation which do not consider potential vorticity have been
reported in abstracts. Buckley et al. (1981) have made long-term
observations over a complex trough-bank system on the Labrador
coast. Blanton et al. (1981) have described and modelled a series
of gyres over a system of nearshore shoals and tidal channels off
the U.S. southeast coast.

The Halten Bank off Norway has depths of about 100 m with a
trench exceeding 500 m between it and the coast (Eide, 1979). The
general flow is to the northeast along the coast. The anticyclonic
vortex trapped by the bank extends to the surface in winter but it
is limited to below the thermocline in summer. In addition, a dome
of cold heavy water develops over the bank which may be important
in bottom water formation.

George's Bank off the U.S. east coast is about 300 km long
and 150 km wide with depths less than 60 m and it borders the Gulf
of Maine (Butman et al., 1982). Deep channels exist on either end
entering the Gulf of Maine and, offshore, it becomes part of the
continental shelf. Two major observations were made. Anticyclonic
(clockwise) circulation occurred around the Bank with year-round
mean speed of 10-20 cm/sec. Unexpectedly strong flow occurred
along the steep northern edge of Bank. Drifter tracks showed that
the circulation can be closed since one made a full loop inside
the vortex. The strong flow over the steep topography is similar
to what is described below on the Alaskan shelf off Kodiak Island.

The continental shelf off Kodiak Island, Alaska, is a complex
bank-trough region (Fig. 5). Banks shoal to 50-100 m and the
troughs extend to depths of 150-200 m. The regional shelf circulation is generally southwestward in the vicinity of Kodiak. A
western boundary current, the Alaska Stream, is seaward of the
shelf break with speeds about 100 cm/sec, also to the southwest.
A seasonal density driven coastal flow in the northern Gulf of
Alaska, the Kenai Current, tends to pass through Shelikof Strait
along the northwestern side of Kodiak Island between it and the
mainland.

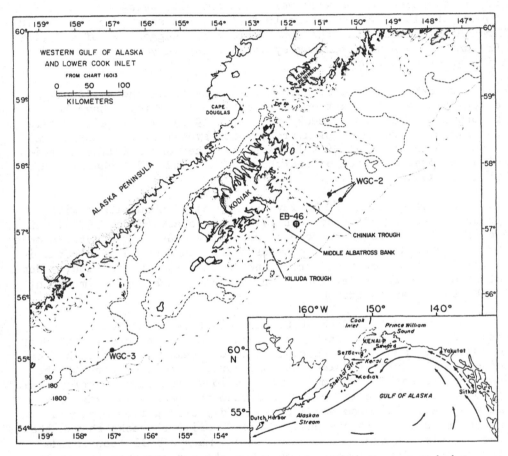

Fig. 5. The Kodiak Island shelf showing Kiliuda and Chiniak
Troughs separated by Middle Albatross Bank and the
locations of the WGC current meter moorings. Depth
is in meters.

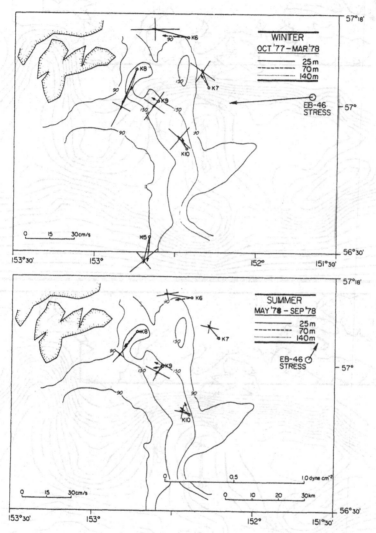

Fig. 6. Mean (4 month) currents near Kiliuda Trough and wind-stress at EB-70 during winter (top) and summer (bottom). Error crosses represent one standard deviation (35 hour low-pass) in the major and minor axes of variance for currents at 25-m depth. Mooring K5 was in place during the same winter period one year earlier and is not con-current with the others. Depth is in meters.

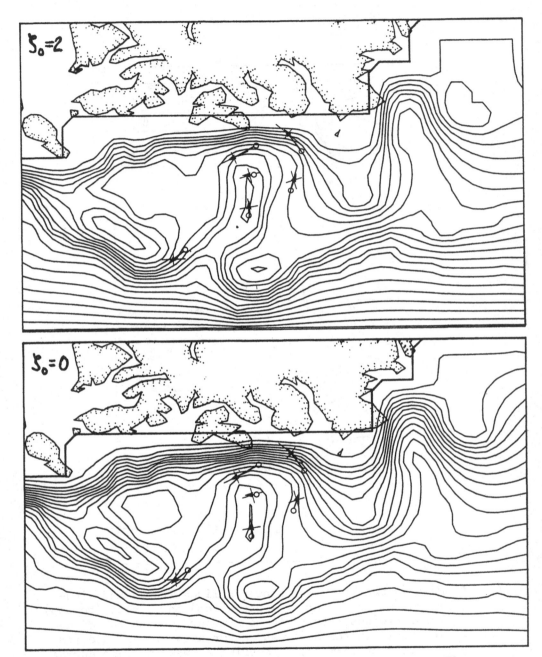

Fig. 7. Streamlines for cases when U is fixed at 5 cm s^{-1} (ε = .02) and ζ_0 is varied between 0 (top) and 2 (bottom). Streamline spacing midway along right hand boundary represents velocity scale. Mean 25-m winter currents are superimposed.

Fig. 6 shows observed mean currents from the Kiliuda Trough and the Mid-Albatross Bank in winter and summer. A steady nearly barotropic flow follows isobaths to form a vortex around the trough. There is no evidence of seasonal change in the mean flow pattern.

A simple steady-state numerical model, based on conservation of potential vorticity (neglecting friction), is in good agreement with observations when the vorticity balance approaches the limit of u · ∇H = 0. This limit is obtained when the Rossby number is small and the scaled topographic variations are o(1), as is the case with the Kodiak Shelf. The effect of increasing the Rossby number or decreasing the topographic relief scale is that the flow is less affected by topography. Fig. 7 shows two examples utilizing a Rossby number ∿ 0.02, based on a velocity scale ∿ 5 cm/sec and a length scale appropriate to the trough width of ∿ 20 km. The first has along-shelf flow uniform along the upstream (right hand) boundary. In the second, the along-shelf current increases seaward from zero at the coast along the upstream boundary.

For small Rossby numbers, flow also tends to be enhanced over areas of greatest topographic steepness. The example above shows

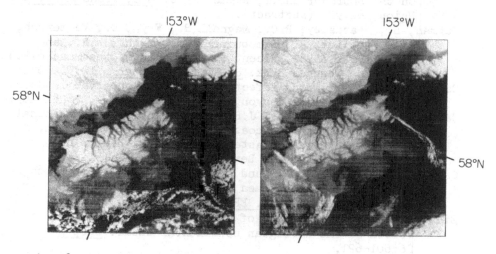

Fig. 8. Sequence of infra-red images taken on February 22 and 26, 1979 by the NIMBUS-7 Coastal Zone Color Scanner. Kodiak Island is the whiter area in the center. White areas indicate cloud or snow cover. Warm water is indicated by darker shades and cool water by lighter shades.

scaled currents near the trough appropriate to those measured
(about 20 cm/sec) even though the scaling velocity (about 5 cm/sec)
was much smaller. This is due to topographic steering whereby
streamlines following isobaths will converge over steeper topography,
and current speeds are roughly proportional to the local topographic
gradient. The strong current along the steep northern edge of
Georges Bank, discussed above, may result from the same phenomenon.

Additional evidence for the steady cyclonic circulation
pattern trapped over the Kiliuda Trough can be seen in various
infra-red satellite images (Fig. 8). Warmer water shown by the
darker shades can be seen extending shoreward over the eastern edge
of the trough. With time this water mass appears to have been
advected around the head of the trough and then seaward over the
western trough edge.

REFERENCES

Beardsley, R.C., and Boicourt, W.C., 1981, On estuarine and
    continental-shelf circulation in the middle Atlantic
    bight. In: "Evolution of Physical Oceanography", B.A.
    Warren and C. Wunsch, eds., MIT Press, Cambridge, 198-233.

Blanton, J.O., Alyea, F.N., and O'Brien, J.J., 1981, Residual
    tidal currents over irregular topography, EOS, Trans.
    Am. Geophys. Union, 62:927 (abstract).

Buckley, J.R., Lemon, D.D., and Fissel, D.B., 1981, Circulation
    on the Labrador shelf, summer 1980, EOS, Trans.Am.Geophys.
    Union, 62:927 (abstract).

Butman, B., Beardsley, R.C., Magnell,B., Frye, D., Vermersch,
    J.A., Schiltz, R., Limeburner, R., Wright, W.R., and
    Noble, M.A., 1982, Recent observations of the mean circu-
    lation on Georges Bank. J. Phys. Oceanogr., 12:569-591.

Cannon, G.A., 1972, Wind effects on currents observed in Juan
    de Fuca submarine canyon, J. Phys. Oceanogr., 2:281-285.

Cannon, G.A., and Holbrook, J.R., 1981, Wind-induced seasonal
    interactions between coastal wind fjord circulation.
    In: "The Norwegian Coastal Current", R. Saetre and
    M. Mork (eds.), Univ. Bergen, Norway, 131-151.

Cannon, G.A., Laird, N.P., and Ryan, T.V., 1972, Currents
    observed in Juan de Fuca submarine canyon and vicinity,
    1971, NOAA Tech. Rept. ERL 252-POL 14, 57 pp.

Eide, L.I., 1979, Evidence of a topographically trapped
    vortex on the Norwegian continental shelf, Deep-Sea Res.,
    26:601-621.

Freeland, H.J., and Denman, K.L., 1983, A topographically
    controlled upwelling centre off southern Vancouver
    Island. (This volume).

Galt, J.A., 1980, A finite element solution procedure for the
    interpolation of current data in complex regions.
    J. Phys. Oceanogr., 10:1984-1997.

Han, G.C., Hansen, D.V., and Galt, J., 1980, Steady state
    diagnostic model of New York Bight. J.Phys.Oceanogr.
    10:1998-2020.

Hickey, B., and Baker, E.T., 1981, Time-dependent particle
    distributions in a moderately broad (∿ 12 km) coastal
    submarine canyon, EOS, Trans. Am. Geophs. Union,
    62:916 (abstract).

Hopkins, T.S., and Dieterle, D.A., 1981, A sea level, shelf
    circulation model, the effects of bathymetry, EOS, Trans.
    Am. Geophys. Union, 62:923 (abstract).

Hsueh, Y., 1980, On the theory of deep flow in the Hudson
    shelf valley, J. Geophys. Res., 85:4913-4918.

Hurlburt, H.E., 1974, The influence of coastline geometry and
    bottom topography on the eastern ocean circulation,
    Ph.D. thesis, Florida State Univ., Tallahassee, 103 pp.

Inman, D.L., Nordstrom, C.E., and Flick, R.E., 1976, Currents
    in submarine canyons: an air-sea interaction,
    Annual Rev. Fluid Mech., 8:275-310.

Keller, G., and Shepard, F.P., 1978, Currents and sedimentary
    processes in submarine canyons off the northeastern
    United States, in: "Sedimentation in Submarine Canyons,
    Fans, and Trenches", D.J. Stanley and G. Kelling (eds.),
    Dowden, Hutchinson and Ross, Inc., Stroudsberg,
    Pennsylvania, 15-32.

Klinck, J.M., O'Brien, J.J., and Svendsen, H., 1981, Simple
    model of fjord and coastal circulation interaction,
    in: "The Norwegian Coastal Current", R. Saetre and
    M. Mork (editors), Univ. Bergen, Norway, 178-214.

Lagerloef, G.S.E., 1983, Topographically induced flow around
    a deep trough transecting the shelf off Kodiak Island,
    Alaska, J.Phys.Oceanogr., 13: (in press).

Lavelle, J.W., Keller, G.H., and Clarke, T.L., 1975, Possible
    bottom current response to surface winds in the Hudson
    shelf channel, J.Geophys.Res., 80:1953-1956.

Mayer, D.A., Han, G.C., and Hansen, D.V., 1982, Circulation
    in the Hudson shelf valley, J.Phys.Oceanogr., 12:
    (in press).

Nelson, T.A., Gadd, P.E., and Clarke, T.L., 1978, Wind-
    induced current flow in the upper Hudson shelf valley,
    J. Geophys. Res., 83:6073-6082.

Peffly, M.B., and O:Brien, J.J., 1976, A three-dimensional
    simulation of coastal upwelling off Oregon, J.Phys.
    Oceanogr., 6:164-180.

Preller, R., and O'Brien,J.J., 1980, The influence of bottom
    topography on upwelling off Peru, J.Phys.Oceanogr.,
    10:1377-1398.

Shaffer, G., 1976, A mesoscale study of coastal upwelling
    variability off NW-Africa, "Meteor" Forsch.-Ergebn.A,
    17:21-72.

Shepard, F.P., Marshall, N.F., McLoughlin, P.A., and
    Sullivan, G.G., 1979, Currents in Submarine Canyons
    and other Seavalleys. Am. Assoc.Petroleum Geologists,
    Tulsa, Oklahoma, Studies in Geology No 8, 173 pp.
Shepard, F.P., Reville, R., and Dietz, R.S., 1939, Ocean-
    bottom currents off the California coast, Science,
    89:488-489
Stetson, H.C., 1937, Current measurements in the Georges Bank
    canyons, Trans. Amer. Geophys. Union, 18:216-2.9.

Contribution 599 from Pacific Marine Environmental Laboratory.

TOPOGRAPHICALLY INDUCED VARIABILITY IN THE BALTIC SEA

A. Aitsam, J. Elken, L. Talsepp and J. Laanemets

Institute of Themophysics and Electrophysics
Academy of Sciences of the Estonian SSR

INTRODUCTION

The low frequency variability of the shallow and strongly stratified Baltic Sea is to a great extent controlled by the bottom topography, whereas the influence of the planetary β-effect is negligible.

The role of bottom topography in the formation of synoptic scale variability depends on the interrelation of bottom slope and stratification of bottom waters, and also on the energy of near bottom currents. In the low-frequency range one can distinguish two different motion regimes. In the case of large bottom slopes, long-term currents across the isobaths cannot exist for motions of isopycnic character, the isopycnal surfaces cannot be sufficiently inclined (with slopes comparable to the bottom slope) because of lack of energy. Transformation of the kinetic energy of upslope currents to potential energy and vice versa is realized in the form of bottom-trapped topographic waves. In the case of moderate bottom slopes the inclination of isopycnic surfaces can exceed the bottom slope and the role of the bottom topography in determining the motion is not so obvious. According to our investigations, in such conditions non-periodic synoptic eddies can exist, whose migration is determined by the bottom topography.

In this paper we discuss the results of measurements of currents, temperature, salinity and density in the deep (more than 100 m) basins of the Baltic Sea. The internal Rossby radius R_d, which determines the lateral scale of baroclinic low-frequency motions, is of the order of 10 km in these areas.

TOPOGRAPHICALLY TRAPPED WAVES IN REGIONS WITH LARGE BOTTOM SLOPES

Topographically trapped waves, similar to low frequency baroclinic shelf waves, appear in the open sea wherever the bottom has
considerable slope. These low-frequency waves are trapped by the
bottom slope: wave amplitudes decrease offslope in the horizontal
direction and increase with depth in the bottom layer. Thus, the
effect of the slope on these trapped waves is similar to that of
the shelf on generation of baroclinic shelf waves.

Observations of low frequency trapped waves were made in
regions where $\alpha\, N_b = 0\ (10^{-4})$, where α is the bottom slope and N_b
the Väisälä frequency in the bottom layer. Current and temperature
were measured by recording current meters on autonomous moorings.
These measurements can be used to show the presence of low-frequency,
slope-trapped topographic waves.

During BOSEX-77, waves with periods of 3 days were seen. As an
example of low frequency oscillations, Fig. 1 shows the low pass filtered time series of the eastern velocity and temperature at 118 m
at station N, located on the steep slope southeast off Gotland. The
waves were travelling along the isobaths with shallower water on the

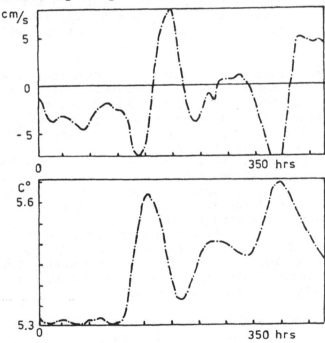

Fig. 1. Time series of the eastern velocity component and of the
        temperature at a depth of 118 m at station N of the
        BOSEX-80 experiment after removal of the high frequency
        oscillation with a 28 hours filter.

right, and wave amplitudes increased with depth in the bottom layer
(Aitsam, Talpsepp, 1981). Absence of these waves from station E,
about 30 km offslope, showed that waves seen at station N were
trapped by the slope. Due to the small number of mooring stations
it was impossible to measure precisely the rate of offslope wave
amplitude decrease.

During the 1980 experiment, clearly distinguishable current
oscillations with periods of 6-8 days at all five mooring stations
were observed. To find the characteristic length scales of dominant
oscillations the frequency wave number spectrum was computed.
Linked to periods of 6-8 days, the wave number spectrum had a dominant peak that corresponded to 22-25 km waves, travelling southwest along the isobaths (Fig. 2). These oscillations were interpreted as unstable waves by using a two-layer channel model with a
large scale steady current along the channel axis (Aitsam, Talpsepp,
1982). In this work a constant slope that differed somewhat from
the real bottom topography was assumed. Real bottom topography
(a section can be seen in Fig. 4b) induced spatial inhomogeneities
of topographic waves. It was found that the waves were more intensive in the region with greater bottom slope (station E) and less
intensive in the region of small slope. The variation of the wave
amplitude can be seen in Fig. 3, where the spectral density S (ω)
of kinetic energy in coordinates ω.S (ω), lg ω (ω = frequency) for
station E (Fig. 3a) and for other stations in the area (Fig. 3b) is
presented. In these plots, the total kinetic energy of the frequency
band is proportional to the area under the curve. It may be seen
that the peak corresponding to periods of 6-8 days in Fig. 3b is
considerably less than that in Fig. 3a. It follows that 6-8 days
oscillations were trapped by the slope.

Next we refer to two models of topographically trapped waves.

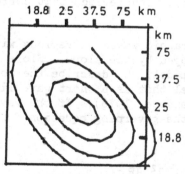

Fig. 2.   Mean wave number spectrum of the BOSEX-80 experiment for
          oscillations with periods of 6-8 days (isolines 95%, 80%,
          60%, 40% of maximum).

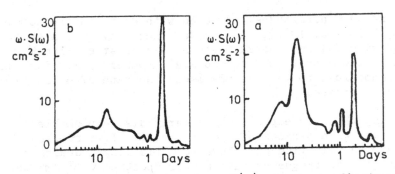

Fig. 3. Kinetic energy spectra in $\omega \cdot S(\omega)$, lg ω coordinates:
a) average over the depths of 41 m and 76 m of station E
b) average over all the depths of stations B, C and D.

Looking for a wave solution in the form $P(z)\,\exp\left[i(kx+ly-\omega t)\right]$ in the case of constant Väisälä frequency (that corresponds to BOSEX-77) and constant bottom slope, Rhines' model (1970) gives the vertical distribution of wave amplitude, P, in the form

$$P = A \cosh\,(\overline{k}Nz/f) \tag{1}$$

and the dispersion relation

$$\omega = -\,\frac{\alpha Ne}{\overline{k}}\,\coth\,(\overline{k}NH_o/f) \tag{2}$$

Here $\overline{k}^2 = k^2 + e^2$, f denotes the Cariolis parameter and H_o denotes the mean depth. Relation (1) gives the increase of wave amplitude with depth. Topographic waves observed during BOSEX-77 were in good agreement with (1), with periods that corresponded to those predicted by (2).

Baroclinic topographic waves in regions with large slopes are also horizontally trapped by the slope. Slope trapping is not described in Rhines' model. To describe the horizontal wave structure we present a model that is a short wave limit of Huthnance's shelf wave model (Huthnance, 1978). If the wave has the form $P(x,z)\,\exp|i(ly+\omega t)|$, the governing equation for the wave amplitude $P(x,z)$ is

$$\frac{\partial^2 P}{\partial x^2} + \frac{l^2-\sigma^2}{S}\,\frac{\partial}{\partial z}\,\frac{l^2}{N^2}\,\frac{\partial P}{\partial z} - l^2 P = 0$$

with boundary conditions

$$\frac{dH}{dx} \left(\frac{\partial P}{\partial x} + \frac{1}{\sigma} P \right) + \frac{1-\sigma^2}{SN^2} \frac{\partial P}{\partial z} = 0, \text{ at } z = -H(x)$$

and

$$\frac{\partial P}{\partial x} = 0, \text{ at } z = 0$$

Here $H(x)$ denotes the sea depth, a function of x only; $S \equiv \max (N^2)$, H^2/f^2L^2; $\tau = \omega/f$; denotes nondimensional frequency; L is characteristic length scale: 1 denotes nondimensional wave number,(supposed large in this limit). After introducing new co-ordinates $\eta \equiv 1(z + H(x))$ and $\xi = (x-x_0)/\sqrt{1}$, it is useful to expand all variables in powers of 1 and thereby obtain the following approximate solution for large 1 in the form (Huthnance, 1978).

$$P = H_m (b\xi) \exp\left[i (b\xi)^2/2\right] \exp\left[-\eta\sqrt{\phi + \overline{H}'^2}\right] ,$$

$$m = 0, 1, 2, \ldots \ldots \tag{3}$$

where H denotes the mth order Hermite's function and $H' = dH/dx$. The solution (3) is trapped by point (x_0, z_0) where NH' has its maximal value. With uniform slope the dispersion relation of this model coincides with that of Rhines (1970). For the stratification and bottom topography that is related to the 1980 experiment, the wave amplitude distribution derived from (3) is presented in Fig. 4. In Fig. 4, $S=2$, $1=6$ and different wave models are denoted by $M=0, 1$ and 2. The function $P(x,z)$ has a maximum at the bottom showing the bottom trapping of waves. The offslope decrease of amplitude is also evident. In the same figure the dispersion curves for different values of the stratification parameter S are presented, showing that wave parameters are greatly dependent on the stratification.

In the latter model the large-scale water motion along the isobaths that was observed in 1980 was not taken into account. In spite of some discrepancies between the data and the modelled prediction, the observed spatial structure of waves is similar.

SYNOPTIC EDDIES IN REGIONS WITH MODERATE BOTTOM SLOPES

The layer of maximum Väisälä frequency N_{max} does not intersect the bottom in these regions. The magnitude of the parameter αN_b ranges up to 10^{-5} sec^{-1}. The spatial structure of any synoptic scale pertubations was studied using repeated CTD-surveys on rectangular grids with 5 miles between vertical casts. The relatively low levels of high-frequency noise in the deep layers enabled us to draw the contour maps of the physical fields and to detect the synoptic eddies from the density data. Two types of eddy (with closed contour lines and stable structure of density pertubations) were identified (Aitsam, Elken, 1982):

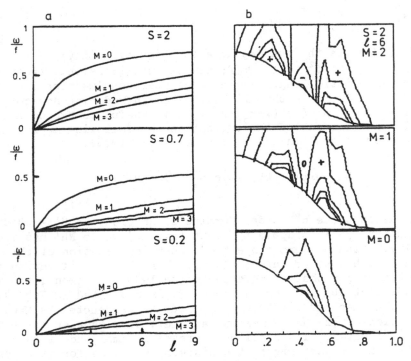

Fig. 4. (a) Dispersion curves for the parameter values S=2, 0.7 and
0.2. (b) Amplitude distribution of waves for l=6 and S=2
in zero (M=0), first (M=1) and second (M=2) mode, versus
nondimensional horizontal length, according to Huthnance's
model.

a) Eddies with raised isopycnals which have a diameter more
than 40 km when the Rossby radius is about 10 km. The vertical
displacement can exceed 20 m (20% of the bottom depth), which
corresponds to a density of available potential energy more than
$1000 \text{ cm}^2/\text{sec}^2$. This kind of eddy appears with remarkably strong
mean currents.

b) Eddies with raised isopycnals in the upper layers of the
halocline and lowered isopycnals in the bottom layers, with
diameters about 20 km. The displacement of isopycnals in the

bottom layers is of the order of 10 m. (downward) and in the upper
layers of the halocline about 5 m (downward). The density of available potential energy is up to 200 cm^2/sec^2. The mean migration of
such eddies is to the "topographic west" (along the isobaths with
shallower water on the right) at speeds of 1 to 3 cm/sec. The generation of two eddies of this type was followed in the aftermath of
disintegration of an intensive eddy of type (a).

The eddies of type (a) have centres with anomalous salinity
values in the upper layers above the halocline, indicating horizontal transport of water by the eddies and increased vertical
mixing. The thermohaline structure of the eddies of type (b) is
more complicated.

Let us give some examples of the synoptic eddies. Maps of the
relative dynamic topography of surveys 18/1-3 are presented in Fig.
5. The eddy of type (a), which had a salty centre in the upper
layers during surveys 18/1-2, had split into a system of two eddies
of type (b) by time of survey 18/3. The eddy with its centre at
station F5 (survey 18/3) had increased salinity in the upper layers
(being the "old" centre), but the eddy with its centre at station
D3 was formed by the convergence in the deep layers. The vertical
displacements of isopycnals are given in Fig. 6.

For surveys 23/1-5 the maps of relative dynamic topography
are presented in Fig. 7. An eddy of type (a), which had reduced
salinity in the upper layers, was observed around station G7 of
survey 23/1. On the map of survey 23/2 a system of two eddies of
type (b) can be seen. During surveys 23/3 and 23/4 the type (b)
eddy stayed with its centre at E4, but by survey 23/5 it had moved
to F5 along the local isobaths to the "topographic west". We also
remark the evolution of a tongue-like positive anomaly of relative
dynamic topography, which migrated to the south along the line of
stations C, to the "topographic west" according to local topography.
When the tongue interacted with the eddy (station E4, surveys 23/3-
4), the eddy did not change its location and shape.

The vertical density structure of the eddies is characterized
in Fig. 8, where sections indicating the depths of the isopycnic
surfaces for surveys 23/1-5 are presented.

It is interesting to compare the parameters of the eddies
with the characteristics of linear quasigeostrophic waves. Taking
into account β-effect and constant bottom slope, two kinds of wave
solutions exist (Rhines, 1977). With Baltic Sea parameters, topographic waves with phase speed to the "topographic west" and with
bottom intensified amplitude are not influenced by the β-effect.
Trapping of motion by the bottom depends on the ratio of Rossby
radius to the lateral scale of the waves. The vertical displacements of isopycnic surfaces all have the same sense. The baroclinic
modes of planetary Rossby waves are modified by a sloping bottom

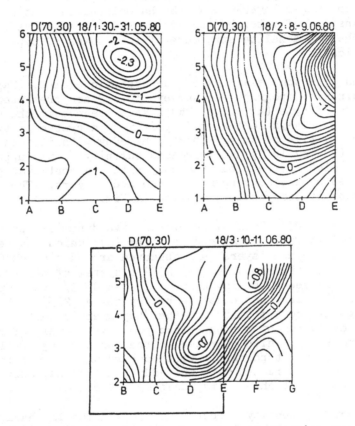

Fig. 5. Maps of relative dynamic topography anomalies for surveys
18/1 (30-31.05.80), 18/2 (8-9.06.80) and 18/3 (10-11.06.80).
The contour interval is 0.2 dyn.cm. for survey 18/1 and 18/3.
Grid scale 5 n miles.

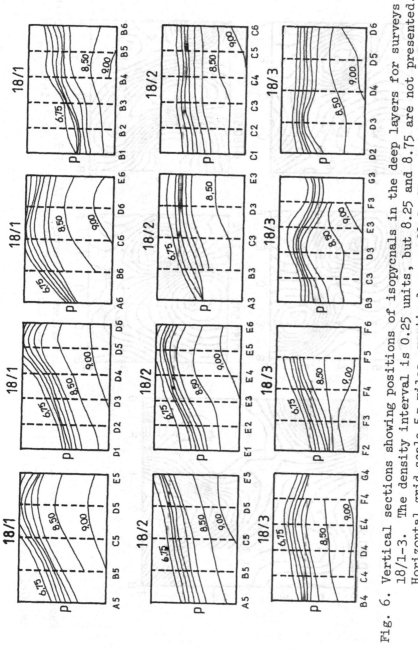

Fig. 6. Vertical sections showing positions of isopycnals in the deep layers for surveys 18/1-3. The density interval is 0.25 units, but 8.25 and 8.75 are not presented. Horizontal grid scale 5 n miles, vertical range 30 decibar.

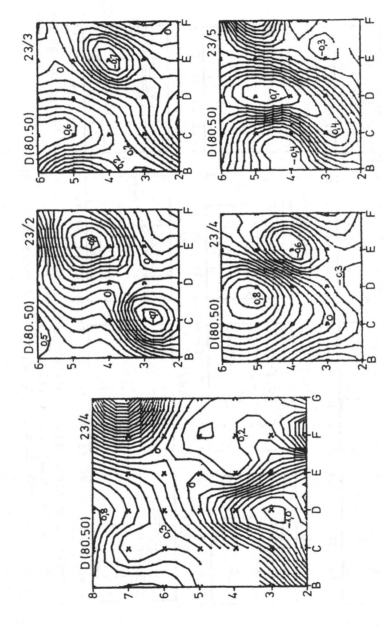

Fig. 7. Maps of relative dynamic topography anomalies for surveys 23/1 (3–5.06.81), 23/2 (15–16.06.81), 23/3 (21–22.06.81), 23/4 (26–27.06.81) and 23/5 (4–5.07.81). The contour interval is 0.1 dyn. cm. Grid scale 5 n miles.

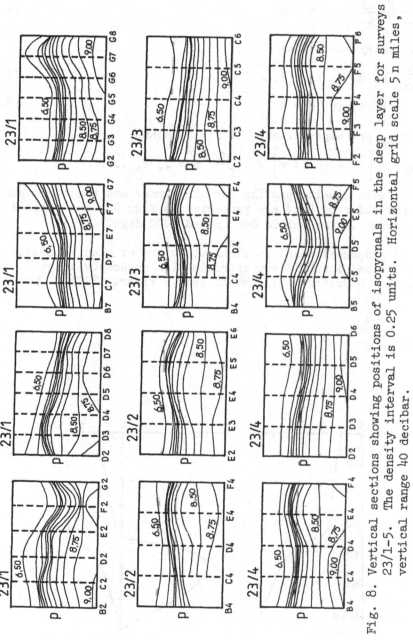

Fig. 8. Vertical sections showing positions of isopycnals in the deep layer for surveys 23/1-5. The density interval is 0.25 units. Horizontal grid scale 5 n miles, vertical range 40 decibar.

and the near-bottom velocity perturbations are close to zero.
Looking for pressure perturbations in the form p' = P(z) exp
i (kx + ly - wt), we get approximately

$$\frac{d}{dz} \frac{f^2}{N^2} \frac{dP}{dz} + \chi^2 P = 0 \qquad (4)$$

with boundary conditions

$$\frac{dP}{dz} = 0 \quad at \quad z = 0, \qquad (5)$$

$$P = 0 \quad at \quad z = H \qquad (6)$$

where $\chi = \chi_n$ are eigenvalues. If N = constant,

$$\chi_n = \left[\frac{\pi}{2} + (n-1) \pi \right] / (\frac{NH}{f}), \, n = 1, \, 2, \, \ldots.$$

The mean density profile for the summer and the corresponding
two first modes of P(z) are presented in Fig. 9. The periods of
Rossby waves in the Baltic Sea exceed 2000 days.

The propagation and vertical structure of oceanic eddies are
well described in the first approximation as Rossby waves
(Koshlyakov & Grachev 1973, McWilliams & Flierl, 1976). The

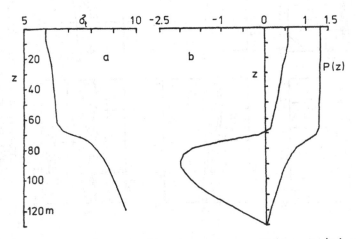

Fig. 9. (a) Mean density profile for survey 23/3 and (b) Normed two
       first modes of vertical structure function (4)-(6).

propagation of the Baltic Sea eddies is in reasonably good accord
with the phase speed of topographic waves, but this interpretation
is refuted by the vertical phase change in isopycnal displacement.
The vertical structure of density perturbations could be identified
with the first and second modes of P(z), but the propagation of the
Baltic eddies cannot be described by Rossby waves. It seems that
the similarity between different properties of the waves and eddies
is insufficient to explain the physical nature of the eddies.

Comparing vertical structure, Rossby number, and the ratio of
lateral scale to the Rossby radius, the Baltic Sea eddies are
similar to those of the ocean (Elken, 1982). The particle rotation
speed can exceed the "phase" speed of the eddies by a factor of
5 or more, so that the eddies are rather stable nonlinear phenomena
and are not obviously influenced by atmospheric forcing. Eddies in
the numerical model of Simons (1982) show similar features.

Consider a simple model of eddy structure with the assumption
(Nelepo, Korotayev, 1979) that the structure of non-linear eddies
is in the first approximation independent of the propagation and
the β-effect. The propagation will be determined by forcing acting
on the eddy as a whole system. Assume that the potential vorticity

$$G\,(\psi) = \Delta\psi + \frac{\partial}{\partial z}\,\frac{f^2}{N^2}\,\frac{\partial\Psi}{\partial z} \tag{7}$$

(Ψ, the streamfunction) is initially concentrated in a circle r<R
(r, distance from the eddy centre) and

$$G\,(\psi) = 0 \tag{8}$$

in the exterior r>R. Looking for axisymmetric eddies, for which
ψ decays faster than arbitrary power of r if $r \to \infty$, the solution
of (8) is

$$\psi = A_o\,K_o\,(\chi_n r)\,(P_n(z) \tag{9}$$

where $K_o\,(\chi_n r)$ is the Bessel function from imaginary argument;
χ_n, $P_n(z)$ are eigenvalues and eigenfunctions of Eqns. (4) – (6).
The solution (9) is singular at r=0, so Eqn. (9) is only valid in
the exterior. To obtain a regular interior solution, which is
continuous and has continuous first derivative with the exterior
solution Eqn. (9) at r = R, we choose the "initial" interior
distribution of the potential vorticity in the form

$$G\,(\psi) = W_o\,(1 - \frac{r}{R})\,P_n\,(z) \tag{10}$$

If G (ψ) is given by Eqn. (7), numerical solutions (Fig. 10) of
Eqns. (8) and (10) depend on the value of the parameter $\chi_n R$. For

our Baltic Sea eddies, $\chi_n R \approx 5$ is a "universal" constant. In this
way the model describes the horizontal and vertical structure of
the eddies and also the lateral scales.

Note that the solution in Fig. 10 is a first approximation to
the solution of the non-linear quasigeostrophic equations in asymptotic expansion for small phase speed relative to water speed, i.e.,
strongly non-linear motion. The propagation of the eddies is determined by the bottom topography but this mechanism is not yet explained.

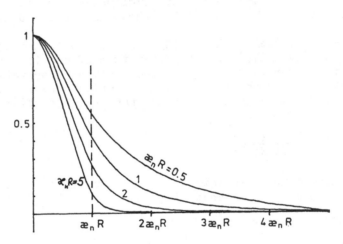

Fig. 10. Nondimensional horizontal eddy structure according to the
        model.

FINESTRUCTURE IN THE SYNOPTIC EDDIES

Finally, the finestructure activity in the synoptic eddies
can be discussed. According to the energy cascade model, finestructure is important, linking synoptic scale processes, internal
waves and turbulence (Woods, 1980). In the centre of an ocean eddy,
vertical mixing signatures have been seen (Fedorov et al., 1981),
as well as increased internal wave amplitudes on the periphery
(Dykeman et al., 1980).

Intrusions and movement of the internal wave field are the
main processes generating finestructure in the deep layers of the

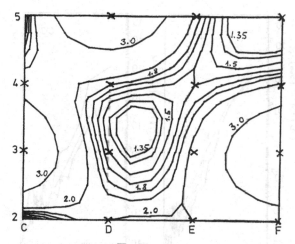

Fig. 11. Distribution of $\overline{\zeta_T}/\overline{\zeta_S}$ on the area of survey 18/3.

Baltic Sea. As a qualitative measure of the role of intrusion in vertical finestructure, the ratio $\overline{\zeta_T}/\overline{\zeta_S}$ was calculated, where $\overline{\zeta}$ and ζ_S are r.m.s. vertical displacements, calculated from temperature and salinity respectively. Mean profiles were calculated using the 5 m cosine-filter. If $\overline{\zeta_T}/\overline{\zeta_S} = 1$ on the scales investigated, temperature and salinity fluctuations are generated by movement of internal waves, but if $\overline{\zeta_T}/\overline{\zeta_S} \neq 1$, certain parts of the structure are formed by intrusive like processes.

These ideas were applied to the CTD surveys so as to descry the intrusive and wave-induced fine structure in the synoptic eddies on scales from 0.5 to 5 m. Eddies seen at D3 and F5 in survey 18/3 (Fig. 5) were in their early development. The distribution of ζ_T/ζ_S is shown in Fig. 11, which shows that ζ_T/ζ_S increased towards the edge of the eddy. This change is also reflected in the spectra of Fig. 12. These temperature and salinity spectra had similar levels and slopes (-3 power) in the centre of the eddy (D3) but while the salinity spectrum kept the same level and slopes at the edge of the eddy (C3), the temperature level increased by one order. It appears that the salinity fluctuations were dominated by internal waves but that intrusive layers increased temperature fluctuations on the eddy edge. The eddy at station E4

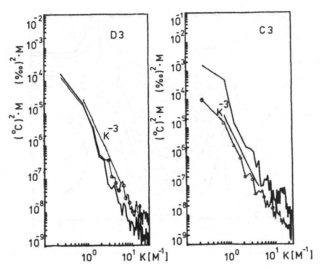

Fig. 12. Spectral densities of the temperature and of the salinity
at stations D3 and C3 of survey 18/3.

(Fig. 7) had been in the area for ten days when a section was
worked through it between surveys 23/3 and 23/4. The distribution
of ζ_T/ζ_S on this section is shown in Fig. 13. In this case the
intrusive finestructure was increasing towards the eddy centre.

REFERENCES

Aitsam, A., Elken, J., 1982, Synoptic scale variability of
        hydrophysical fields in the Baltic Proper on the basis
        of CTD measurements. In:"Hydronamics of semi-enclosed
        seas", Amsterdam, Elsevier, pp. 433-468.
Aitsam, A., Talpsepp, L., 1981, Analysis of bottom trapped
        topographic waves in the Baltic Sea. In: "The Investi-
        gation and Modelling of Processes in the Baltic Sea".
        Tallinn, pp. 58-70.
Aitsam, A., Talpsepp, L., 1982, Synoptic variability of currents
        in the Baltic Proper. In: "Hydronamics of Semi-
        enclosed Seas", Amsterdam, Elsevier, pp. 469-488.

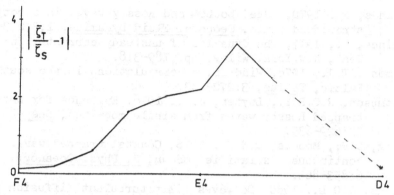

Fig. 13. Distribution of $|\bar\zeta_T/\bar\zeta_S-1|$ at the section of the eddy F4-D4, measured between surveys 23/3 and 23/4.

Bryden, H., 1976, Horizontal advection of temperature for low-frequency motions, Deep-Sea Res., 1165-1174.

Dykman, V.Z., Korotayev, G.K., Panteleyev, N.A., Slepyshev, A.A., 1980, Local balance of energy in the synoptic eddy (in Russian). In:"Structure, kinematics and dynamics of synoptic eddies", Sevastopol, pp. 81-92.

Elken, J., 1981, Comparison of some aspects of the Baltic Sea and ocean synoptic variability (in Russian), Okeanologichesky Issledovanya (in press).

Fedorov, K.N., Ginzburg, A.J., Zatsepin, A.G., 1981, Thermo-haline structure and tracks of mixing in synoptic eddies and Gulfstream rings (in Russian), Okeanologya, XXI.1: 25-29.

Huthnance, J.M., 1978, On coastal trapped waves: analysis and numerical calculation by inverse iteration, J. Phys. Oceanogr., 8, 1:74-92.

Konyaev, K.V., 1981, Spectral analysis of random oceanological fields (in Russian), Leningrad, Gidrometeoizdat, 207 pp.

Koshlyakov, M.N., Grachev, Y.M., 1973, Mesoscale currents at a hydrophysical polygon in the Tropical Atlantic, Deep-Sea Res., 23:285-300.

McWilliams, J., Flierl, G., 1976, Optimal quasigeostrophic wave analysis of MODE array data, Deep-Sea Res., 23:285-300.

Nelepo, J., Kopotayev, G.K., 1979, Structure of synoptic variability on the basis of hydrological surveys on the POLYMODE area (in Russian), Morskie gidrofizicheskie issledovanya, 2:142-151.

Rhines, P., 1970, Edge, bottom and Rossby waves in a rotating
    stratified fluid, Geophys. Fluid Dynamics, 1,3:273-302.
Rhines, P., 1977, The dynamics of unsteady currents, In:"The
    Sea", New York, Wiley, pp. 189-318.
Simons, T.J., 1978, Wind-driven circulations in the southwest
    Baltic, Tellus, 3:272-283.
Thompson, R.O.R.Y., Luyten, J.R., 1976, Evidence for bottom-
    trapped Rossby waves from single moorings, Deep-Sea Res.,
    23:629-635.
Wang, D.P., Mooers, C.N.K., 1976, Coastal trapped waves in a
    continuously stratified ocean, J. Phys. Oceanogr.,
    6:853-863.
Woods, J.D.L., 1980, Do waves limit turbulent diffusion in
    the ocean? Nature, 288:219-224.

AN ADDITIONAL ANALYSIS OF INERTIAL OSCILLATIONS ON

THE CONTINENTAL SHELF

Claude Millot and Catherine Moulin

Antenne du Laboratoire d'Océanographie
Physique du Muséum
La Seyne, France

ABSTRACT

    Inertial oscillations on the continental shelf have been
studied with a theoretical model and spectral analysis of numerous
time series (Millot and Crepon, 1981); the mean temperature over a
specific depth interval including the thermocline was supposed to
be representative of the amplitude of the internal waves in the
inertial frequency band. To estimate better the vertical displacements due to these waves, we take into account the vertical temperature gradient. Such an analysis improves the accuracy of some of
the results already obtained, the method is extended to measurements
in almost homogeneous layers, and the reliability of the theory is
increased.

INTRODUCTION

    Millot and Crepon (1981), henceforth MC, presented observations
and theory of inertial waves induced by the wind on the continental
shelf. The theoretical work consisted in an analytical model of a
two-layer ocean which is submitted to a wind impulse. The main
result was that inertial oscillations are probably associated with
the following processes : horizontal motions with inertial frequency
(f_i) are mainly locally wind generated, and internal waves whose
frequency is $\geqslant f_i$ propagate from discontinuities. These results are
consistent with those from the analysis of long and numerous time
series obtained during summer on the continental shelf of the Gulf
of Lions.

    A typical mooring measured the current in the surface and
bottom layers and the temperature over the whole depth. The thermo-

271

cline was subject to a seasonal deepening and to wind induced up- and down welling (Millot),1979), temperature signals over the whole depth were roughly in phase for frequencies up to 0.1 cph. A time series which represents the oscillations of the thermocline in the inertial frequency band, whatever its mean level, is computed (MC) by summing the various signals obtained within a depth inter- val including the thermocline.

The ensemble averaged spectral estimates of the depth averaged temperatures, the surface currents and the bottom currents are drawn in Fig. 1. As predicted by the model, spectral analysis of the data showed that currents mainly oscillated at f_i although measured tem- perature oscillations at frequencies above f_i were detected. Bottom currents, relatively energetic, were in this frequency band. Also, currents measured in the surface layer up to 43 km apart were coh- erent below f_i and currents measured in each layer at the same pos- ition were coherent at frequencies above f_i. The depth averaged tem- perature and the currents at the same point were coherent mainly at f_i. In MC, it was shown that the direction of propagation of the baroclinic waves could be computed from current and temperature measurements at one point only. The computed directions were roughly normal to the nearest shore and make the theory quite reliable.

Although the observations and the theory presented in MC fit very well, the hypothesis involving depth averaged representation of the oscillations of the thermocline allows a major criticism: under varying conditions of stratification, the mean temperature within a depth interval is not proportional to the mean amplitude of the vertical displacement. It is the purpose of this paper to test the MC model with computed time series of the vertical velo- city.

THE DEPTH AVERAGED VERTICAL VELOCITY

Firstly, we suppose that heat exchange with the atmosphere and solar radiation do not significantly disturb the temperature measurements. This is valid because this perturbation has a daily frequency and is detectable in a few metres near the surface only. Secondly. we neglect the advective terms in comparison with the convective ones in the conservation equation. This is admissible in the inertial frequency band because spatial gradients of $1^{\circ}C/10km$ in the horizontal and $1^{\circ}C/m$ (at the thermocline level) in the ver- tical are associated with horizontal displacements of some km and vertical displacements of some m: there is a factor of about 10 between the two effects.
These hypotheses lead to:

$$\frac{\partial T}{\partial t} \, dt + \frac{\partial T}{\partial z} \, dz = 0$$

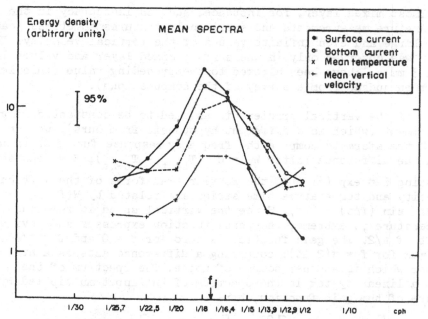

Fig. 1. Each spectrum is an average of 5 to 7 spectra associated
         with time series of about 3 months in length, obtained at
         different places and during different summers on the con-
         tinental shelf of the Gulf of Lions. Let us emphasize that
         typical frequencies of energy density in the vertical are
         greater than similar ones in the horizontal.

where T is the temperature measured at time t and depth z. Then the
vertical velocity is $W = \dfrac{dz}{dt} = -\dfrac{\partial T}{\partial t} \cdot \left(\dfrac{\partial T}{\partial z}\right)^{-1}$, positive upwards.
The first and second terms in this product are computed respectively
at a constant level and time.

    For the deep ocean, many authors have used a constant tempera-
ture gradient in the vertical which is but a rough approximation in
the surface layer. Therefore we take into account the instantaneous
vertical gradient : Δt and Δz being the time and depth intervals
of the time series $T(t,z)$, the vertical velocity at level z_j and
time t_i is defined by

$$W_{ij} = -\frac{\Delta z}{\Delta t} \cdot \frac{T(t_{i+1},\, z_j) - T(t_{i-1},\, z_j)}{T(t_i,\, z_{j+1}) - T(t_i,\, z_{j-1})}$$

We have considered time series for which $\Delta z = 5$ or 2.2 m and
$\Delta t = .5$ or 1 hour when the temperature is defined to 0.025°C. In

an almost mixed layer, for instance, such an inaccuracy in the
temperature measurements and small-scale features sometimes leads to
very large and even infinite values of the vertical velocity. This
error was observed only in the surface mixed layer and values larger
than 3 mm s^{-1} have been altered to the preceding value (this level
has been judged from a survey of the computations).

If the vertical gradient is assumed to be constant during the
experiment (which is a different hypothesis from ours), we can use
the Z transform to compute the frequency response function associated
with the difference filter $W_t = C (T_{t+\Delta t} - T_{t-\Delta t})$; C = constant.

Denoting $Z = \exp (jf \Delta t)$, the Fourier transforms of the vertical
velocity and temperature time series are related by W(f) =
C 2i sin ($f\Delta t$) · T(f). Since the vertical speed is zero when the
temperature is extreme, the phase function expresses and obvious
shift of $\pi/2$. The gain function is zero for $f = 0$ and $f = \pi/\Delta t$, and
maximum for $f = \pi(2 \Delta t)$: computing a difference acts as a high-pass
filter which is another source of noise. The spectrum of the output
from a linear system is the spectrum of the input multiplied by the
square of the gain function, here

$$\left[\frac{2\Delta z}{T_{z+1} - T_{z-1}} \right]^2 \cdot \left[\frac{\sin f\Delta t}{\Delta t} \right]^2$$

In the vicinity of $f = f_j = 0.57$ cph, Δt of the order of one hour
leads to $\sin f\Delta t / \Delta t \sim f$ and thence to a transfer function proportional
to f^2 (the same as for a pure sinusoidal wave) and independent of
Δt. Despite using a more complex algorithm, we have found that an
f^2 relationship roughly exists between the vertical velocity and
temperature spectra at a specific depth. Although the spectra in
Fig. 1 represent depth-and-ensemble averaged parameters, such a
relationship may be suspected. So the vertical velocity time series
are more noisy than the measured temperature ones. Nevertheless,
the amplitudes of the largest oscillations on the two curves are
not proportional. Then it is possible to get information from temperature time series taken in an almost homogeneous layer : a set
of rather smooth signals may reveal large vertical displacements
provided that the measurements have sufficient accuracy. When
considering the noise in the vertical velocity time series computed
at specific levels, what time series represents the thermocline
oscillations the best?

Coherence analyses in the f_j band show that the vertical
velocities computed at different levels are coherent with a zero
phase lag and it follows that oscillations mainly occur in the
first baroclinic mode. Around mid depth, the vertical velocities
are more similar than the temperature. We have computed a depth
averaged vertical velocity time series by summing the time series

from eight different levels. This procedure acts as a low-pass
filter and defines a parameter which we have found to be the best
representation of the amplitude of the oscillations for the following
reasons : in the f_i band, the coherence between the horizontal
currents and the depth averaged vertical velocity is larger than the
coherence between the horizontal currents and the vertical velocity
computed at any level ; also, the vertical velocities at specific
levels are coherent with the depth averaged vertical velocity. A
depth averaged vertical velocity is drawn in Fig. 2 together with
the depth averaged temperature within roughly the same depth interval.

Seven depth averaged vertical velocity time series (see Table
I in MC) of about 3 months, namely 75 B, 75 C, 77 B, 77 C, 78 A,
78 B and 78 E have been specially analysed. Regarding the depth
averaged temperature and the surface and bottom currents, we are
not able to explain all the differences observed between the various
spectra. However, in order to reach some statistical aspect of the
vertical velocity in the f_i band, we have computed an ensemble
averaged spectrum (MC) which is drawn in Fig. 1. This spectrum shows
that the energy density at f_i ($\sim$ 1/18 cph) is of the order of those
at 1/16.4 and 1/15 cph. This is similar to the result obtained with
the temperature.

THE DIRECTION OF PROPAGATION OF THE INTERNAL WAVES

The phase relationship between the current and the vertical
velocity at f_i was established in MC ; also, the representation of
a scalar by horizontal polarized currents, in order to use the spectral analysis of two-dimension vectors (Gonella, 1972), is explained.
Both surface and bottom currents are coherent with the depth averaged
vertical velocity only for those moorings located less than about
20 km offshore. The maximum coherence is observed at f_i but we
emphasize that the bottom current and the vertical velocity are
coherent for some frequencies $\geq f_i$. Coherence values obtained with
the temperature (MC) were not so large and this must be considered
as an improvement in the analysis. As a matter of fact, Fig. 1 shows
some similarities between the bottom current and vertical velocity
spectra. Also, theoretical results indicate that the current (at
f_i) due to the local wind is created in the surface layer only so
that baroclinic currents in the bottom layer are less masked than
in the surface layer and the coherence analysis is more accurate.
Let us mention that, when significant, the phase lags for different
estimates are similar for each pair (current-vertical velocity) and
that, if ϕ is the phase lag between the vertical velocity and the
surface current, the phase lag between the vertical velocity and
the bottom current is $\sim \phi + \pi$. When the points are more than 40
km from shore, the coherence values are not significant.

The directions of propagation computed from measurements at
one point only are shown in Fig. 3. At all the coastal points, the

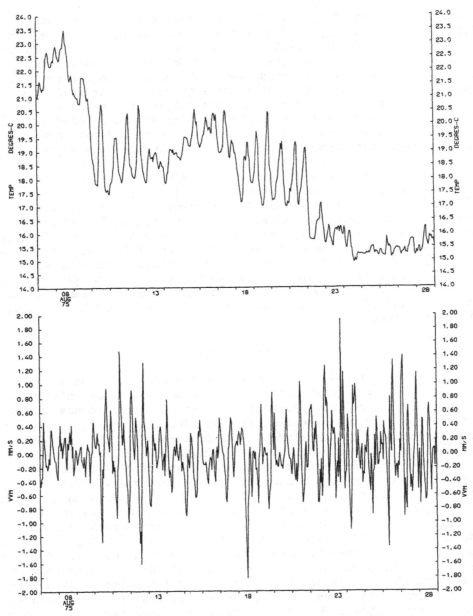

Fig. 2. Depth averaged temperature and mean vertical velocity between
about 2.5 m and 18 m at 75 B (see Fig. 3). When the stratification is smooth (10th to 13th), oscillations appear on the
two curves. When the stratification increases (20th to 21st),
a large temperature signal does not correspond to significant
amplitudes for the internal waves. Conversely, when the stratification decreases (in the surface mixed layer before the upwelling event - 9th to 10th - and from the 23rd when the whole
column is homogeneous), actual oscillations are not detected
with one temperature sensor.

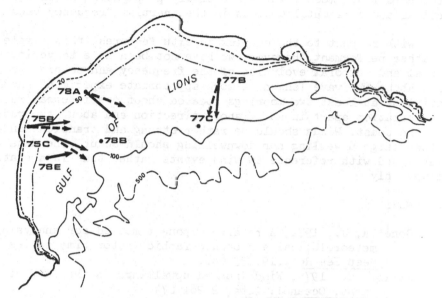

Fig. 3. The directions of propagation drawn from the phase-
lags (when significant) between the horizontal
currents and the mean vertical velocity show that
the internal waves are created at the shore. No direction is dominant at the offshore points. These directions are roughly similar to the ones computed with
the depth averaged temperatures and strongly support
the theoretical results presented in MC.

internal waves are propagating from the nearest coast.

DISCUSSION

Taking into account the vertical temperature gradient and
computing a vertical velocity make it possible to represent the
amplitude of the internal waves in a more accurate manner than with
the depth averaged temperature. The main improvement with respect
to the results presented in MC is that bottom currents have been
found to be coherent with vertical motions at frequencies greater
than or equal to the inertial frequency.

The spectral analysis of the whole data set supports most of
the MC conclusions from the model. It is very probable that the
current induced by the local wind in the surface layer and a barotropic mode generated at the coast combine to make the two layers
oscillate at the inertial frequency with an out of phase relation-

ship. Baroclinic modes, which are slowly propagating seaward, induce
vertical and horizontal motions in the inertial frequency band.

With respect to the complex in situ features, this simple model
is rather performant. However, we have not been able to verify the
spatial and temporal evolution of the frequency and nothing is
known about the wave length. A more appropriate experiment should
consist of at least two moorings located about one internal radius
of deformation apart in an offshore direction and about two radii
from the coast. Winds should be rather strong and transient and
neither large upwelling nor downwelling should occur. The data should
be analysed with reference to wind events rather than in a statistical way only.

REFERENCES

Gonella, J., 1972, A rotary-component method for analysing
    meteorological and oceanographic vector time series,
    Deep Sea Res.,19:833-846.
Millot, C., 1979, Wind induced upwellings in the Gulf of
    Lions, Oceanol. Acta, 2:261-274.
Millot, C., and Crepon, M., 1981, Inertial oscillations on
    the continental shelf of the Gulf of Lions.
    Observations and theory, J. Phys. Oceanogr., 11:639-657.

COASTAL UPWELLING, CYCLOGENESIS AND SQUID FISHING NEAR

CAPE FAREWELL, NEW ZEALAND

M.J.Bowman, S.M.Chiswell, P.L.Lapennas and R.A.Murtagh

Marine Sciences Research Center
State University of New York
Stony Brook, NY 11794, USA

B.A.Foster, V.Wilkinson and W.Battaerd

Biology Departments
University of Auckland
New Zealand

INTRODUCTION

Many observations have shown that coastal upwelling is an important feature of eastern boundary currents. It has been noted that upwelling is often localized near capes and promontories, or above submarine coastal features such as seamounts. Examples include California (Bernstein et al., 1977; Traganza et al., 1981), Baja California (Barton and Argote, 1980), Chile (Johnson et al., 1980), Peru (Preller and O'Brien, 1980), South West Africa (Bang and Andrews, 1974) and New Zealand (Bowman et al., 1983a).

Examples of numerical simulations of coastal upwelling dynamics include the f-plane model of O'Brien and Hurlburt (1972), the β plane model of Hurlburt and Thompson (1973) and the work of Hurlburt (1974), Johnson (1975), and Peffley and O'Brien (1976) who studied the effects on coastal upwelling of coastline geometry and bottom topography. Other investigators have concentrated on using sea level (vertically integrated) numerical models to study the barotropic response of coastal currents to the effects of headlands, seamounts and islands (e.g., Pingree, 1978; Pingree and Maddock, 1979, 1980). Although such models exclude baroclinic effects, they are useful in investigating such phenomena as the generation, advection and dissipation of vorticity due to coastline irregularities and topographic features. They also explain the creation of stagnation

279

regions in the lee of islands and promontories, and the physics of
offshore sand bank formation due to bottom stress variations.
Scouring and redeposition of sediments may result in a grading of
material according to the distribution of maximum bottom stress.
Coriolis effects have been invoked to explain the assymetric
formation of such sandbanks in regions of cyclonic as opposed to
anticyclonic circulation (Pingree, 1978).

Hsueh and O'Brien (1971), and Hsueh and Ou (1975) also pointed
to the importance of Ekman turning in the bottom boundary layer to
coastal upwelling. Curvature of the flow in the presence of bottom
friction can significantly enhance upwelling through centrifugal
effects (Garrett and Loucks, 1976). These effects may occur in flow
around headlands or within cyclonic vortices, but were largely
ignored in Arthur's (1965) study of upwelling generated by vorticity
accelerations near headlands in eastern boundary currents. Robinson
(1981) carefully isolated barotropic vorticity generation mechanisms
in a fluctuating flow by estimating the relative contributions of
stretching of planetary vorticity, nonlinear friction, advection, and
topographic production.

Here, we present results obtained from an interdisciplinary
sampling cruise during the austral summer of 1980-81 in a coastal
upwelling region near Cook Strait on the west coast of central New
Zealand (Fig. 1). Satellite thermal infra-red imagery (Fig. 2) has
shown that the region is one of persistent cyclogenesis. Cold core
eddies drift from upwelling centers equatorward into the western
approaches to Cook Strait, or towards Cape Egmont, depending on local
winds. From a physical point of view, the region is of interest due
to the apparently persistent nature of the upwelling, even in the
presence of downwelling favorable winds, and by its confinement to a
region of convex curvature in the coastline at the northwestern
corner of the South Island. The upwelling and eddy shedding zone is
also biologically productive. A major Japanese squid fishing fleet
exploits what appears to be a short and tightly coupled food chain
and enjoys particularly profitable fishing for New Zealand arrow
squid (Notodarus solani) around the frontal boundaries of senescent
eddies (~10 days old) which have drifted into the western approaches.

This energetic region was a good location to study topographic
forcing of cyclogenesis and upwelling as well as initiating studies
directed towards food chain interactions. We also present preliminary
results of numerical simulations directed towards understanding the
roles of coastal curvature and abrupt topographic features in
generating and modifying vorticity in the coastal boundary zone.
These computations have provided some insight into the various
torques and their effects on a coastal current flowing over a
topographically undulating bottom in the presence of rotation and
nonlinear bottom friction.

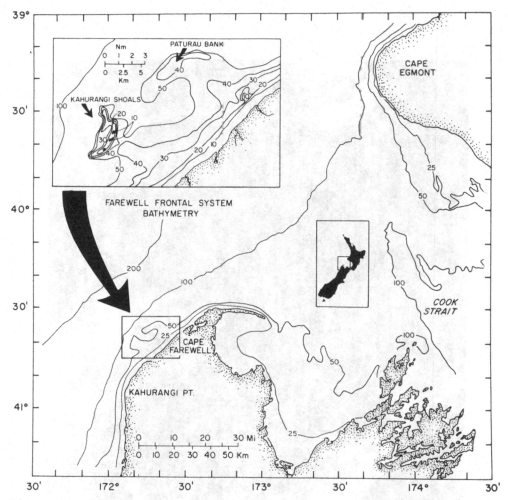

Figure 1.   Locator map of the western shelf of central New Zealand.

OCEANOGRAPHY OF THE STUDY REGION

    The prevailing westerlies at these latitudes (39-42°S) are
modified by a more or less regular progression of cyclones and
anticyclones throughout the year. Local winds are highly variable,
being strongly influenced by coastal mountains and in fair weather by
sea breezes.

    Strong surface fronts associated with upwelling in the Cape
Farewell region have been observed on several occasions during the
last decade (Stanton, 1971, 1976; Bowman et al., 1983 a,b) and are

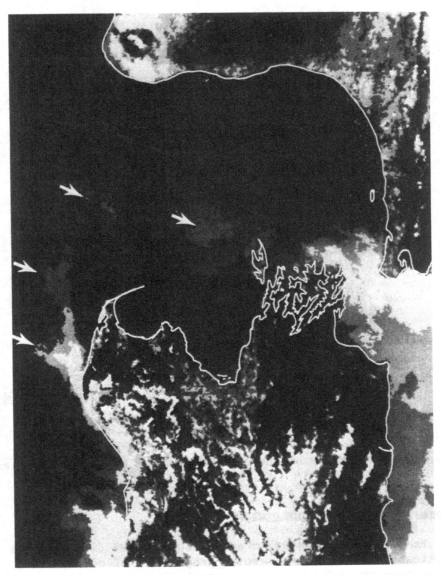

Figure 2. Sea surface temperature, Feb. 9 1982. Upwelling centers can be seen at Cape Foulwind (lower left) and Kahurangi Point. Three detached eddies are identified by arrows. Evidence of strong tidal mixing can be seen in Cook Strait.

clearly evident in thermal IR imagery (Fig. 2). The upwelling is apparently associated with fluctuations in the Westland Current, which is the coastal component of the equatorward flowing eastern boundary current of the Tasman Sea. Some basic characteristics of the coastal circulation have been derived from drift card measurements and dynamic height calculations (Brodie, 1960; Garner, 1961). The current appears quite variable; it usually sweeps around Farewell Spit and into Cook Strait, where it becomes known as the D'Urville Current. Stanton (1976) suggested a strong influence by local winds, with the Westland Current disappearing or reversing during periods of low wind stress. Under strong southerly wind forcing, the D'Urville Current diminishes or ceases and the Westland Current continues northward past Cape Egmont. Eulerian current data from the western approaches to Cook Strait analyzed by Heath (1978) also showed strong meteorologically forced barotropic currents in the region.

Cook Strait possesses strong and variable tides (Bowman et al., 1980); summer stratification is controlled by spatial variations in tidal mixing (Bowman et al., 1983a) although it is considerably modified by wind stress and advective intrusions (Bowman et al., 1983b). The western approaches to Cook Strait (South Taranaki Bight) contain oligotrophic water of subtropical origin derived from the surface layers of the Tasman Sea. Tidal streams in the Cape Farewell study area, derived from a numerical model (Bowman et al., 1980) lie in the range $0.1 - 0.2$ m s^{-1}, with a small region of flow intensification close to shore, just north of the tip of Farewell Spit where currents reach 0.35 m s^{-1} (Bowman et al., 1983b). The ratio of the M_2/S_2 tides is about 3.0 on the west coast (Ministry of Transport, 1981) which leads to a doubling in spring tidal streams over neaps and thus a fourfold increase in maximum tidal bottom stress.

The bottom topography on the west coast is characterized by a steeply sloped (~3%), but relatively featureless shelf down to about 50 m followed by a more gradual drop off (~0.5%) to the shelf break about 45 km offshore (Fig. 1). Northeast of Cape Farewell the shelf is broad and flat, forming a saddle known as the Egmont Terrace, located between the North and South Islands, the Tasman Sea to the east and the Cook Strait canyon to the west. On the inner shelf two prominent features, Kahurangi Shoals and Paturau Bank, rise approximately 50 m and 25 m from a surrounding depth of about 60 m.

EXPERIMENTAL DESIGN AND TECHNIQUES

Since tidal mixing variations were originally suspected to play some role in modifying the upwelling intensity in the Cape Farewell region, sampling cruises aboard HMNZS "Tui" were designed to cover both spring and neap tides. Rapid surface mapping of temperature,

salinity, nutrients (NO_3+NO_2, $Si(OH)_4$) and chlorophyll a
fluorescence along zig-zag tracks allowed us to locate, survey, and
resurvey regions of upwelling and cyclogenesis.

Surface temperature and salinity were obtained with an
Interocean 550 CTD placed in a deck bucket of running sea water.
Nutrients were measured with a Technicon Mark II Autoanalyzer.
Chlorophyll a (chl a) was determined from fluorescence measured with
a Turners Associates fluorometer calibrated every 30 minutes against
extracted chl a (Yentsch and Menzel, 1963; Holm-Hansen et al., 1965).
These instruments were supplied continuously with sea water pumped
from 2 m depth. Hydrocast stations were made along several transects
across the upwelling zone as well as across detached eddies to
determine vertical structure. Zooplankton (0.1 and 0.3 mm mesh) casts
were taken at each station together with the hydrobottle and CTD
casts.

Radio tracked drifters, drogued at 5 m depth, were deployed at
appropriate times to measure mean circulation as well as to tag
eddies for subsequent remapping. Winds were recorded from the ship's
anemometer.

INITIAL SURVEY OF GREATER COOK STRAIT

An initial survey of the greater Cook Strait region including
the Cape Farewell upwelling zone was made over a 9 day period (Jan
22-30, 1981).

It is not intended to discuss these results in detail, but only
to briefly describe some important features of the region (Figs. 3-6).
Further details are given in Bowman (1983b).
Region I, known as the South Taranaki Bight, consisted mainly of
Tasman Sea surface water of subtropical origin. In summer
these surface waters were warm (T >19 C), moderately stratified
(ΔT~4 C), of high salinity (S >35), low nutrient NO_3+NO_2 <0.25μM
and low in chl a.
Region II is Tasman Bay, identified by very warm (T > 20 C),
highly stratified (ΔT > 6 C) and nutrient depleted coastal water.
Tasman Bay is well sheltered from strong winds from both the
southwestern and southeastern quarters.
Region III is the tidally mixed frontal zone around Marlborough
Sounds, the dendritic system of drowned river valley estuaries on the
northeastern tip of the South Island. Inshore waters were depressed
in temperature (T < 15 C) but high in nutrients (NO_3+NO_2< 4μM)
and chl a (> 1 mg m^{-3}), indicating a favorable growth regime for
phytoplankton within these marginally stratified waters (ΔT~1 C).
Region IV is the narrows of Cook Strait, where both intense
tidal mixing and significant advective influences existed. An
intrusion of nutrient depleted South Taranaki Bight water can clearly

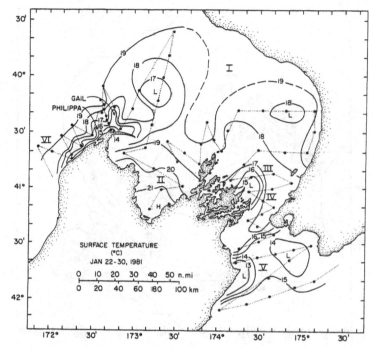

Figure 3. Surface temperature Jan. 22-30, 1981.

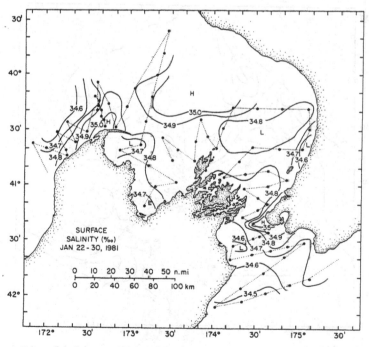

Figure 4. Surface salinity Jan. 22-30, 1981.

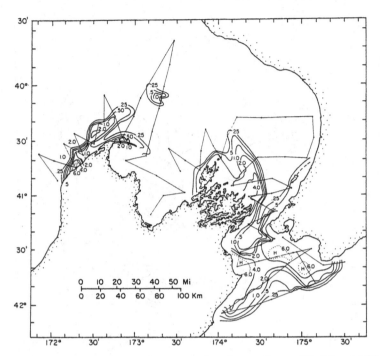

Figure 5.   Surface nitrate-nitrite, Jan. 22-30, 1981.

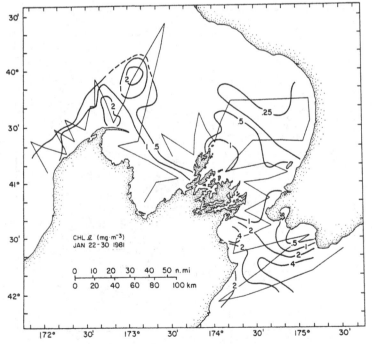

Figure 6.   Surface chlorophyll a, Jan. 22-30, 1981.

be identified in the surface $NO_3 + NO_2$ distribution (Fig. 5).

Region V is southeastern Cook Strait, characterized by complex hydrographic patterns representing the mixing of Cook Strait, East Cape Current, and Southland Current waters. Evidence of upwelling along both North and South Island shorelines can be noted, particularly in the temperature and nutrient maps (Figs. 3 & 5).

Region VI is the Cape Farewell upwelling zone which forms the focus of this paper.

SURVEYS OF THE FAREWELL FRONT DURING NEAP TIDES

Six surface mapping surveys were made of the Cape Farewell region between successive neap and spring tides.

Surveys I and II (Jan 29 & 30, 31) were made during neap tides and generally light (~5-8 m s^{-1}) downwelling favorable wind conditions. During these surveys we were able to identify localized sources of upwelling. Also we made some initial estimates of the rate of detachment and reformation of cyclonic eddies arising from instabilities in the upwelling zone.

Strongest upwelling on survey I (Jan 29) occurred near the coast at the base of Farewell Spit, as evidenced by lowest temperature (~14 C) and highest salinity (~35.0) (Figs. 3, 4), although elevated nutrient levels indicated an active upwelling cell further south towards Kahurangi Point (Fig. 5). Two meanders were located and labelled Phillipa and Gail (Fig. 3); the presence of an enclosed region of depressed temperatures, elevated nutrients and chl a (Figs. 5, 6), 60 km NNE of the tip of Farewell Spit suggested that this was an eddy previously shed from the upwelling zone. The shape of the streamer had a close similarity to the temperature distribution in the Feb 9, 1982 satellite thermal IR image (Fig. 2).

Survey II (Jan 30,31), made immediately following the first, provided an opportunity to study the development and propagation of the frontal meanders. Meander Phillipa detached from the frontal zone and propagated northwards at an approximate speed of 0.18 m s^{-1} to be replaced by meander Gail about 48 hours later (Fig. 7-10 inserts). Surface nitrate-nitrite levels were below instrument resolution (<0.25 μM) within Phillipa, presumably due to phytoplankton uptake, with only a trace left within Gail. A patch of chl a with peak concentrations > 3 mg m^{-3} appeared in Gail (Fig. 10 insert), suggesting rapid growth in the marginally stratified frontal waters.

A section was taken across the frontal zone on January 30 (stations 74-80; Fig.7 insert). Density closely tracked temperature with evidence of strong upwelling at the coast. Surface water appeared to be originating from a depth of about 100 m.

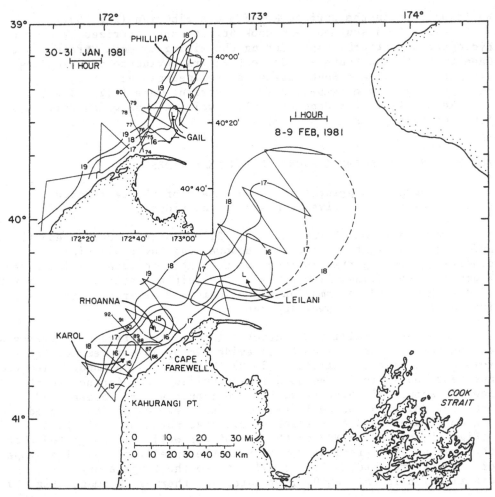

Figure 7. Sea surface temperature, Jan. 30-31, 1981 (insert) and
Feb. 8-9, 1981

SURVEYS OF THE FAREWELL FRONT DURING SPRING TIDES

     Sampling of the frontal zone was repeated in survey III a week
later during spring tides (Feb. 8-9). The upwelling zone had increased
markedly in size and encompassed a significant fraction of the South
Taranaki Bight (Figs. 7-10). During this and subsequent surveys,
winds were light (0-7 m s^{-1}) and in the 36 hours prior to survey
III, were generally upwelling favorable (from the south), but were
again downwelling favorable for the remainder of the cruise. The
source of the upwelling was now concentrated near the 45° bend in

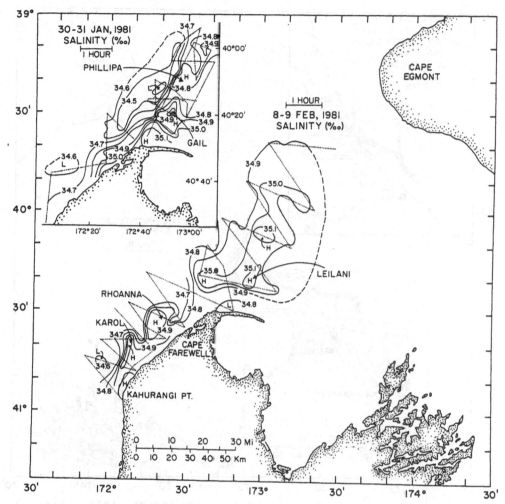

Figure 8.   Sea surface salinity, Jan. 30-31, 1981 (insert) and Feb. 8-9, 1981.

the coastline at Kahurangi Point. Three cold core eddies (Leilani, Rhoanna and Karol) were embedded in a wedge of recently-upwelled water centered over the 100 m isobath; a fourth eddy appeared to be developing near Kahurangi Point. Highest nutrients were located in the offing of the Point (Fig. 9); chl a was highest within Leilani (>2.5 mg m^{-3}), with a smaller patch inshore of eddy Karol. In contrast to the neaps survey, a band of warm, nutrient depleted surface water was located along the shore adjacent to Cape Farewell (Fig. 7).

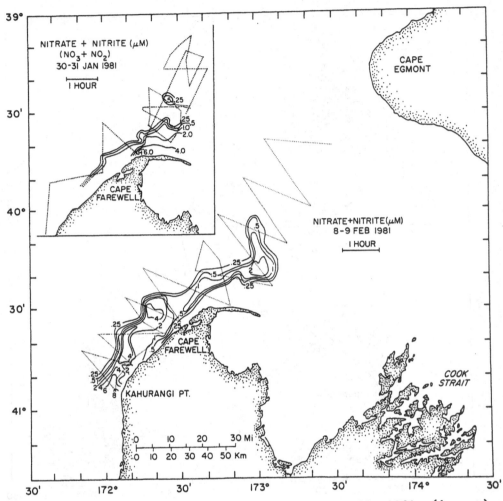

Figure 9. Sea surface nitrate–nitrite, Jan. 30–31, 1981. (insert)
and Feb. 8–9 1981.

Survey IV, made on February 9 and 10, included the coastal
region immediately south of Kahurangi Point (Fig. 11). Mapping
confirmed that upwelling was localized near to the Point. During this
period, eddy Rhoanna stalled and reversed its northeastward drift,
and coalesced with eddy Karol to form Karolanna located directly over
Kahurangi Shoals. The sharp horizontal density gradients observed
through Rhoanna (stations 93–99; Fig. 12) could have induced both
strong cyclonic rotation and upwelling. Maximum upwelling was
apparently situated about 10 km offshore in the eddy center.
Significant convergence at, and shear across, the surface fronts of

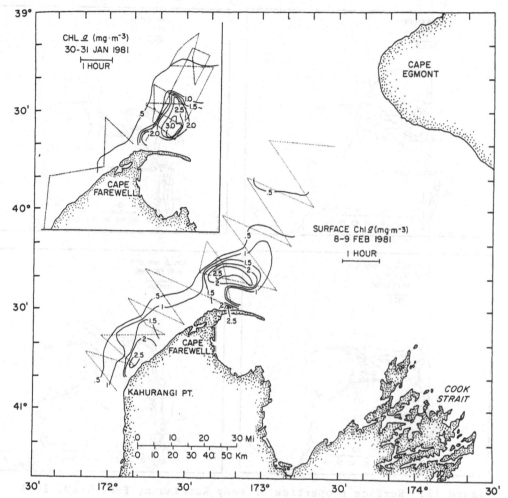

Figure 10.   Sea surface chlorophyll a, Jan. 30-31, 1981. (insert) and
Feb. 8-9, 1981.

Karolanna were accompanied by flotsam accumulations, such as seaweed
and tree trunks, flocks of feeding seabirds and sudden changes in
ship headings experienced during the mapping exercises as the ship
slowly crossed the front. Maximum tangential speeds were estimated
from the isopycnal slopes as 0.8 m s^{-1}, representing a rotational
period of about 22 hours for an eddy of 20 km diameter.

Eddy Karolanna was again surveyed (#V) on February 11. It had
remained located over the Shoals overnight; the only significant
change appeared in chl a concentrations which had ordered themselves

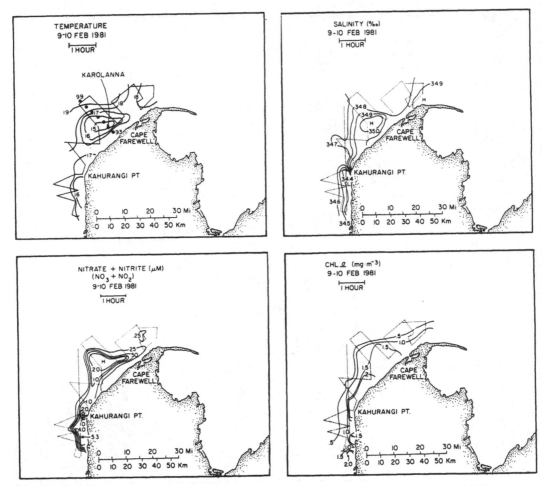

Figure 11. Surface properties of eddy Karolanna, Feb. 9-10, 1981.

into a more circular pattern. Eddy Leilani was also relocated and remapped on February 11. It had migrated westward about 20 km within the D'Urville Current at a mean speed of .06 m s^{-1} between the 8th and 11th. Nitrate-nitrite was completely exhausted in Leilani (i.e., <0.25 μM) by the 11th.

A final survey (#VI) of eddy Leilani was made overnight during February 11-12 (Fig. 13). Although nitrate-nitrite was depleted, silicate (Si(OH)$_4$) was still a useful water mass tracer (Murtagh, 1982). Twenty four Japanese squid fishing ships from a 50 vessel fleet were observed squid jigging around the fronts of eddy Leilani during this survey (Fig. 13a). A vertical section across Leilani (Fig. 14) showed that it was in a senescent stage, as evidenced by a slumping of isopleths and fragmentation of the eddy. Presumably

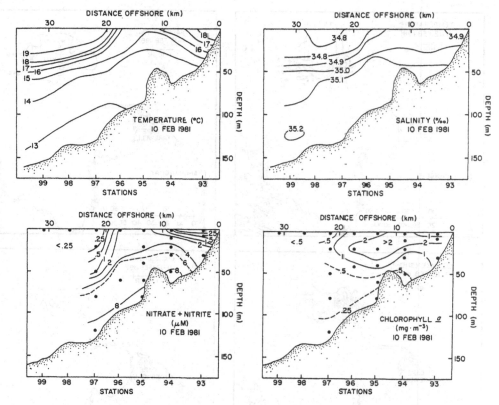

Figure 12.   Vertical section across Karolanna, Feb. 10, 1981. Station positions are given in Fig. 11.

frictional stresses dissipated vorticity and disintegrated the physical structure of the eddy. These repeated surveys on Leilani suggested an eddy lifetime of about 10-15 days from formation at Kahurangi Point to dissipation in Cook Strait.

DYNAMICS AND NUMERICAL SIMULATIONS

Since winds were light and downwelling favorable during much of the survey, our results suggest that the upwelling was driven by onshore bottom Ekman transport reinforced by centrifugal and topographic effects. In order to gain some preliminary insight into these mechanisms, we applied a sea level (vertically integrated) nonlinear numerical model to simulate the barotropic circulation over the continental shelf in the region of interest. We synthesized a steady, barotropic current between the coast and the 500 m isobath of

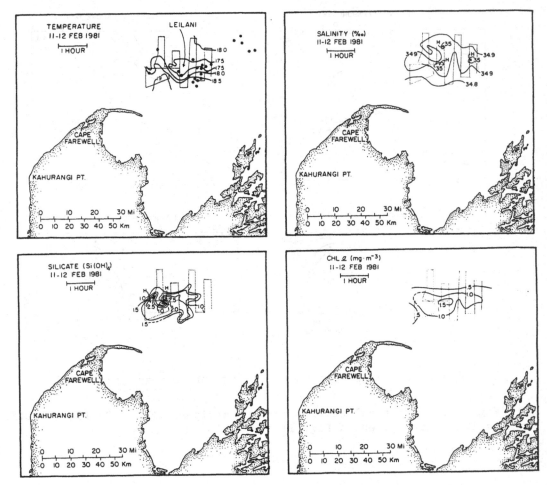

Figure 13. Surface properties of eddy Leilani, Feb. 12, 1981.

total transport 10^6 m^{3}s^{-1} whose velocities, spatial extent,
and general flow patterns were consistent with our observed drogue
trajectories and historical hydrographic and satellite observations.
This current, driven by along and cross shore pressure gradients (no
winds) was swept around Kahurangi Point and Cape Farewell, firstly on
a 4 n mile grid. The area covered (6913 km^3) is shown in Fig. 15.
The model was then rerun on 1 n mile and 1/3 n mile grids for the
sub-areas shown in Fig. 15, to obtain sufficient resolution over the
Kahurangi Shoals, using the results of the larger models sequentially
to obtain boundary conditions for the higher resolution simulations.

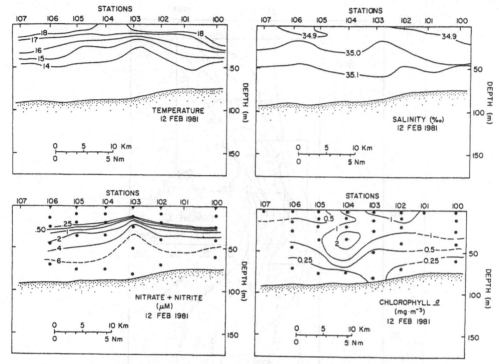

Figure 14.   Vertical section across Leilani, Feb. 12, 1981.

Relative vorticity

The distribution of relative vorticity was derived from the computed flow field to study Ekman pumping of bottom water into the quasi-geostrophic interior. We also investigated the various torques that contribute to the production and dissipation of relative vorticity ie., by stretching and squashing of water columns and bottom friction.

Figure 16 shows three regions of elevated negative vorticity. The first, at the base of Farewell Spit is a region of strong negative vorticity (~f) whose location corresponds to the upwelling center noted during survey I (Fig. 3). The second region lies close inshore of Kahurangi Shoals, and the third lies over the outer flank of the shoals. Again, both have magnitude ~f. Insight into the mechanisms producing this vorticity distribution was obtained by computing the generation and dissipation terms of eqn (A4) along mass transport streamlines (see appendix).

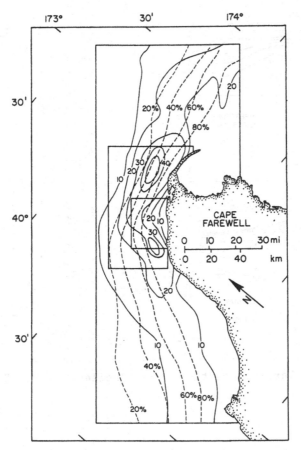

Figure 15. Domains of 4, 1, and 1/3 n mile models. Selected
streamlines (labelled as % of total transport between 500m isobath
and contour), and isotachs (cm s^{-1}).

    Four selected streamlines are superimposed on the depth contours
in Fig. 17. These illustrate how the flow intensifies over the sill
separating Kahurangi Shoals and the Point (maximum speeds 0.50 m s^{-1})
before weakening over the Paturau sandbank (minimum speeds
~ 0.15 m s^{-1}). The flow intensifies again as the bottom
gradient steepens at the right of the figure and currents peak at
about 0.60 m s^{-1} 15 km offshore of the base of Farewell Spit (Fig
15).

    Coriolis effects (viz., stretching of planetary vorticity) will
act upstream or downstream of topographic features where the flow is
directed up and down the bottom gradient (i.e., where streamlines

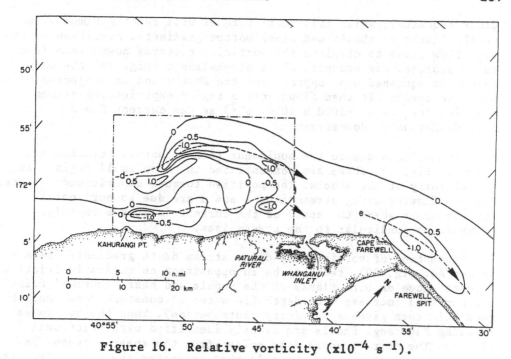

Figure 16. Relative vorticity ($\times 10^{-4}$ s^{-1}).

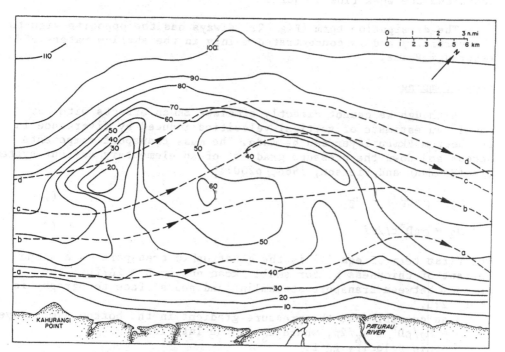

Figure 17. Bathymetry (m) and selected streamlines.

cross isobaths), while frictional torques will reach maximum values
on the flanks of shoals and steep bottom gradients. Advection by the
mean flow tends to displace the vorticity patterns downstream from
their sources. For example, along streamline d (Fig. 18) the water
column is squashed upon approaching the Shoals and is subjected to a
positive torque. It then flows into a region experiencing strong
negative torques ($< -1000 \times 10^{-11}$ s^{-2}) as the current flows
into deeper water downstream.

The effects due to the nonlinear form of bottom friction are
shown in Fig. 19. These are concentrated in three small regions in
the vicinity of the shoals. The positive torques experienced by water
columns flowing along streamlines a and d are due to horizontal
current shear where the speed of the current increases rapidly
offshore perpendicular to the streamlines.

Production of vorticity by cross stream depth gradients (term c
in eqn A4; Fig. 20), tends to be in opposition to quadratic friction
effects since on both flanks of the shoals and near Kahurangi Point
the currents increase with depth (in water of constant depth only the
quadratic term gives a vorticity contribution). Thus the two lobes
flanking Kahurangi Shoals are clearly identified with frictional
effects. The third inshore lobe arises from the shear between the
current accelerating over the sill that separates the Shoals from the
coast and the weak flow inshore.

The dissipation term (Fig. 21) always has the opposite sign to
the vorticity and is concentrated mainly in the shallow waters off
Kahurangi Point.

Ekman Pumping

Although we cannot directly simulate Ekman effects with our
model, an estimate of vertical velocities induced by divergence in
the bottom Ekman layer can be made. The mass transport components
across and down the pressure gradient of an elementary current system
are (Neumann and Pierson, 1966, p200):

$$M_1 = \rho g h \beta / f - M_2 \tag{1}$$

$$\text{and } M_2 = \rho g D^* \beta / 2\pi f \tag{2}$$

The first term of eqn (1) is the geostrophic transport of a slope
current of thickness h. For small Ekman number $E = 2A/fh^2$, the
angle θ between transport streamline and sea surface topography is
$\theta = D^*/2\pi h$.
The mean current down the pressure gradient in the bottom Ekman layer
is $u_2 = M_2/\rho D^* = u_g/2\pi$, where u_g is the interior
geostrophic velocity $g\beta/f$.
With linearized friction, the vertically integrated equations give

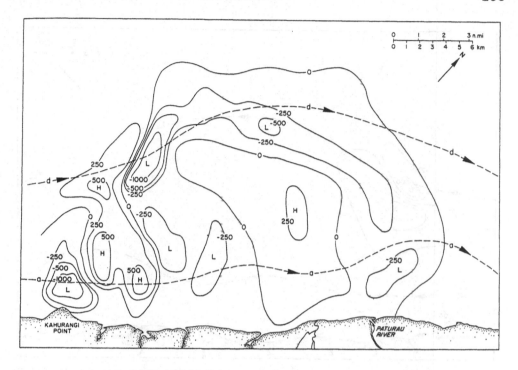

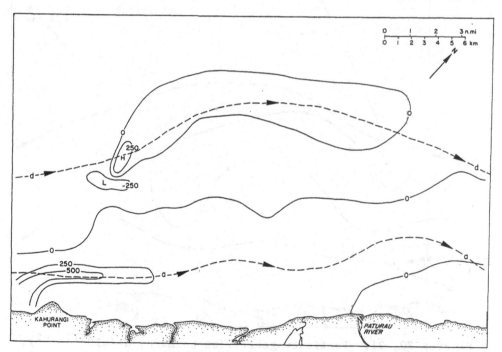

Figures 18, 19. Vorticity generation terms a, b (x10^{-11} s^{-2}).

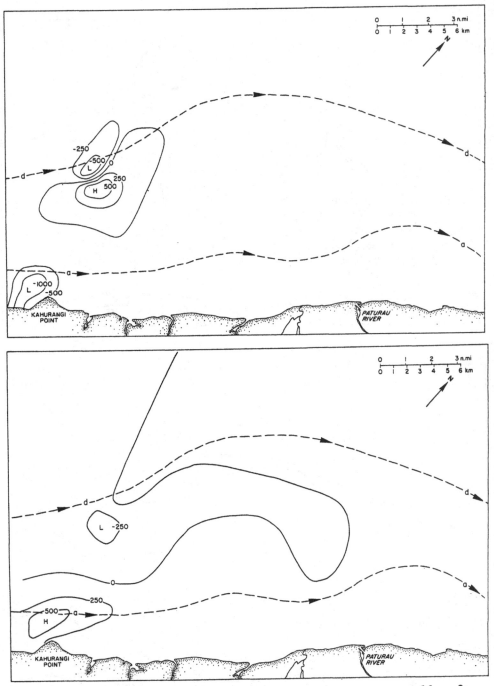

Figures 20, 21.   Vorticity generation terms c, d (x10^{-11} s^{-2}).

$$\theta' = k'/fh \tag{3}$$

(For example, for $k' = 5.4 \times 10^{-4}$ m s^{-1}, $f = 10^{-4}$ s^{-1} and $h = 75$ m, $\theta' \sim 4°$).
Equating the two transports, i.e., setting $\theta = \theta'$, gives

$$k' = (Af/2)^{1/2} \tag{4}$$

Also $D^* = \pi(2A/f)^{1/2}$ \hfill (5)

The Ekman pumping out of the bottom layer into the geostrophic interior is (Pedlosky, 1979, p198)

$$w = (A/2f)^{1/2} \, \nabla_\wedge \vec{u}_g$$

Since, from eqn (3), θ' is independent of u, $\nabla_\wedge \vec{u} = \nabla_\wedge \vec{u}_g$, so

$$w = (A/2f)^{1/2} \, \nabla_\wedge \vec{u} \tag{6}$$

From eqns (4 & 5) we estimate $A \sim 60$ cm^2 s^{-1} and D* ~ 35 m. The uplift z of a parcel of water

$$\int_0^t w \, dt$$

during transit time t through each of the three regions of high negative vorticity shown in Fig. 16 were thus estimated by integrating ζ along streamlines a, d, and e using eqn (6).
For streamline a, z = 50 m over 15 hr,
     d, z = 90 m over 50 hr,
     e, z = 10 m over 15 hr.
We conclude that midshelf Ekman pumping to be significant near the Kahurangi Shoals, less so near Cape Farewell. However strong currents in the offing of the Cape may induce upwelling at the coast (see next section).

Centrifugal effects

     In regions where streamlines curve sharply, centrifugal effects can augment Ekman flow through increasing sea surface slopes in the radial direction. A first order estimate of this (landward) radial bottom current u_r is (Garrett & Loucks, 1976)

$$u_r = h_0 u_g (f + u_g /r)/k' \tag{7}$$

where h_0 is the depth of a bottom frictional layer, assumed to be some fraction of D^*. The first term in eqn (7) is identified with the Ekman flow down-gradient in the absence of curvature. The second term is the contribution due to centrifugal effects. For streamline

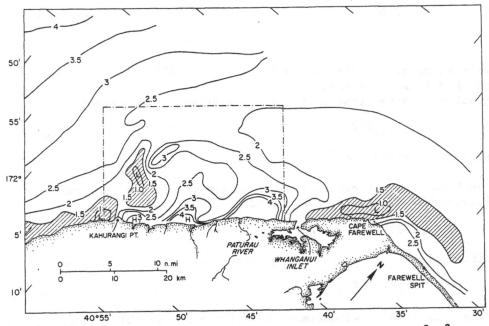

Figure 22. Stratification index, s, on log scale ($cm^{-2} s^3$).

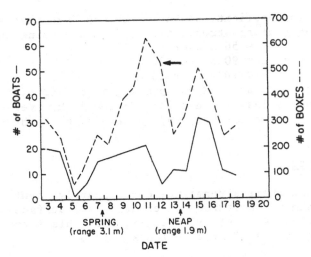

Figure 23. Fleet count (solid line), plus daily squid catch per boat (dashed line; 1 box contains 28 squid). The best yield for the month was 88 boxes/boat (Feb. 12; see arrow).

e, we estimate r = 30 km near Cape Farewell.
Taking a mean alongshore current u_g = 0.3 m s^{-1} and equating
u_r = u_g/2π gives a value for h_o = 1 m.
Hence u_r = (5 + 0.5) cm s^{-1} Taking the width of the upwelling
zone L to be 1 km (see Fig 3), then w ~ $u_y h_o$/L = 5 m
day^{-1}. We thus conclude that centifugal effects are negligible in
the Cape Farewell upwelling center.

Vertical mixing

We estimated the importance of vertical mixing arising from
turbulent energy dissipation within the Westland Current by computing
the Simpson-Hunter (1974) stratification index
s = $\log_{10}$(h/ku^3) (Fig. 22). Breakdown of summer stratification
in temperate latitude, tidally energetic shelf seas often occur for a
critical value of s ~ 1.5 in cgs units. Two areas of low s (< 1.5),
shaded in Fig. 22, lie over the Kahurangi Shoals and off Cape
Farewell. In addition, wind and swell induced stirring over the
Shoals (the top of the Shoals reaches to within 10 m of the sea
surface) will be important. Spring to neap modulations of h/ku^3
are expected to be ~8. We conclude that vertical mixing over the
Shoals and off Cape Farewell should be important mechanisms
contributing to the observed appearance of the cold water. Figures 3
and 7 support the hypothesis of strong neap to spring modulations of
upwelling; however more current measurements are needed to confirm
these ideas.

Cyclogenesis

From our neap surveys, we estimated meander wavelengths ~28 km,
phase velocities ~0.4 m s^{-1} and periods ~20 hours. This is close
to the inertial period 18.4 hours, suggesting that southerly wind
gusts lasting a few days modulate the intensity of the Westland
Current and generate transient meandering and upwellings at inertial
frequencies. The closeness of the eddy and the inertial periods also
suggests a resonant interaction which reinforces cyclogenesis.

Eddy Spindown

We estimated the spin down time τ of a quasi-geostrophic eddy as
(Pedlosky, 1979, p205):

$$\tau = 1/fE^{1/2} = 20 \text{ hr for } A = 58 \text{ cm}^2 \text{ s}^{-1} \text{ and } h = 75 \text{ m.}$$

Our surveys suggest an eddy lifetime of 10- 15 days; this suggests
that the eddies are deriving significant kinetic energy from the
potential energy of the surrounding, deeper, stratified water.

NOTES ON BIOLOGICAL FINDINGS AND FISHING STRATEGIES

The rates of phytoplankton production and biomass in the Farewell upwelling zone will be discussed in detail in Lapennas et al. (in prep.). The results presented there show that surface chl a concentrations in the upwelling zone were increased some 4 to 6 times background levels farther offshore. These patches usually were located near to, but not necessarily coincidental to the core of the upwelling cell. Presumably, production was greatest in the frontal boundaries where marginal stratification coupled with an adequate light regime led to favorable growth conditions.Copepod abundance was at least four times higher within and around the shoreward perimeter of the eddies than offshore. Larval stages of euphausiids were present in increased concentrations at the southern edge of eddy Leilani on February 12. Increased concentrations of fecal pellets from grazing copepods and sinking phytoplankton in frontal boundaries of senescent eddies are thought to attract the krill, and as the krill move towards the surface at night the squid follow.

Fishing at night is more effective if only because of the shorter distances of jigging. The fishermen concentrate their efforts at these frontal boundaries. They use local weather observations, sonar and synoptic sea surface temperature maps derived from daily fleet reports to locate schools of squid.The fishermen told us that catches were poor when the temperature was less than 15 C and the water of the wrong color (presumably upwelled water). Visibility is sometimes restricted because of sea fog in such areas. Fishing is best when a cold atmospheric front approaches, but weather conditions are too rough for good catches at times of low barometric pressure. The captains interpret and share their daily catches to guide their search for further concentrations of squid. They can distinguish between fish, zooplankton and squid on their sonars. Thus a combination of fishing experience, local knowledge, sharing of physical oceanographic data and fishing catches, coupled with fleet mobility and flexibility to move to areas where fishing is good is used to maximize catches and keep the industry profitable.

Analysis of the fleet logs kept by the fishermen showed that the fleet moved into the Cape Farewell region subsequent to spring tides on February 8. Perhaps they were anticipating increased eddy shedding and migration after spring tides. Fishing catches peaked on February 11 (Fig. 23), after which the ships dispersed as tides slackened towards neaps. It can also be seen that maximum and minimum fishing efforts were phased about 2-3 days following spring and neap tides, respectively.

SUMMARY

Although there are a great many references in the literature to wind driven ("classical") coastal upwelling, it seems that insufficient attention has been paid to the role of bottom frictional driven upwelling in shallow seas. No doubt this is partly due to the difficulty of measuring currents with sufficient resolution in regions of tortuous bathymetry and partly due to the masking of such effects by wind driven upwelling itself.

Results from a sampling cruise in the austral summer of 1980-81 near Cape Farewell have confirmed the existence of a persistent upwelling zone that appears to have significant variations between neap and spring tides. Some of these modulations in the intensity of upwelling may also be related to fluctuations in the Westland Current and local winds. Numerical simulations have identified localized regions where frictional upwelling and vertical mixing may be significant; these regions are proximate to upwelling centers mapped during the surveys.

It seems that a fortuitous combination of physical characteristics exists that produces an environment favorable to the generation of a more or less continuous stream of eddies from the northwestern tip of the South Island during summer. The Kahurangi Shoals are prominent midshelf obstructions lying in the path of a fluctuating but persistent coastal current. Mesoscale cyclonic eddies are shed in their wake, bringing nutrients to the surface as they drift onto the broad, unobstructed flat plateau of the Egmont Terrace. Here they are advected within the prevailing currents before being mixed away in the turbulent waters of Cook Strait to the east, by strong winds, or through viscous dissipation on the western shelf. The small diameter and vigorous circulation of the eddies suggests that centrifugal effects may be important in augmenting their secondary circulation since the earth's and the eddy rotation are in the same sense.

Cyclogenesis seems to be important in cross frontal mixing across the Cape Farewell upwelling zone, and in providing a favorable environment for enhanced biological productivity in the South Taranaki Bight. Preliminary calculations on a nitrate- nitrite budget for the stratified waters of the South Taranaki Bight indicate that these eddies may contribute up to 35% of the total nitrogen demand by phytoplankton.

The apparent relationships between upwelling, cyclogenesis and squid fishing activities offers the prospect of managing this resource more effectively with a better understanding of the physical and biological interactions that are operative in the region.

ACKNOWLEDGMENTS

    This project was a major component of the Maui Development
Environmental Study, carried out by the University of Auckland
(Coordinator, Prof. A. C. Kibblewhite),and commissioned by Shell BP
and Todd Oil Services Ltd (acting on behalf of Maui Development Ltd)
as part of the development of the Maui gas field off the Taranaki
coast. Support was also provided by the National Science Foundation
(NSF-INT grant 78-01159,NSF-OCE grants 77-26970 and 81-8283). We also
wish to acknowledge the help of Mr. B.H. Olsson, Director of the
Defence Scientific Establishment for the use of the HMNZS "Tui", to
Lt. Commander G.C. Wright, RNZN, for assistance at sea, to Ms V.
Abolins for cartographic assistance and to Mr G. Carroll for
computing advice. The research program was facilitated by a
Memorandum of Understanding in Marine Sciences between the State
University of New York at Stony Brook and the University of Auckland.
Contribution 343 of the Marine Sciences Research Center (MSRC).

APPENDIX

Model equations

    The model is based on the finite difference semi-implicit
staggered grid scheme of Leenderste (1967). The vertically integrated
equations of momentum and continuity are

$$\frac{\partial u}{\partial t} + u\frac{\partial u}{\partial x} + v\frac{\partial u}{\partial y} = -g\frac{\partial \eta}{\partial x} + fv - \frac{ku(u^2+v^2)^{1/2}}{h} \tag{A1}$$

$$\frac{\partial v}{\partial t} + u\frac{\partial v}{\partial x} + v\frac{\partial v}{\partial y} = -g\frac{\partial \eta}{\partial y} - fu - \frac{kv(u^2+v^2)^{1/2}}{h} \tag{A2}$$

$$\frac{\partial \eta}{\partial t} + \frac{\partial}{\partial x}(Hu) + \frac{\partial}{\partial y}(Hv) = 0 \qquad \text{where } H = h + \eta \tag{A3}$$

In an oscillatory flow with non zero mean u, the time averaged bottom
stress can be linearized and approximated as $ku_ou = k'u$, where
u_o is the amplitude of the oscillatory component (Saunders,
1977). Thus the linearized equations are more appropiate for the
study of residual flows in the prescence of significant tidal
currents, while the quadratic form should be used in simulating
steady flows or the oscillatory currents explicitly. Shelf currents
near Cape Farewell have not been well studied but are generally
expected to be in the range of 0.1- 0.2 m s^{-1}. We chose to use
the quadratic formulation in studying vorticity effects near
Kahurangi Shoals; but the results obtained with linearized friction

are not expected to be significantly altered.

The dimensionless quadratic drag coefficient k was set equal to

$k = gN^2/h^{1/3}$, where the Manning coefficient $N = .028$ m$^{-1/3}$ s.
For example, for h = 25 m, k = .0026
                  and h = 75 m, k = .0018.
The model was run with a constant value of $f = -0.95 \times 10^{-4}$ s^{-1}.

Vorticity Equation

The Lagrangian rate of change of relative vorticity, for $h \gg \eta$ is
(Robinson, 1981)

$$\frac{d\zeta}{dt} = \frac{(\zeta+f)}{h}\frac{dh}{dt} + \frac{k\vec{u}}{h} \wedge \nabla|\vec{u}| - \frac{k|\vec{u}|\vec{u}}{h^2} \wedge \nabla h - \frac{k|\vec{u}|\zeta}{h} \tag{A4}$$

$\qquad\qquad$ (a) $\qquad\qquad$ (b) $\qquad\qquad$ (c) $\qquad\qquad$ (d)

Term (a) represents a torque arising from changes in water depth,
i.e., the potential vorticity effect. The second term (b) generates
vorticity when there is lateral shear in the flow. This term is due
to the quadratic frictional law and generates a torque even when the
depth is constant. It is not present in the linearized friction
formulation. Term (c) arises when there is a depth gradient normal to
the velocity. Then there is a shear in the depth averaged frictional
force which generates vorticity. Term (d) represents vorticity
dissipation by bottom friction.

NOTATION INDEX

A	vertical eddy viscosity
D^*	thickness of bottom Ekman layer
E	vertical Ekman number
f	Coriolis parameter
g	acceleration due to gravity
h	water depth
h_0	depth of bottom frictional layer
k	quadratic drag cofficient
k	linear drag coefficient
L	width of upwelling zone
M	mass transport
N	Manning coefficient
r	radius of curvature of flow
s	stratification index
t	time
u, v	depth mean currents in x, y directions
u_g	geostrophic current
u_r	radial component of bottom flow
w	upwelling velocity

η   sea surface elevation
β   sea surface slope
ζ   vertical component of vorticity
ρ   seawater density
θ   angle between transport streamline and sea surface topography
            as calculated from Ekman theory
θ´   angle between transport streamline and sea surface topography
            as calculated from the vertically integrated equations
τ   eddy spin down time

REFERENCES

Arthur, R. S. 1965. On the circulation of vertical motion in eastern
        boundary currents from determinations of horizontal motion.
        J. Geophys. Res. 70:2799-2803.
Bang, N. D. and W. R. H. Andrews. 1974. Direct current measurements
        of shelf-edge frontal jet in the southern Benguela system.
        J. Mar. Res. 32:405-417.
Barton, E. D. and M. L. Argote. 1980. Hydrographic variability in an
        upwelling area off northern Baja, California in June, 1976.
        J. Mar. Res. 38:631-649.
Bernstein, R. L., L. Breaker and R. Whritner. 1977. California
        current eddy formation: ship, air and satellite results.
        Science. 195:353-359.
Bowman, M. J., A. C. Kibblewhite and D. E. Ash. 1980. M2 tidal
        effects in greater Cook Strait, New Zealand. J. Geophys.
        Res. 85:2728-2742.
Bowman, M. J., A. C. Kibblewhite, S. M. Chiswell and R. A. Murtagh.
        1983a. Shelf fronts and tidal stirring in greater Cook Strait,
        New Zealand. Oceanologica Acta. (In press).
Bowman, M. J., A. C. Kibblewhite, S. M. Chiswell, R. A. Murtagh and
        B. G. Sanderson. 1983b. Circulation and mixing in greater Cook
        Strait, New Zealand. Oceanologica Acta. (In press.)
Brodie, J. W. 1960. Coastal surface currents around New Zealand.
        N.Z. J. Geol. Geophys. 3:235-252.
Garner, D. M. 1961. Hydrology of New Zealand waters,1955. N.Z. Dep.
        Scient. Ind. Res. Bull. 138.
Garrett, C. J. R. and R. H. Loucks. 1976. Upwelling along the
        Yarmouth shore of Nova Scotia. J. Fish. Res. Board Can.
        33:116-117.
Heath, R. A. 1978. Atmospherically induced water motions off the west
        coast of New Zealand. N.Z. J. Mar. Freshw. Res. 12:381-390.
Holm-Hansen, O., C. J. Lorenzen, R. W. Holmes and J. D. H.
        Strickland. 1965. Fluorometric determination of chlorophyll.
        J. Cons. Perm. Int. Explor. Mer. 30:3-15.
Hsueh, Y and J. J. O´Brien. 1971. Steady coastal upwelling induced by
        an along-shore current. J. Phys. Oceanogr. 1:180-186.

Hsueh, Y. and H. Ou. 1975. On the possibilities of coastal, mid-shelf, and shelf break upwelling. J. Phys. Oceanogr. 5:670-682.

Hurlburt, H. E. 1974. The influence of coastline geometry and bottom topography on the eastern ocean circulation. Ph.D. thesis. Florida State Univ. 103pp.

Hurlburt, H. E. and J. D. Thompson. 1973. Coastal upwelling on a beta plane. J. Phys. Oceanogr. 3:16-32.

Johnson, D. R., T. Fonesca and H. Sievers. 1980. Upwelling in the Humboldt coastal current near Valparaiso, Chile. J. Mar. Res. 38:1-16.

Johnson, J. A. 1975. Upwelling over a curved continental shelf. Mem. Soc. Roy. des Sciences de Liege. 6th series. VII:93-103.

Lapennas, P. P., V. Wilkinson, M. J. Bowman and R. A. Murtagh. Phytoplankton production in relation to upwelling and cyclogenesis in coastal waters off New Zealand. In prep.

Leendertse, J. J. 1967. Aspects of a computational model for long period water wave propagation. Mem. RM 5294-PR, Rand Corp. Santa Monica, Calif.

Ministry of Transport. 1981. New Zealand Tide Tables. Marine Div. Wellington, New Zealand.

Murtagh, R. A. 1982. Summer nutrients in greater Cook Strait, New Zealand. MS thesis. Marine Sciences Reseach Center.

Neumann, G. and W. J. Pierson. 1966. "Principles of physical oceanography." Prentice-Hall, Englewood Cliffs.

O'Brien, J. J. and H. E. Hurlburt. 1972. A numerical model of coastal upwelling. J. Phys. Oceanogr. 2:14-26.

Pedlosky, J. 1979. "Geophysical Fluid Dynamics." Springer Verlag, New York.

Peffley, M. B. and J. J. O'Brien. 1976. A three dimensional simulation of coastal upwelling off Oregon. J. Phys. Oceanogr. 6:164-180.

Pingree, R. D. 1978. The formation of the Shambles and other banks by tidal stirring of the seas. J. Mar. Biol. Assn. U.K. 58:211-226.

Pingree, R. D. and L. Maddock. 1979. The tidal physics of headland flows and offshore tidal bank formation. Mar. Geol. 32:269-289.

Pingree, R. D. and L. Maddock. 1980. The effects of bottom friction and earth's rotation on an island's wake. J. Mar. Biol. Assn. U.K. 60:499-508.

Preller, R. and J. J. O'Brien. 1980. The influence of bottom topography on upwelling off Peru. J. Phys. Oceanogr. 10:1377-1398.

Robinson, J. S. 1981. Tidal vorticity and residual circulation. Deep Sea Res. 28A:195-212.

Saunders, P. M. 1977. Average drag in an oscillatory flow. Deep Sea Res. 24:381-384

Simpson, J. H. and J. R. Hunter. 1974. Fronts in the Irish Sea.
        Nature. 250:404-406
Stanton, B. R. 1971. The hydrology of the Karamea Bight.
        N.Z. J. Mar. Freshw. Res. 5:141-163.
Stanton, B. R. 1976. Circulation and hydrology off the west coast of
        the South Island, New Zealand. N.Z. J. Mar. Freshw. Res.
        10:445-467.
Traganza, E. D., J. C. Conrad and L. C. Breaker. 1981. Satellite
        observations of a cyclonic upwelling system and giant plume in
        the Californian Current. in "Coastal Upwelling." ed. F. A.
        Richards. Amer. Geophys. Union. Wash. D.C. 519pp.
Yentsch, C. M. and D. W. Menzel. 1963. A method for the determination
        of phytoplankton chlorophyll and phaeophytin by fluorescence.
        Deep-Sea Res. 10:221-231.

WHIRLS IN THE NORWEGIAN COASTAL CURRENT

T.A. McClimans and J.H. Nilsen

Norwegian Hydrodynamic Laboratories
Trondheim, Norway

ABSTRACT

Laboratory measurements of instabilities in a baroclinic coastal current (Vinger, et.al., 1981) have been scaled to the conditions of the Norwegian Coastal Current. Large whirls on the order of 50-100 km in diameter show maximum local current speeds up to 1 m/s. Comparison with the NORSEX field study (Mork and Johannessen 1979) and other field measurements is favourable.

Available satellite images covering the 500 km long coast from Lista to Stadt imply that there are 4-5 large whirls 9 months of the year and 0-3 whirls the remaining time. The development of whirls northward from the convex coast at Lista suggests a growing, finite-amplitude wave. Within a whirl, the coastal jet (0.5-1.0 m/s) detaches from the coast, penetrating 70-100 km seaward.

INTRODUCTION

The Norwegian Coastal Current is a density stratified flow driven by the outflow of fresh water from northern Europe and confined to the continent by the Coriolis force. Similar flows occur along most continental coasts. Although most current maps depict these baroclinic flows as stationary residual currents, it has been long known that they may exhibit instability in which large waves grow into meanders, whirls and rings – a so-called baroclinic instability (see Greenspan, 1969).

Variability in the Norwegian Coastal Current has been observed for decades (Eggvin, 1940; Sælen, 1959). Meterologic variability has often been cited as the primary cause. The development of off-

shore petroleum reserves has led to an urgent demand for more
precise answers to causes and effects of this variability.

Early thermal mapping of the coastal waters from satellites
showed rather turbulent chaotic flows. Enhanced thermal images
have revealed time and again the type of structures shown in Fig.
1. Similar structures have been modelled in the laboratory in the
absence of wind forcing as shown in Fig. 2. The results, produced
in a 5 m diameter rotating basin, are scaled to a typical condition
off the west coast of Norway. Current velocities within the whirls
scale to about 70 cm/s in nature.

Clearly, these large scale structures contain enough momentum
to be predictable on a time scale of several days and are probably
unaffected by weak or moderate winds. This is important for the
planning of offshore operations and clean up of oilspills.

The State Pollution Control Authority commissioned NHL to
estimate the size and occurrence of whirls along the west coast of
Norway and their effects on the transport of offshore oil spills
(McClimans & Nilsen, 1982). The results showed that, on the average,
a large wave, meander or whirl about 50-100 km in size occurs about
every 125 km along the coast. The shoreward wind-drift of an oil
patch could vary several 10's of kilometers depending on its initial
location with respect to a whirl.

In the following, some results from remote sensing, laboratory
simulation and direct field measurements are presented to show the
existence of and some effects of such large flow structures. The
causes are not fully understood of but the flow has the essential
features of a hydrodynamic instability for which phase speeds and
growth rates may be determined.

LABORATORY SIMULATIONS

The results in Fig. 2 show an essential feature of a baroclinic coastal current whose velocity is less than the phase speed
of an internal Kelvin wave. In particular, waves grow to meanders
in which the coastal jet separates from the coast. This development will be referred to as a whirl. As the initial current speed
increases, the distance to flow separation increases and, for large
speeds, the instability resembles the Kelvin-Helmholz instability.
This suggests a barotropic instability, although it is most likely
that all the observed wave growths are mixed instabilities. All
flows observed in the laboratory are close to geostrophic.

The waves observed in the laboratory originate from the source
of the current. To interpret them for a natural flow requires a
source in nature. Alternatively, the phase speeds and growth rates
may be interpreted in terms of already developed waves regardless

Fig. 1. Enhanced thermal image of the Norwegian Coastal Current
(NOAA-8 data Courtesy Tromsø Telemetristasjon).

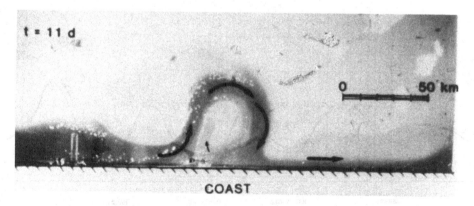

Fig. 2. Laboratory simulation of a baroclinic coastal current.
Length scale 1:53200; time scale 1:3230 (Vinger &
McClimans, 1980).

of their source. For this, the stage of development in terms of
amplitude and steepness, or separation, must be determined. The
scaled phase speed of the whirl in Fig. 2 (15 + 1 km/day) agrees
favourably with that observed by Johannessen & Mork (1979). Newer
laboratory simulations to determine the limits of scaling
(McClimans & Green, 1982) give sufficient data for estimating both
phase speeds and growth rates for various initial conditions. Topographic effects have yet to be tested in the large basin.

The laboratory results provide the general flow structure as
well as the wave kinematics and can be used for quantitative
comparisons with field data. This provides a basis for understanding the dynamics.

SATELLITE THERMAL IMAGES

Fig. 1 contains an example of an enhanced thermal image from
NOAA-8. The large structures, of which there appear many, are
easily identified in terms of size and position.

A typical picture of the seaward front of the Norwegian
Coastal Current from the southern tip of Norway at Lista to Stadt,
500 km to the north, is given in Fig. 3. This picture is based on
27 satellite images from 1981 of which several contained only
partial coverage or no contrast (when the coastal water differed
only a few tenths of a degree centigrade from the oceanic water).
The number of individual waves used in constructing the shape of
each of 5 waves is given on the curve. An arrow shows the wave
propagation to the north. The results suggest that waves are formed
at or east of Lista, grow and become steeper toward Stadt. The
"front" depicted in Fig. 3 is quite similar to that on an enhanced
satellite image shown by Johannessen & Mork (1979).

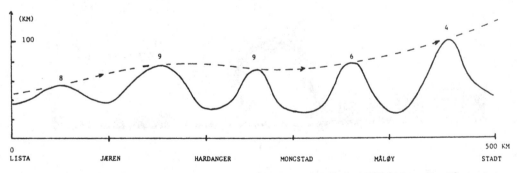

Fig. 3. Average amplitude and length of coastal whirls at five
        stages along the western coast of Norway. (McClimans &
        Nilsen, 1982).

This is the extent of the information derived from thermal
imagery. The similarity between the laboratory whirl and the
natural whirls is apparent. The flow field must de determined by
the laboratory results or, preferably, direct field measurements.

FIELD MEASUREMENTS

Both satellite imagery and laboratory simulation are indirect
means of looking at the phenomena at hand. The limitations caused
by the scale effects are not well documented for such large scale
processes and there is some question as to the ability of surface
temperature to reflect the important dynamics below. Thus direct
measurements of currents and hydrography in a vertical section are
of great importance.

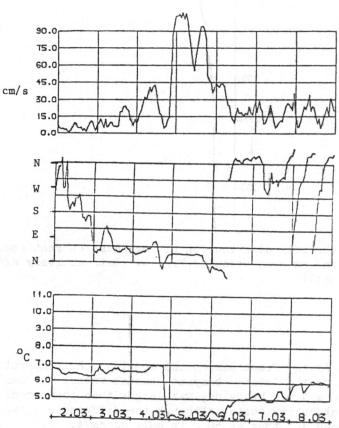

Fig. 4. Current speed, direction and temperature during an extreme
event ca 70 km off the coast of Norway. (See cross on Fig.
1.) (Lønseth et al., 1982; data courtesy A/S Norske Shell).

Some measurements have been made about 70 km from shore
(Lønseth et al., 1982). On several occasions, abnormally large
current speeds were observed toward the northeast as shown in Fig.
4. The measurements were made using a SIMRAD triaxial acoustic
current meter 2 m below the surface. The shoreline runs approximately north. The changing current direction from east to north-
west reveals the outer edge of a whirl propagating northwards along

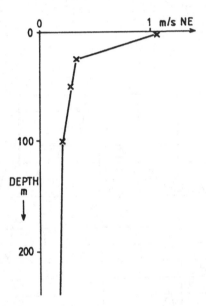

Fig. 5. Vertical distribution of northeast current speed during
the extreme event of Fig. 4. (Data courtesy A/S Norske
Shell)

the coast. As the front passes the mooring, the temperature drops
3°C revealing the fresher buoyant coastal water. The maximum current,
measured in a 4 m/s headwind, was larger than that predicted on the
basis of laboratory tests, although the difference between the
surface speed and that at 50 m is only 80 cm/s.

It is possible that the laboratory simulations smooth out the
details of the structure. A vertical distribution of the current
in the northeast direction shown in Fig. 5 shows that the current

is indeed thin 70 km from shore – less than a centimeter in the
distorted laboratory model and most likely affected by viscous
smoothing. It should be noted that the local surface flow in Fig. 5
may be faster than the phase speed of interfacial waves. Newer
measurements should resolve this issue.

WIND AND WHIRLS

There is little indication that weak or moderate winds affect
the general nature of a meandering baroclinic coastal current. A
daily breeze may move the uppermost meter of the water column a few
kilometers. The wind induced vertical mixing causes the water to
take on the character of the water below. Thus on the scale of
coastal whirls, daily breezes will have little effect on the structure seen on the thermal images. What, then, is the effect of wind
and can it explain some of the variability observed in nature?

Sustained longshore winds have been known to cause upwelling
of deep water along a coast. A north wind W along the western
coast (northern hemisphere) may produce a seaward transport Q up
to

$$Q = \frac{\rho_a}{\rho} \frac{c_D W^2}{f} \tag{1}$$

per unit length of coast. Here ρ_a/ρ is the ratio of air density to
water density ($\sim 10^{-3}$), c_D is an air-sea drag coefficient ($\sim 10^{-3}$)
and f is the socalled Coriolis parameter (twice the local rotation
rate $\sim 10^{-4}$ s^{-1}).

The seaward transport in a whirl is estimated to be on the
order of 0.2 Sv[*] over a width of about 20 km. This has the equivalent (local) effect of a sustained north wind of about 30 m/s but
occupies only a small portion of the coast. According to the preliminary survey of satellite images and field measurements, it appears
that this effect occurs about twice a month and increases in strength
northwards along the west coast of Norway.

In Fig. 6 the daily average longshore wind is compared with
current observations at 15 m depth in a fjord near Stavanger.
(see Fig. 1). A north wind is expected to drive this layer out of
the fjord (Svendsen, 1980). Two particular events of the opposite
character are noted with question marks. These may be the results
of whirls in the coastal current. Clearly, a more systematic coverage of whirls and velocity measurements would be valuable.

---

[*] 1 Sv. = 10^6 m^3/s.

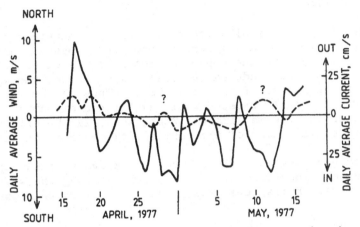

Fig. 6. A comparison between longshore wind speed (——) and seaward transport (---) at 15 m depth near Stavanger.
(Data courtesy of Regionplankontoret for Jæren)

COMMENTS ON SCALING

        Figs. 1 and 2 show similar dynamics over length scales from
1 to 10^5m. Scale effects due to viscosity were implied but the
general flow pattern appears to be scaled. Thus, the results
presented here should apply to the Keweenaw Current in Lake
Superior (length scale $\sim 10^4$m) for which there exists well-
documented coastal upwelling (Niebauer et.al., 1977). In particular,
a longshore 6 m/s wind was able to move the deeper water along the
coast to the surface within a day. The density difference was 1
kg/m^3 or about half that of the Norwegian Coastal Current. Over a
period of 3 days, the wind moved the coastal current 1 km from
land.

According to Froude's law the equivalent dynamics can be scaled in terms of the following ratios:

$$L_r = \frac{\text{Length in nature}}{\text{Length in model}}$$

$$\varepsilon_r = \frac{\text{Density difference in nature}}{\text{Density difference in model}} \tag{2}$$

$$T_r = \frac{\text{Time in nature}}{\text{Time in model}} = (L_r/\varepsilon_r)^{\frac{1}{2}}$$

$$U_r = \frac{\text{Shear velocity in nature}}{\text{Shear velocity in model}} = (L_r \varepsilon_r)^{\frac{1}{2}}$$

The results of the Keweenaw study may be used to model up-welling in the Norwegian Coastal Current, in which case $\varepsilon_r \cong 2$ and $L_r \cong 10$. Thus $U_r = W_r \cong 4.5$ and $T_r \cong 2.2$. A 27 m/s north wind lasting 6.6 days would drive the Norwegian Coastal Current 10 km offshore. Incipient surface upwelling is expected to occur within about 2 days with a sustained whole gale.

We conclude that the dynamics of the Norwegian Coastal Current are little affected by weak to moderate winds and may therefore be quantitatively simulated in the laboratory without wind stress.

ACKNOWLEDGEMENTS

We thank A/S Norske Shell and their partners in Block 31/2 and Regionplankontoret for Jæren for the use of their data. NOAA-8 thermal images were provided by Tromsø Telemetristasjon.

REFERENCES

Eggvin, J., 1940. The movements of a cold water front. Rep. Norw. Fishery mar. Invest. 7:1-151.

Greenspan, H.P., 1969. "The theory of rotating fluids". Cambridge University Press.

Johannessen, O.M. and Mork, M., 1979. Remote sensing experiment in the Norwegian Coastal Waters. Spring 1979. Samarbeids-prosjektet den Norske kyststrøm. Report 3/79. Geophysical Institute, University of Bergen.

Lønseth, L., Haver, S., Mathisen, J.P. and Tryggestad, S., 1982. Environmental conditions at Block 31/2: Analysis of METOCEAN data from June 1980 to June 1981. OTTER Report STF88 F82009 to A/S Norske Shell. (Proprietary).

Mathisen, J.P., Nittve, A., Sægrov, S. and Thendrup, A., 1977.
    Recipient investigations near the Stavanger peninsula.
    Physical marine evaluations of effluent discharge in By-
    fjorden and Gandsfjorden. River and Harbour Laboratory
    Report STF60 F78004 to Regionalplankontoret for Jæren.
    (Proprietary, in Norwegian).

McClimans, T.A. and Green, T., 1982. Phase speed and growth of
    whirls in a baroclinic coastal current. River and Harbour
    Laboratory Report. (In preparation).

McClimans, T.A. and Nilsen, J.H., 1982. Whirls in the coastal
    current and their importance to the forcast of oil spills.
    River and Harbour Laboratory Report STF60 A82029.
    (In Norwegian).

Niebauer, H.J., Green, T. and Ragotzkie, 1977. Coastal upwelling/
    downwelling cycles in southern Lake Superior. J.Phys. Oceanogr.
    7:918-927.

Svendsen, H., 1980. Exchange processes above sill level between
    fjords and coastal water, in "Fjord Oceanography" H.J. Freeland,
    D.M. Farmer and C.D. Levings, eds. Plenum Publishing Corp.,
    New York.

Sælen, O.H., 1959. Studies in the Norwegian Atlantic Current.
    Part 1: The Sognefjord section. Geofys. Pub., 20(13) 1-28.

Vinger, Å. and McClimans, T.A., 1980. Laboratory studies of
    baroclinic coastal currents along a straight, vertical coast-
    line. River and Harbour Laboratory Report SFT60 A80081.

OBSERVATIONS OF INSTABILITIES OF A GREAT LAKES COASTAL CURRENT

Theodore Green

University of Wisconsin
Madison, Wisconsin, USA

ABSTRACT

The small and mesoscale features of the very sharp front
bounding a coastal current off the south shore of Lake Superior
have been studied, using moored current meters, towed thermistors,
aerial photography, and an airborne thermal scanner. Estimates are
made of frontal slope, the motion and evolution of meanders along
the front, the (surface) velocity shear and convergence near the
front, and the along-front coherence of surface-temperature fine
structure. The meanders are related to laboratory and theoretical
studies of similar phenomena by others.

INTRODUCTION

Coastal currents in the Great Lakes of North America have been
the object of considerable study over the last two decades. The
broad outline of their behavior is now reasonably clear (e.g.,
Csanady, 1975). However, the available data, while hinting at
meanders associated with such currents, and on the same scale, have
been inconclusive on this point. Measurements of growth and propagation speed are also lacking.

We attempted to verify the existence of such meanders in a
preliminary set of experiments in the summer of 1979. The data
provide far more convincing evidence of instabilities of coastal
currents than has been available, and the time sequence allows some
estimates of growth rates and phase speeds. They also suggest one
mechanism for entrainment of central lake water into the current.
These data are outlined below, following a brief introduction to
the overall character of the current.

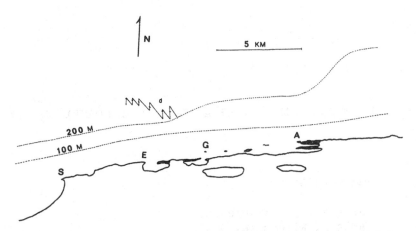

Fig. 1. The geography of the area on the south shore of Lake
        Superior where the eastward flowing Keweenaw Current has
        been studied. S (Sand Point), E (Eagle Harbor), G (Gull
        Island), and A (Agate Harbor) are referred to in the text.
        See also Fig. 2.

BACKGROUND

        The Keweenaw Current runs eastward along the south shore of
Lake Superior (Figs. 1 and 2), and is probably associated with
springtime warming of the large, shallow area of the Lake upstream,
to the west, in conjunction with some river input. The transmissivity of the nearshore water composing the current is about half that
of the central-lake water. In the study area, the bottom contours
are quite straight and parallel to shore. The bottom slope is one
in ten, one of the highest anywhere in the Great Lakes. The current
is rather shallow (perhaps 20 to 30 m deep), and is unlikely to be
affected by the bottom "bump" near Agate Harbor. The central lake
water adjacent to the current.has a velocity which is very small:
often below the threshold of moored current meters (Niebauer et al.,
1977).

        When steady, the current is in approximate geostrophic balance
(Yeske and Green, 1975) with a typical speed of 40 cm/sec. Upwell-
ing/downwelling cycles are superimposed on this mean state
(Niebauer et al., loc cit), much as those better described in Lake
Ontario (e.g., Csanady, 1974). The outer edge is sharply defined
by a strong surface front, which moves slowly outward from shore
with an increase in the amount of warm, nearshore water until full

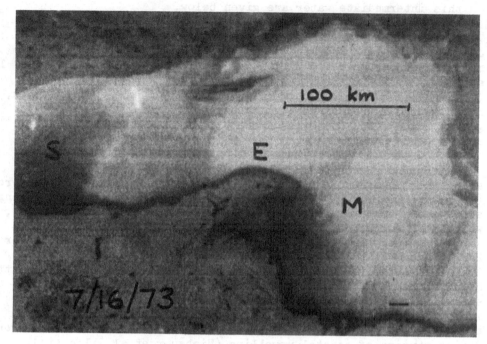

Fig. 2. A NOAA VHRR satellite thermal image of most of Lake
Superior, showing the Keweenaw Peninsula on the south
shore, Eagle Harbor (E), the warm water to the west (S),
and meanders downstream of the Peninsula (M). Here, dark
tones denote warm water, and light tones cold water.

stratification occurs sometime in September. Aerial photogrammetry
studies usually indicate a surface convergence at this front on the
order of 10^{-4} s (Yeske and Green, loc cit). The front will be used
below as an indicator of coherent meanders associated with the
current.

The relation of this front to the "thermal bar", itself the
subject of several studies, is not always clear. Our measurements
sometimes show them to coincide: the sharp temperature change
includes the $4^{\circ}C$ isotherm. At other times (and places along the
front, at the same time), the thermal bar was found a few hundred
meters offshore of the main front, and the "intermediate water"
(my term) between the bar and the front had a quite uniform temperature and turbidity. Examples are shown in Green and Terrell
(1978). Rodgers and Sato (1979) and Noble and Anderson (1968) also
refer to the stepped nature of temperature and turbidity changes
in the direction normal to shore. New data and some conjectures

on this intermediate water are given below.

The character of the current changes seasonally. In early July, 1979 (during the study described below), the overall densimetric Froude number $F \equiv U/\sqrt{g'h} \sim 1$, the Rossby number $R \equiv U/fL \sim 1$, and the internal deformation radius $D = \sqrt{g'h}/f \sim 2$ km. Here, U (20 cm/s) is a measured current speed, L (2 km) the current width, h (20 m) the current depth, g' ($g\Delta\rho/\rho = 0.3$ cm/s^2) the reduced gravity, and f the Coriolis parameter. These estimates are, of course, quite approximate.

Airborne radiometer data taken in early studies show rapid along-front changes in both front character and distance from shore (Green and Terrell, 1978), and hint at current meanders with a long-shore space scale of a few km, and a (growth) time scale of a day, which move downstream at about 10 cm/s. The dead-reckoning navigation limited the spacing of the north-south flight lines to 1 or 2 km, which was too coarse to resolve such features accurately. Available satellite data, with ~ 1 km resolution, were also of little help, although eddies downstream of the point where the current separated from the coast are sometimes seen (Fig. 2). The above-mentioned aerial photogrammery studies also suggested meanders on the scale of the current width, as did airborne thermal scans taken during studies of coastal upwelling (Niebauer et al., loc cit). Current-meter data taken during the summer of 1973 show, in addition to the expected near-internal-period activity at about 16 hours, well defined periods of from 9 to 14 hours, significantly below the inertial and Poincaré wave periods. An example is shown in Figure 3. (Note that the lowest barotropic mode of Lake Superior has a period of about 8 hours).

None of these studies, however, was directed primarily at Current meanders. All the above-mentioned indications were at best intriguing. It should also be noted that other evidence of Great Lakes coastal current meanders exists, such as the measurements of Boyce (1977), the baroclinic channel instability theory of Rao and Doughty (1981), and a few (isolated) satellite images (see, e.g., Rao and Doughty).

THE PRESENT STUDY

Procedure

In each experiment, a DC-3 aircraft equipped with a Texas Instruments RS18A far-infrared line scanner having a 2.5 μ-radian detector made straight (as possible), approximately shore-parallel passes about each 15 minutes over the area from Sand Point (S) to well beyond Agate Harbor (A), both of which are shown on Fig. 1. The aircraft altitude was limited by the lack of oxygen to 10.000

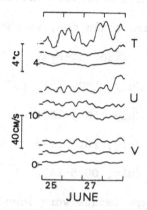

Fig. 3. Temperature and current data taken about 3 km north of
Gull Island, at (top to bottom) depths of 10 m, 35 m,
and 60 m, in June, 1973. The curves are offset in the
vertical: the horizontal marks on the left are at 4°C
(for T), 10 cm/s (for the eastward velocity component, U),
and 0 cm/s (for the northward velocity component, V).
The data have been smoothed with a Gaussian filter set to
pass 10% of the variance at a 4-hour period. Note the
9-11 hour periods indicated. Temperature rises usually
correspond to northward and eastward moving water, in
accord with meanders in a warm, high-momentum coastal
current.

ft, so that the scan spot on the lake surface was 25 ft wide. At
this altitude, the scanner "sees" a swath about 20.000 ft wide,
which very fortunately usually encompassed both front and shore.
This is crucial for accurately transforming the front imagery to
earth coordinates. Aerial photographs were taken on the first and
last pass of the experiments, each of which lasted about three
hours.

At the same time, a Boston whaler was towing a near-surface thermistor back and forth across the front in a zig zag pattern, taking hydrographic data with a mechanical bathythermograph, measuring radiometric lake surface temperatures with a Barnes PRT-5 radiometer, and noting positions of surface slicks. The boat study area ran from Gull Island (G on Figure 1) to Eagle Harbor (E). Positioning was done with a Motorola Miniranger III radar.

Four experiments were conducted, as shown below.

Experiment	Time Interval	No. Passes
(a)	1 July, 1524–1812 EDT	11
(b)	2 July, 1045–1251	8
(c)	2 July, 1613–1748	7
(d)	3 July, 0945–1151	9

Conditions during the experiments were ideal for our work: waves were less than one foot high, winds measured with a cup anemometer on G were under 2 m/s, except for a few hours during the night of July 1–2, the sky was almost cloud-free. Hydrographic data taken off Gull Island are shown in Fig. 4. Although there was little wind, the cross section thermal structure varied greatly from one day to the next, probably for reasons given below. Note the isolated blob of warm water on 29 June, which will also be discussed below.

A drogue set in the middle of the current on 2 July showed a speed of 20 cm/s. The central lake water just outside the current was almost stagnant.

The digitally recorded infrared radiance data obtained were first processed to remove the tangential distortion associated with the uniform rotation rate of the scanner mirror, and adjusted using the mean aircraft speed to make shore features conform to their known shapes in an approximate fashion. Film was made of this "first-cut" correction.

A subset of the data was further stretched by different amounts in the long and cross-flight directions in an iterative manner, to give shoreline features which corresponded to within 5 % of their known shapes. This was done using a Stanford Technology Corp. Model 70 Graphics Processor. The radiance values were then enhanced, and 35 mm photographs taken of the resulting imagery appearing on the display. Selected frontal features were enlarged and their evolution over time investigated.

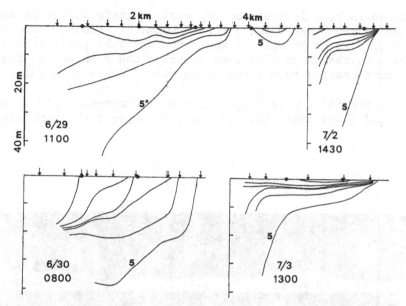

Fig. 4. Hydrographic data taken during the experiments.
Observation points are shown by arrows. Isotherms
of integer values of temperature (°C) are shown;
the warmer water is always nearer the surface.

Large Scale Instabilities

The thermal image from each experiment having the broadest
coverage is shown in Fig. 5. Light tones denote high surface
temperatures, dark tones low. The boat radiometer showed the very
uniform central lake water to have a surface temperature slightly
below 4°C, and the main mass of the coastal current to have surface
temperatures from 8°C to 10°C. These were both within 0.05°C of the
subsurface (10 cm) temperature. The surface temperature varied
slowly away from shore, except near the front. Here, the boat thermistor data suggest entrainment into the current (blobs of cold
water in the warm water, and not the opposite) and the above-
mentioned intermediate water, sometimes contaminated with internal
waves with a scale of 10 m.

Wavelike features with lengths of a few km appear along the
entire front. The shapes and positions of these features change
little over the course of one experiment, probably less than the
position errors on the imagery caused by slight unrecorded changes
in aircraft speed and direction. The geometric accuracy of the
imagery in Fig. 5 can be judged by comparing the shoreline with

that shown in Fig. 1. The accuracy east of Agate Harbor is sometimes decreased because the aircraft began its (counterclockwise)
turn. Fig. 6 is an example of the more carefully processed imagery.
It shows, at the expense of some coverage and resolution, that
significant changes have taken place over 7 hours on 2 July.

When the imagery is compared on a distance-time plot, and
Figs. 5 and 6 are taken into account, a rather consistent downstream

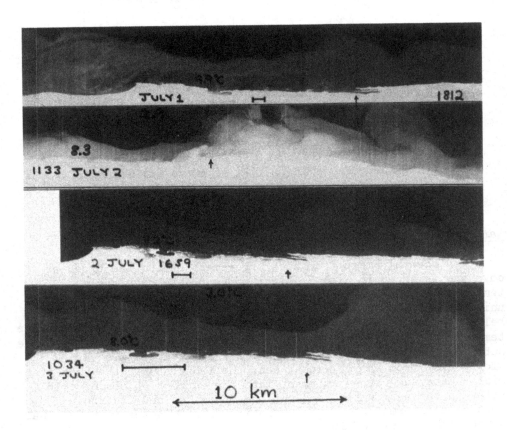

Fig. 5. Thermal imagery showing the evolution and progression
        of instabilities of the Keweenaw Current in Lake
        Superior. Light tones denote warm water, dark tones
        cold water. These images have been selected on the
        basis of area coverage; the longshore scales and
        tone-temperature correlation vary from one image to
        the next. Arrows denote the position of Agate Harbor
        (see Fig. 1).

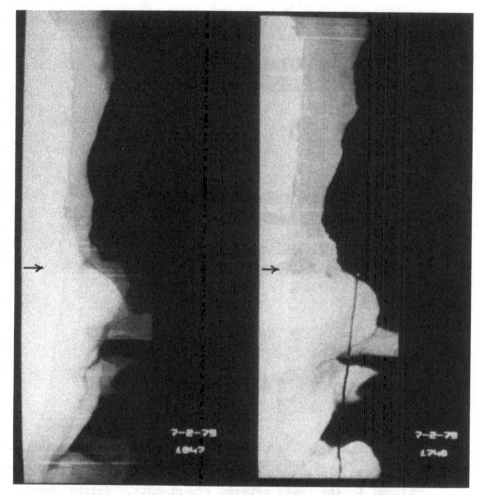

Fig. 6. Two carefully mapped thermal images, showing the
changes in surface temperature features between
1047 (left) and 1748 (right) on July 2nd.

propagation of features is evident (Fig. 7). The "backward breaking"
wave of experiment (a) becomes the large meander of (b), which
continues to move downstream in (c). Other meanders exist, further
downstream. A new wave comes into view upstream in (c), and is
breaking (backwards) in (d). A smaller wave grows and propagates
near E in (b) and (c), and is breaking in (d). The phase of all
these features is within 20 % of 4.5 cm/s, but seems to diminish
downstream of A, as shown by the apparent bunching up of meanders.
The observed wavelengths and phase speeds are loosely consistent
with the 1973 current meter data and with the meander seen by Green
and Terrell (loc cit). The waves also grow: the amplitude doubling

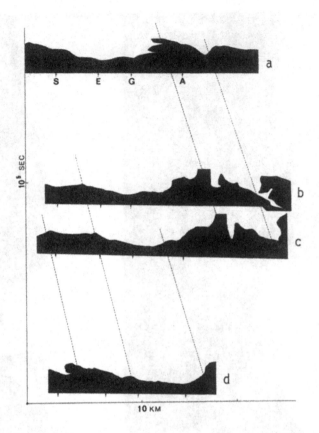

Fig. 7. Blacked-in versions of the coastal current imagery shown
         in Fig. 5, plotted very closely to the same horizontal
         (longshore) scale, and with vertical spacings propor-
         tional to the times between experiments. Arrows show
         the positions of points S,E,G,A on each image. Dashed
         lines connect what seems to be the same feature on each
         image. The actual shorelines have been replaced by
         straight lines, for clarity.

times are all on the order of a day, also in accord with the earlier estimate of Green and Terrell.

     The breaking waves seem to produce a mixed water type of
fairly uniform surface temperature (see the thermal-bar discussion
above). This intermediate water is quite obvious in the meander

near A in experiment (b). This water also has an intermediate
transmissivity, as is evident in aerial photographs taken during
(b), such as Fig. 8. Detailed subsurface temperature measurements
suggest that this water is only a few meters deep; some of the color
change may be due to this shallowness. The intermediate water also
appears on several thermistor tows, though not on all of them. The
breaking waves also offer an explanation of the detached blob of
warm water seen in the hydrography of 29 June, and of similar blobs
seen elsewhere (e.g., Csanady, 1974).

These waves are very likely due to an instability of the geostrophic mean flow. Similar waves have been seen in experiments of
(1) Stern (1980), (2) Vinger et al.(1980), and (3) Griffiths and
Linden (1980). The phase speed to current speed ratios found in (2)
and (3) are similar to those in our field observations, as are the

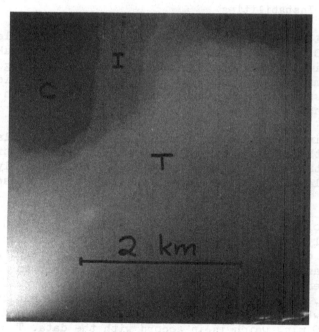

Fig. 8.   An aerial photograph of the large meander about 2 km
          east of Agate Harbor (see also Figs. 1 and 6), showing
          the sharp color fronts between the nearshore, turbid
          water (T), the intermediate water (I), and the offshore,
          clear water (C).

wavelength to current width ratios in (3). The latter are about
twice ours. The scaled amplitude doubling times found in (2) and
in more recent work with the same apparatus are about five times
ours, at the same F, but very close to ours for F ~ 0.6. Our field
estimates are much less secure than those made in the laboratory,
and I expect that agreement to better than perhaps a factor of two
may well be fortuitous. The experiments of (2) showed backward
breaking waves only for F > 1. In our case F ~ 1 and a detailed
comparison with (2) awaits the analysis of the more recent work.

The theory of Stern assumes zero total vorticity and wavelengths
much greater than L and predicts a backward breaking behavior when
F,L, and D have values close to those observed in the field. Whereas
we have no vorticity determinations in 1979, the aerial photogrammetry in 1973 suggests that it varies substantially across the current.
Further, less restrictive calculations along the lines suggested
by Stern may be in order. The theory of Griffiths and Linden is a
channel model, in some ways similar to that of Rao and Doughty
(loc cit). Although there is some agreement with experiment, the
absence of an unconstrained surface front is unsettling. Our
observed wavelengths are much shorter than those predicted by (3).

Small Scale Instabilities

Frontal instabilities also occur on the much smaller scales of
100 m and 30 minutes. Here, most of the (backward) breaking process
could sometime be monitored. An example is shown in Fig. 9. The
process seems similar to that of the larger instabilities, but
somewhat less convoluted; this could be associated with the scan
spot size, however.

These instabilities are clearly local phenomena, largely independent of the rotation of the earth. Their space and time scales
imply that they depend on the local shear and are likely barotropic.
There is a large gap between these wavelengths and those discussed
above.

It is of interest to know if their growth rate and wave length
λ are loosely compatible with stability theory. Our ignorance of
the details of the near front velocity structure permits us without
much embarrassment to consider the simplest model: an infinite
shear between a stagnant fluid and one of velocity u. Then stability
theory (e.g., Haurwitz and Panofsky, 1950) shows that the time scale
of wave growth should be $\tau = \lambda/\pi u$. Taking $\lambda = 200$ m and u = 10 cm/s
gives $\tau \sim 30$ min, which is in accord with the data. The phase speed
should be about 5 cm/s, for the same model. This speed cannot be
estimated from the data because of aircraft navigation errors and
the short time of existence of a wave. However, it is not incompa-
tible with the data.

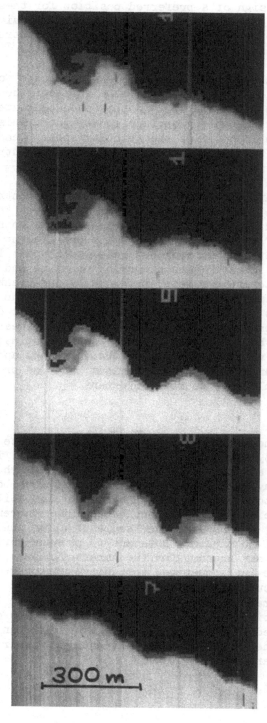

Fig. 9. Thermal images of a small scale instability observed in experiment (d) near Agate Harbor. Light tones represent surface water temperatures of about 8⁰C, dark tones those of about 3⁰C. The earliest image is at the bottom; the times are 1100, 1147, 1159, 1207, and 1215.

There was no sign of a preferred position for these small waves, with respect to the large waves. They are almost certainly a major cause of the lack of correlation of the temperature structure along the front which was seen in the boat thermistor data.

These waves also cause intermediate water to be formed. However, the water does not persist at the surface; the front remains quite sharp. It seems that in this case the intermediate water is removed by the strong surface convergence. This convergence probably also intensifies the velocity shear, leading to further breaking. It seems that this cycle leads to entrainment of central lake water into the current, at depth.

A first cut attempt to estimate the amount of entrainment associated with the small scale instabilities assumes that the production of intermediate water at the surface equals both its removal from the surface by the convergence and its entrainment by the mean flow. Then the fractional entrainment rate is FE = N · I/V, where N is the number of instabilities occurring per time and distance, I the volume of intermediate water produced by one breaking wave, and V is the cross sectional area of the current. Crude estimates of these quantities from the imagery, the hydrographic data, and some exploratory measurements of current depth very near the front give FE ~ 10^{-4} per hour. This is probably an overestimate as some of the downwelling water might well <u>not</u> be entrained and is also very small. Entrainment by other mechanisms, such as breaking internal waves, could well be more important.

SUMMARY

Sequential thermal imagery has shown the presence of two types of instability of a Great Lakes coastal current. The larger instabilities have a space scale near the current width, a time scale of about a day, and seem loosely in accord with both theory and various laboratory experiments. The smaller instabilities are confined to the vicinity of the front between the current and the central lake water, and have a space scale of a few hundred meters, and a time scale of 30 minutes. Both act to produce horizontally mixed water, which may account for the numerous observations of a stepped temperature (and sometimes turbidity) structure near the outer edge of the current, and which must be important to the transport of water normal to shore. Further work is needed to clearly document the conditions under which such instabilities occur (e.g., when are they masked by other processes?), the importance of these instabilities to the dynamics of the current, and their interaction with the thermal bar.

ACKNOWLEDGEMENTS

I wish to thank Drs. F.L. Scarpace and L. Fisher for their
assistance in gathering the infrared imagery. This work was founded
by a grant from the NASA Office of University Affairs (NGL-50-002-
127) and by Office of Naval Research Contract N00014-79-C-0066.
The manuscript was mostly prepared while I was a Visiting Scientist
at the Norwegian Hydrodynamic Laboratories, Trondheim. While there,
I benefitted from several discussions with Dr. T.A. McClimans about
his experiments.

REFERENCES

Boyce, F.M., 1977, Response of the coastal boundary layer on the
    north shore of Lake Ontario to a fall storm, J.Physical
    Oceanography, 7:719.

Csanady, G.T., 1975, Hydrodynamics of large lakes, Annual Review
    of Fluid Mechanics, 7:357.

Csanady, G.T., 1974, Spring thermocline behavior in Lake
    Ontario during IFYGL, J. Physical Oceanography, 4:425.

Green, T., and Terrell, R.E., 1978, The surface temperature
    structure associated with the Keweenaw Current in Lake
    Superior, J. Geophysical Research, 83:419.

Griffiths, R.W., and Linden, P.F., The stability of buoyancy-
    driven coastal currents, Dyn. Atm. and Oceans, 5:281.

Haurwitz, B., and Panofsky, H.A., 1950, Stability and meandering
    of the Gulf Stream, Trans. Amer. Geophys. Union, 31:723.

Niebauer, H.J., Green, T., and Ragotzkie, R.A., 1977, Coastal
    upwelling/downwelling cycles in southern Lake Superior,
    J. Physical Oceanography, 7:918.

Noble, V.E., and Anderson, R.F., 1968, Temperature and current
    in Grand Haven, Michigan, vicinity during thermal bar
    conditions, Proc. 11th Conf. Great Lakes Res., 470.

Rao, D.B., and Doughty, B.C., 1981, Instability of coastal
    currents in the Great Lakes, Arch. Met. Geoph. Biokl.,
    Ser. A, 30:145.

Rodgers, G.K., and Sato, G.K., 1970, Factors affecting the
    progress of the thermal bar of spring in Lake Ontario,
    Proc. 13th Conf. Great Lakes Res., 942.

Stern, M.E., 1980, Geostrophic fronts, bores, breaking and
    blocking waves, J. Fluid Mech., 99:687.

Vinger, A., McClimans, T.A., and Tryggestad, S., 1980, Labo-
    ratory observations of instabilities in a straight
    coastal current, in: "Stratified Flows", Vol. 1
    (Eds. T. Carstens and T.A. McClimans), TAPIR.

Yeske, L.A., and Green, T., 1975, Short-period variations in
    a Great Lakes coastal current by aerial photogrammetry,
    J. Physical Oceanography, 5:125.

STRATIFIED FLOW OVER SILLS

D.M. Farmer

Institute of Ocean Sciences
P.O. Box 6000
Sidney, B.C., V8L 1K1

INTRODUCTION

A feature of many coastal waters is the presence of topographic irregularities, such as ridges, banks or sills, that inhibit or at least modify the movement of water past them. An important example occurs in the flow over sills in fjords; here the water is often strongly stratified due to river discharge, it may have a net shear due to gravitational circulation and the flow will be time dependent due to meteorological and tidal forcing. We review this topic from the viewpoint of fjord oceanography, but the problem is much more general than this and has its origins in atmospheric studies such as airflow over mountains, as well as in certain aspects of classical hydraulics.

The study of such flows is important to oceanographers because constrictions tend to control the movement of water and also, by increasing local flow speeds and shears, to enhance the opportunity for turbulent mixing. Thus they may exercise a disproportionate influence on the local oceanography. However, attempts to measure water movement and properties near a sill need to be based on a clear understanding of the fluid dynamics. This leads us to a consideration of relevant scales and of a classification for sill flows, and also to a brief discussion of theoretical approaches. We conclude our review with an examination of some measurement techniques and a recent example of observations of sill flows on the British Columbia coast.

In reviewing the relevant dynamics we shall concentrate on tidal forcing; however it should be kept in mind that meteorological effects or exchanges due to modification of offshore density

337

structure, may often predominate on coasts that are not subject to
strong tides. Nevertheless many of the processes occurring in the
latter context also occur for tidal forcing. An excellent review
of stratified flow over obstacles has been presented by Huppert
(1981). We also refer here to a useful discussion of the linear
theory by Lee (1972), Bell (1975) and others. The present discussion
draws from these earlier papers, but will emphasize the application
to fjords.

In fjords the sills may be either bed-rock ridges that
survived glacial carving, or they may be remnant moraines.
Typically a fjord will have at least one sill and often, two or
more. Figure 1 shows a plot of depth along the axis of Knight
Inlet in British Columbia which has a prominent sill 75 km from the
inlet head, together with a plot of the tidal kinetic energy U_T^2
defined as

$$U_T^2(x) = (A^{-1} \frac{d\zeta}{dt} \int_0^x Bdx')^2, \tag{1}$$

where $A(x)$ and $B(x)$ are the local cross-sectional area and channel
breadth respectively and ζ is the tidal elevation. (The assumption
implicit here is that $\frac{d\zeta}{dx}$ is negligible). This figure emphasises the
greatly enhanced kinetic energy associated with the tide near the
main sill. Flows over shallower sills in the outer reaches of a
fjord have disproportionate tidal kinetic energies and are therefore
the most likely sites for hydraulic phenomena and turbulence
generation.

In addition to the release of tidal energy for mixing, sills
also form the barrier which inhibits ventilation of the deep basins
in fjords (c.f. Gade & Edwards, 1980) and provide the conditions
for hydraulic control of estuarine circulation discussed in various
theoretical models of fjords (i.e. Stigebrandt, 1981). Mining
operations in which the mine waste is released into the fjord have
provided additional motivation for studying such flows. Figure 2
shows a sketch of one proposed scheme in which the mine tailings
are projected to fill an inner basin and spill down the sill into
the outer basin. Prediction of the behaviour of tailings in this
example would require a thorough understanding of the dynamics of
stratified flow over the sill. In addition to the physical and
engineering consequences of sill flows, there appear to be
intriguing biological implications. Apart from the larger scale
biological effects of basin ventilation and estuarine circulation,
the strong currents over some sills also provide a unique ecological
niche for certain benthic communities. One remarkable recent
discovery is that corals growing on boulders on the sill crest
eventually reach a size that allows the current to dislodge the

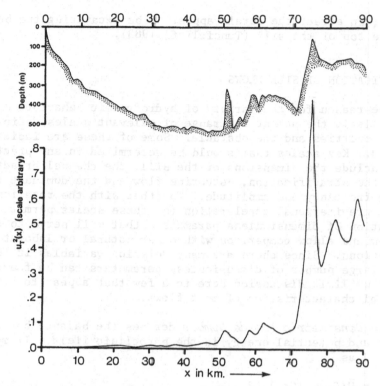

Fig.1.   Bottom depth and tidal kinetic energy for Knight Inlet,
British Columbia, showing dominant effect of sill.

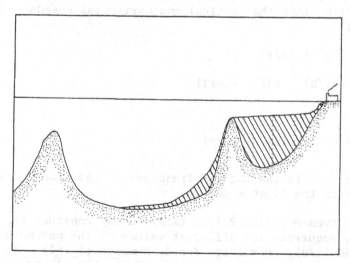

Fig.2.   Example of a proposed mine tailing disposal scheme
illustrating significance of sills to tailing dispersal problems.

boulder; in effect the corals appear to be excavating the boulders
from the top of the sill (Tunnicliff, 1983).

CLASSIFICATION OF SILL FLOWS

    One reason for the variety of hydrodynamic behaviour observed
near sills is the number and range of relevant scales defining the
water properties and the obstacle. Some of these are indicated in
Figure 3. Key scales that should be determined in any practical
study include the dimensions of the sill, the channel breadth and
depth, the stratification, estuarine flow and the dominant tidal
forcing frequency and amplitude. Together with the rotation rate
(Ω) and gravitational acceleration (g) these scales permit
calculation of dimensionless parameters that will serve to classify
the flow and allow comparison with other natural or laboratory
observations. Since there are many defining variables it follows
that a large number of dimensionless parameters can be found;
however we limit discussion here to a few that appear to determine
essential characteristics of most flows.

    The *Densimetric Froude Number* defines the balance between
kinetic and potential energy in the baroclinic field. It may be
expressed as

$$F_i = \bar{U}/C_i, \quad i = 1, 2 \ldots \tag{2}$$

where $\bar{U}$ is the depth mean flow and C_i the long internal wave speed
of mode i. Let L_T be the length of the tidal excursion, λ the
wavelength, H the total depth, A the obstacle height and T the
tidal period, then the vertical and horizontal speeds W and U may
be scaled as

$$W \sim (L_T/\lambda) \, (A/T)$$

$$U \sim W(\lambda/H) \sim (A/H) \, (L_T/T)$$

and

$$F_1 = U/C_1 \sim (L_T/H) \, (\omega/N) \tag{3}$$

where $\omega = 2\pi/T$ is the angular frequency of the tide and we restrict
attention to the first mode.

    For given *Relative Height* (A/H) we may consider the physical
processes occurring for different values of the parameters
(L_T/H) and (ω/N). For large Froude Numbers ($F_i \gg 1$), the kinetic
energy exceeds the potential energy of the baroclinic field, thus
inhibiting development of wave-like behaviour of mode i. For

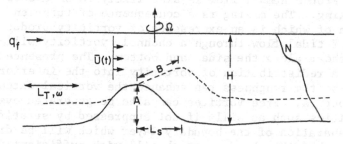

Fig.3.   Schematic diagram illustrating various scales that
determine the kind of flows occurring over sills. The sill is
defined by at least 3 scales: breadth (B), height (A) and length (Ls).
The flow field is defined by total depth (H), stratification N(z),
mean speed $\bar{U}(t)$ which has a tidal component of frequency (ω) and
excursion (L_T).   A fresh-water transport (q_f) may be imposed on the
flow and for wide sills the rotation (Ω) will be relevant.

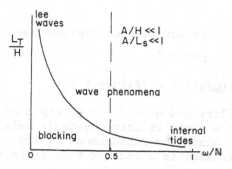

Fig.4.   Schematic diagram illustrating different types of flow
applicable for harmonic forcing over a low, smooth obstacle. At
one extreme is the 'acoustic' limit for which internal tides are
generated; lee waves occur in the quasi-steady limit, while
blocking results for sufficiently small values of ω/N and L_T/H.
The curve represents a line of constant Froude Number.

$F_1 \gg 1$ no wave-like behaviour of any mode is possible and the
density field behaves like a passive contaminant, although as
discussed below, intermediate cases where $F_2 > 1 > F_1$ can also occur.

High Froude Number flow is also likely to be associated with
strong mixing. The mixing is a consequence of turbulence, the
generation of which is an expression of vorticity production. In
the case of tidal flow through a channel, vorticity will arise from
boundary shear along the sides and bottom. The presence of a sill
can cause a redistribution of vorticity into the interior of the
fluid and bottom roughness can enhance the vertical extent over
which it occurs. Free vortices can also be generated over bottom
irregularities such as sills if not suppressed by stratification,
through separation of the boundary layer which will be discussed
below. On the other hand a smooth sill with sufficiently small
Aspect Ratio (A/L_s), where L_s is the sill length, may lead to something approximating potential flow at large Froude Numbers.

At small Froude numbers ($F_i \ll 1$) blocking will occur, since
there will then be insufficient kinetic energy in the upstream flow
to raise the deep fluid up to the sill crest. However, separation
of different responses into categories as implied by a Froude Number
scaling is often less clear in nature than the theoretical arguments
suggest. For example, observations indicate that for some sill flows
blocking invariably occurs at depth, simultaneously with the generation of wave-like phenomena nearer the sill crest. This is most
likely to be the case when the water is deep on either side of the
sill.

This scheme can be extended to include a *Bulk Richardson Number*

$$\bar{R}_i = N^2 / \left(\frac{dU}{dZ}\right)^2 \sim N^2 H^2 / U^2$$

$$\sim [(A/H)(L/H)(\omega/H)]^{-2} \sim [(A/H)F_i]^{-2}. \tag{4}$$

Shear flow instability and mixing can be expected for sufficiently
small $\bar{R}_i$; thus for a given relative obstacle height a large Froude
Number may lead to mixing by enhancement of the vertical shear,
which has its expression in a reduced Bulk Richardson Number.

Similar arguments can be used to develop classification schemes
on the basis of various other dimensionless parameters, but we draw
attention to two of special significance. First, an *Estuarine
Froude Number* F_e can be formed relating the fresh water discharge
q_f for a semi-enclosed basin such as a fjord, to the stratification
N and water depth h over the sill crest:

$$F_e = q_f / N h^2. \tag{5}$$

This parameter is related to the Estuarine Froude Number

proposed for two-layer flow by Stigebrandt (1981) and it plays a
central role in certain fjord circulation models. For sufficiently
large values of F_e the estuarine circulation is hydraulically limited
in the sense of "overmixing" as originally defined by Stommel &
Farmer (1952). Second, rotation Ω and sill breadth B can be used to
form an *Internal Rossby Number* R_B:

$$R_B = NH/\Omega B \tag{6}$$

determining the significance of rotation to the flow. Although we
limit our discussion to sill flows for which rotation is not
important, R_B as well as a related parameter (see Sambuco & Whitehead,
1976) defining the significance of rotation to hydraulic control,
should be checked in any observational program.

   While a multiparameter classification based on the above could
be described it is more useful to consider special cases. Figure 4
shows a schematic diagram indicating the response to tidal forcing
for different values of the parameters (L_T/H) and (ω/N), for small
Relative Height (A/H) and small *Aspect Ratio* (A/L_s). The solid
curve separates blocking effects from wave generation, while the
wave generation is further divided into two regimes having limiting
examples of lee waves and internal tides. Bell (1975) has discussed
the linear theory applicable to the wave generation due to harmonic
flow over a sill in an infinitely deep fluid. The governing
equation is

$$D^2\nabla^2 W + N^2 W_{xx} = 0 \tag{7}$$

where $D = \dfrac{\partial}{\partial t} + U\dfrac{\partial}{\partial x}$

and where the bottom boundary condition is taken as

$$W = U\frac{dh}{dx} . \tag{8}$$

This linear condition implies that the slope of the rays is much
greater than the slope of the bottom. The essential characteristics
of the limiting solutions are given in Table I. Bell discusses the
implications of sill shape and shows how the spatial derivatives
defining the obstacle determine relative contributions to the
corresponding internal modes. Neither the description in terms of
modes, nor that in terms of rays is able to explain fully the
observed internal response to such flows. A thorough analysis of
the linear problem with application of both approaches to observed
currents is described elsewhere in this volume (Cushman-Roisin and
Svendsen, 1982) and will not be further discussed here.

TABLE I

	Lee Wave Limit	Acoustic Limit (i.e. Internal Tides)
Wave frequency	U/L_s	ω
Slope of rays	U/NL_s	ω/N
Linearity constraints	$\dfrac{NH}{U} \ll 1$	$\dfrac{NH}{\omega L_s} \ll 1$

The generation and propagation of internal tides based on
linear theory has been widely discussed by Baines (1982) and others
and the corresponding linear, inviscid theory for harmonically
forced lee waves in fluid of finite depth has been treated by
Lee (1972) and Bell (1975). We mention two results from the finite
depth lee wave theory that are useful in interpreting observations
of sill flows. The length λ of the wave is

$$\lambda = 2HF_i(1-F_i^2)^{-\frac{1}{2}} \tag{9}$$

and the speed C_E at which energy is radiated away from the sill is

$$C_E = C_p - C_g = U(1-F_i^2) \tag{10}$$

where C_p is the phase speed and C_g the group velocity. For slow
flows such as might occur shortly after slack water or over sills
near which the tidal currents are relatively weak, the wave-length
is large. As the flow speed increases the wave-length decreases as
also does C_E. Thus energy accumulates on the lee face of the sill
in the form of lee waves of increasing amplitude.

FINITE AMPLITUDE EFFECTS

The assumptions used in deriving the linearised equations are
violated for flow over large obstacles. Two quite different
approaches have been used to investigate such flows. For the case
of layered stratification, the flow has been treated within the
framework of internal hydraulics for which the hydrostatic
approximation is assumed. Continuously stratified flows over
finite obstacles have been studied through a transformation of the
two-dimensional nonlinear equations which under certain conditions

can be reduced to the linear Helmholtz equation. Each of these approaches sheds some light on observed flows over fjord sills. The hydrostatic approximation will only be valid over sills that are long so that vertical accelerations can be neglected.

The theory of steady 2-layer flow yields solutions for several different types of response which depend on the numbers $\bar{F}_1$, $\bar{F}_2$ for each layer (see Long, 1954, 1970, 1972, 1974, Houghton & Kasahara, 1968, Armi, 1974). $\bar{F}_1$ and $\bar{F}_2$ are the respective densimetric Froude Numbers for each layer. (i.e. $\bar{F}_1 = U_1/\sqrt{g'h_1}$, where U_1 and h_1 are the mean velocity and layer depth and g' the reduced gravity.) If a transition between super-critical and sub-critical conditions occurs a hydraulic jump is expected downstream of the obstacle and evidence of such jumps has been observed near the sills of tidally forced fjords (i.e. Farmer & Smith, 1980a). In the steady solution for which a smooth transition occurs between subcritical flow upstream of the crest and super-critical flow downstream, an adjustment of the upstream condition is required to allow the appropriate critical condition to occur at the crest. This is most easily interpreted as the transient response generated when the flow is suddenly switched on and its manifestation in a single layer fluid is illustrated in Figure 5a. The adjustment is shown here as a turbulent bore or shock wave.

In coastal waters the situation is somewhat different in that the flow is modulated by the ebb and flood of the tide so that the gradual increase in speed may not result in an upstream bore. However Farmer & Smith (1980a) observed a bore propagating upstream from the sill in Knight Inlet which would be consistent with this interpretation. The observation is indicated schematically in Figure 5b; the observed bore was undular rather than turbulent and was generated before the maximum ebb on a spring tide, when the acceleration of the flow was relatively high. In contrast to the single layer example of Figure 5a, the interfacial adjustment in this case resulted in a deepening of the interface, as required to maintain a higher interfacial wave speed.

Within the jump itself turbulence is intense and entrainment and mixing occurs. As yet there have been few attempts to measure the amount of mixing that takes place, but it is clear from laboratory measurements (i.e. Macagno and Macagno, 1975) that hydraulic jumps are associated with high entrainment rates.

The theory of hydraulic flows and transitions has been extended to account for shear at the interface and more general upstream conditions (Armi, 1974). Su (1976) and Lee and Su (1977) considered the problem of multiple layers as an approach to representing internal hydraulic jumps in continuously stratified fluids and showed that for an exponential density stratification jumps could only occur over an extremely narrow range of flow speeds.

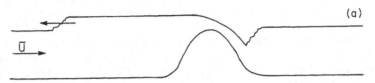

Fig.5a. Schematic diagram showing transient response of single
layer fluid to a flow suddenly started. Critical conditions occur
over the sill crest with supercritical flow downstream and subcritical flow upstream. The upstream adjustment occurs as a
travelling bore or shock wave which leaves a deeper layer behind
it. (Adapted from Houghton & Kasahara, 1968).

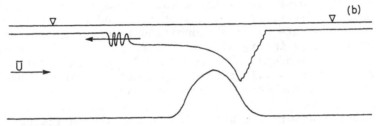

Fig.5b. Upstream response in a two-layer fluid to an increased
flow, with critical conditions over the sill crest. Here the
travelling bore, which is undular, results in an increased
surface layer thickness. This effect was observed in Knight
Inlet. (Farmer & Smith, 1980a).

However, as pointed out by Yih (1980), it is questionable whether
such multiple layer models which specifically exclude turbulence
and mixing, can adequately represent hydraulic jumps in real flows.

Attempts to compare solutions of Lee and Su's (1977) model with
flows observed in nature have proved useful under certain restricted
conditions (Gardner & Smith, 1980). Experience shows however (J.
Smith, personal communication) that while the nonlinear model may
reproduce the initial distortion of the flow quite well, the
inviscid assumption prevents the modelling of realistic flow
speeds on the lee face of the obstacle. In effect, failure to
account for drag on the bottom, as well as mass and momentum
exchange between layers, leads to unrealistically high velocities.

Thus the dynamics of the inviscid solutions are understood fairly well; the next step, namely the incorporation of a satisfactory representation of mixing processes into these simpler models, guided by appropriate field observations, has yet to be taken.

The theory of continuously stratified flow over obstacles of finite amplitude is largely based on an exact first integral of the equations of motion derived by Dubreil-Jacotin (1937) and Long (1953) which becomes linear for special choices of stratification and velocity. An unfortunate limitation of this approach is that the flow is only specified along streamlines that originate upstream and there are also difficulties associated with the problem of upstream influence (Baines, 1977). Nevertheless the theory appears to describe many features of observed flows including the formation of jets and eddies, which Long (1955) demonstrated using a series of laboratory and theoretical comparisons. Although the form of the differential equation is determined upstream and thus the flow is not defined within closed streamlines, Long's solutions still bear remarkable fidelity to the laboratory observations of rotors and eddies. Yih (1980) questions whether ruling out the validity of solutions with closed streamlines may not entail the sacrifice of solutions that have some validity for more interesting flows. An alternative approach to modelling stratified flow around finite amplitude obstacles, which avoids the problem of upstream influence encountered in Long's model for subcritical conditions, is obtained through use of the inviscid unsteady Oseen equations (Janowitz, 1968, 1981).

Finite amplitude effects also have a measurable influence on the internal tides generated over sills. Blackford (1978) described a non-linear mechanism for generating internal tidal oscillations in a pycnocline that lies deeper than the sill depth. The analysis is instructive though the model is restricted to a rather limited class of inlets. He points out that Bernoulli effects close to the sill will generate an even harmonic of the forcing frequency and a d.c. signal, and friction will generate odd harmonics (see also the discussion by Stigebrandt, 1980). He considers only the response to a single forcing frequency, presumably the same mechanism would predict a rich crop of exotic tidal lines if forcing at several tidal lines was allowed. In Knight Inlet, current observations at 170m depth seaward of the sill do indeed exhibit a rich tidal spectrum with large amplitudes at non-linear tidal lines (i.e. MSf, MSN_2 and many ter-, tetra- and penta-diurnal lines). As discussed above, a large Froude Number flow over the sill generates jets, lee waves and hydraulic jumps. However, no matter what the detailed mechanism of the sill flow in these cases it would seem inevitable that in the farfield, internal tides including some harmonic components, will always be found.

It was pointed out by Freeland and Farmer (1980) that more
energy will be extracted from the barotropic tide at spring tides
than at neap tides. The hydraulic transitions that enable large
amounts of energy to be extracted do not occur on every tide, but
are more common near springs than near neaps. Hence, more turbulence
is available for mixing near springs than neaps and we expect the
general circulation to be modulated at the beat period of M_2 and S_2
which is the MSf tide of period 14.7d. Unequivocal observations of
large, purely internal, MSf tides are shown and it seems likely
that a large MSf internal tide will be a signature of any inlet
whose turbulence is derived from the tide. The highly non-linear
events observed around the sills of fjords subject to large tides
force the large scale stratification at the dominant frequencies
and must inevitably produce a wide range of internal signals by
non-linear interaction between the astronomical components. At
higher frequencies these become observable in the surface elevation
signal and thus we see, in Knight Inlet for example, a rich tidal
spectrum.

In the case of tidal forcing through a long channel in which
extensive mixing occurs gravitationally induced circulation can be
modulated by the spring-neap cycle. This is a more extreme
manifestation of tidal effects than the generation of an internal
MSf tide described above and has been observed by Geyer & Cannon
(1982) in Puget Sound.

BOUNDARY LAYER SEPARATION

An important series of experiments together with a linear
analysis, carried out by Brighton (1977), demonstrate the role played
by lee waves in controlling the separation of the boundary layer from
the obstacle. This work has been complemented by Sykes' (1978)
nonlinear calculations of flow over a ridge, which accurately predict
the suppression and onset of separation for a range of Froude Numbers,
and experiments carried out on three dimensional hills both by
Brighton (1977) and Hunt & Snyder (1980).

The lee waves can either promote or suppress separation
depending upon the ratio of wavelength to the half length of the
sill. For sills with gentle slopes and lee waves with length λ
(equation 9), separation occurs under a rotor at a distance of
approximately λ downstream of the sill. It appears that the down-
slope acceleration produced by the internal wave effectively
suppresses separation on the sill while promoting it downstream.
With increased Froude Number λ may drop to the scale length of the
sill ($\sim 2 L_s$), in which case separation occurs on the downstream
slope of the sill, this time as a consequence of the lee wave
rather than the boundary layer.

For large Froude Numbers, separation is controlled by the
boundary layer flow; the wave length of the lee waves, if they
occur at all, is too large to suppress the separation. These ideas
have been extended to the study of boundary layer separation in
hydraulic flow (Huppert, 1980).

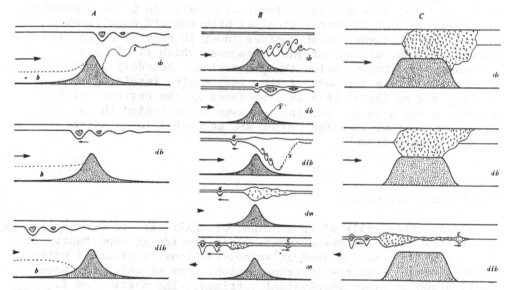

Fig.6.   Schematic diagram illustrating 3 different types of
response to tidally forced stratified flow over a sill.
A: (i) Lee waves form at the fresh-water/salt-water interface.
Beneath the trough of the first lee wave the boundary layer
separates (S). Upstream of the sill the deeper water is blocked.
(ii) and (iii) As the tide slackens the lee waves move over the
crest and advance upstream. B: (i) In the early stages of the
cycle the boundary layer separates from the crest forming a
growing shear layer. (ii) At maximum flow, mode 2 lee waves
form, suppressing the boundary layer separation, but a critical
layer with growing instabilities runs down the lee face of the
sill (iii). As slack water approaches (iv) the lee wave collapses
forming a layer of mixed water which intrudes upstream accompanied
by internal waves (v). C: Over a long sill mixing occurs through
most of the tidal cycle (i) and (ii), collapsing to form an
intrusion near slack water (iii).

Figure 6 is a schematic diagram indicating some of the internal responses that can occur due to tidal forcing of stratified water over a sill. The mechanisms displayed in 6A and 6B are to be expected over short sills and have been observed both in Knight Inlet and Observatory Inlet, B.C. Figure 6C shows some effects that can occur over a sill that is long enough to allow considerable mixing. The mixed water spreads out as an intrusion accompanied by nonlinear internal waves.

The first two examples in Figure 6 (A & B), illustrate differences in the modal structure of the internal response. 6A corresponds to the weaker stratification encountered during the low run-off season. Lee waves of first internal mode are formed during maximum ebb (or flood) and are then released to form a train of first mode nonlinear internal waves. In 6B we represent the response that is more typical of high run-off conditions. In this case the very fresh surface layer is only slightly distorted and most of the internal response, which is of second internal mode, occurs below the pycnocline. Boundary layer separation occurs for part of the tidal cycle, resulting in a growing mixing layer; this is suppressed as the current reaches a maximum and a large breaking lee wave forms behind the sill. Such observations in Knight Inlet are described in detail by Farmer & Smith (1980b).

MEASUREMENT APPROACHES

Flows over sills offer a number of challenges to oceanographers interested in direct measurement. They provide an opportunity for developing and testing innovative observational approaches using existing or perhaps new instrumentation, some of which may have application in other geophysical settings. The motivation for developing new techniques arises from the range of space and time scales associated with typical sill flows. For example a typical sill length might be 1-5 km, but the spatial variability of the flow will usually be such as to require a much smaller horizontal scale of measurement. For example internal lee waves may require measurement with a horizontal scale of 50m and vertical scale of 5m for adequate resolution. The assumption of statistical homogeneity in the horizontal plane, which can be made with some justification in the deep ocean, does not apply to small scale topographic forcing. Similar arguments apply to the relevant time scales. Even a quasi-steady response to semi diurnal tidal forcing can only be considered stationary for 2-3 hours.

These considerations point towards the need for remote sensing techniques and, as shown below, these have proved most useful. However these remote sensing approaches are complementary to standard

profile measurements by CTD and to recording current meter deploy-
ments. An optimum approach to measuring sill flows appears to
include the deployment of a modest array of recording instruments,
preferably current meters, for an extended period, together with a
program of ship-board measurements both from anchor stations to
provide high vertical resolution time series and also from slow
traverses across the sill using both towed and remote sensing
instruments.

     The problem of interpretation of current meter measurements in
the absence of a larger picture of the sill induced flow, can be
illustrated by the following hypothetical example. Suppose a
current meter is deployed near a sill for the purpose of
establishing mean (i.e. tidally averaged) transports, in such a
way that it resides within the turbulent rotor of a lee wave
during an ebb tide. Then during the ebb the measured current will
be close to zero but during the flood it will be directed landwards.
Thus although a tidally averaged current can be found for that
particular location, no reliable conclusions can be drawn about the
tidally averaged transport. This is an extreme case, but it points
to a general difficulty in the measurement of topographically
controlled flows subject to strong inertial effects.

     Two remote sensing approaches have proved useful for the
measurement of such flows, both from moving and anchored ships.
The first and simplest technique is echo-sounding. High
frequency echo-sounders operating in the range 50–200 KH$_z$ can track
horizontally coherent structures. The sound is scattered by
plankton and nekton and also by temperature micro-structure. It is
not easy to separate the contributions from these two sources, but
this is not a great disadvantage if the scattering layers tend to
occur in horizontal planes, as often seems to be true. Then
perturbations of these layers, such as may be caused by flow over
a sill, can be interpreted as perturbations in the streamlines.
If the flow becomes turbulent the echo assumes a distinctive
signature resulting in a diffuse patch on a facsimile recording.

     An echo-sounding can also be used to locate a profiling instru-
ment since the instrument itself forms a much stronger target than
biota or temperature micro-structure and may appear on the image
even if it is outside the main lobe. This greatly facilitates
interpretation of simultaneous measurements from, for example, a
CTD or current profiler. In effect the echo-sounding allows flow
visualization while the profiling instrument acquires quantitative
data.

     An echo-sounding is a reproduction of the amplitude envelope of
the acoustic signal and it is left to the human eye to integrate the
resulting image and infer the water movements. An alternative
technique makes use of the phase information in the signal in order

to remotely sense velocity. Such Doppler systems are still new to
oceanographers and a number of difficulties, such as adequate correction for ship motion, remain to be solved. Speed is only sensed along
the axis of the acoustic beam so that three beams are required to
resolve the motion completely; however in the two-dimensional flow
often encountered over sills in narrow channels a single beam Doppler
probe can provide a detailed view of the velocity field. In the
protected waters of fjords, the absence of significant wave motion
provides almost ideal conditions for remote sensing using this
technique.

One scheme embodying these techniques which has proved useful
for the study of tidally forced flows in the fjords of the N.E.
Pacific coast is shown in Figure 7. The ship is accurately positioned
by microwave, using two shore stations. The echo-sounder, Doppler
beam and towed CTD and current-meter operate simultaneously as the
ship slowly traverses the sill. Recording current meters, and perhaps
thermistor chains, can provide a longer time series at discrete depths.

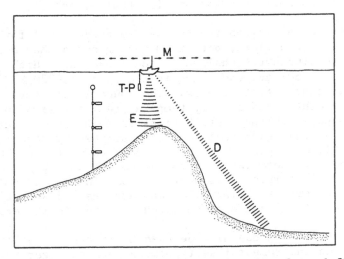

Fig.7.   A scheme for measuring topographically forced flows in the
coastal waters of British Columbia, which are subject to strong tidal
forcing. A towed profiler (T-P) obtains 3 components of current and
also conductivity and temperature at one depth as the ship moves
slowly over the sill. A downwards pointing echo-sounder (E) records
perturbations to the flow observable as scatterers (biota, micro-
structure) are moved vertically beneath the vessel. A narrow beam
range-gated Doppler probe (D) is directed forwards to profile the
longitudinal component of velocity. The ship is positioned with
microwave equipment (M). Recording current meters obtain time series
records to assist in the interpretation of horizontally profiled data.

OBSERVATIONS

Observations of many of the types of sill flows described above have been presented in earlier papers (Haury, Briscoe & Orr, 1979, Farmer & Smith, 1980a, b, Smith & Farmer, 1980). Here we take the opportunity of discussing a recent example of measurements obtained in Observatory Inlet, B.C. using a specially designed range-gated Doppler probe. The observations illustrate certain features of lee wave response discussed above as well as demonstrating the potential of acoustic remote sensing techniques in the study of topographically controlled coastal flows.

The sill in Observatory Inlet is quite shallow (30-60m) and is located just up-inlet of the mouth of the Nass River (Figure 8a), the maximum discharge of which takes place in May and June. There is also appreciable run-off at this time of year from streams and rivers entering the upper reaches and side-arms of the inlet so that the surface is typically capped by a layer of fresh water, beneath which the stratification decreases with increasing depth (Figure 8b). With a tidal range of up to 5m, currents in excess of 1 ms^{-1} are common near the sill crest.

The relative height (A/H) of the sill in this example is about 0.8 and the Aspect Ratio (A/L_s) is approximately 0.3. Thus it is no surprise that the resulting flow is highly nonlinear and at certain times leads to the shedding of vortices from the crest as the boundary layer separates at certain stages of the tide. The frequency ratio (ω/N) is of order 10^{-2} to 10^{-3} and the parameter $U/\omega L_s$, used by Bell (1975) to determine the relationship of the flow to the acoustic limit, is 1.2. The low value of ω/N places this flow in the quasi-static or lee wave limit, although in contrast to the example of Knight Inlet discussed by Farmer & Smith (1980a), the scaling does not unambiguously remove it from the acoustic limit. We might therefore expect both lee waves and internal tides to be generated over the Observatory sill. However rotation is not important here since $R_B \gg 1$. As indicated above the dominant discharge from the Nass River is down-inlet of the sill, so that although the stratification is modified by the large mass of fresh water that this river produces, we do not expect a very strong estuarine circulation in the sense implied by large values of F_e (~1). A more likely scenario is the periodic over-riding of critical conditions that might otherwise inhibit exchange over the sill, as discussed in detail by Stigebrandt (1977).

During June 1982 a number of slow traverses were made over the Observatory sill in the research vessel VECTOR. Simultaneous measurements were made with a 102 KH$_z$ echo-sounder pointing straight down and a narrow beam (1°) 215 KH$_z$ Doppler system pointing forwards, inclined at an angle of 30° to the vertical as shown in Figure 7.

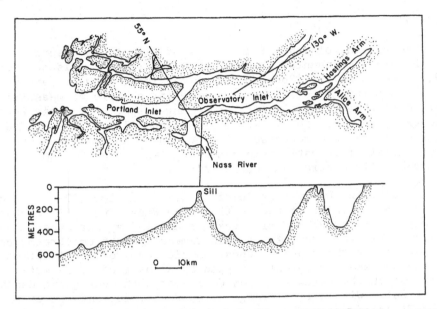

Fig.8a.   Map showing location and depth profile of Observatory
Inlet, B.C.   The observations shown in Figure 9 were obtained near
the sill.   Note that the principal fresh-water influx enters the
inlet from the Nass River, just downstream of the sill.

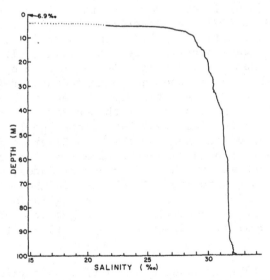

Fig.8b.   Salinity profile taken near the sill at the same time
as the data shown in Figure 9 were collected.   A sharp pycnocline
at 5m separates an almost fresh surface layer from the weakly
stratified, deeper water.

The Doppler system was triggered by the echo-sounder so as to avoid mutual interference. Acoustic returns from the Doppler were transformed into speed using the covariance estimator described by Rummler (1972). The pulse length was 10 milliseconds which determined the range resolution (7.3m along the beam, 6.4m in the vertical) and the pulse was repeated at intervals of 0.5s. Each estimate was averaged over a 15s interval and the resulting speed estimates resolved into horizontal components. Simultaneous vessel position information derived from microwave equipment was used to recover ship speed which was then removed from the Doppler speed estimates.

Figure 9a shows the echo-sounding image and 9b the corresponding Doppler measurements. The echo-sounding illustrates the movement of water up the slope to the crest of the sill, followed by a lee wave downstream of the crest. A prominent feature of the acoustic image is the diffuse and irregular patch, just downstream of the crest, in the trough of the lee wave. This appears similar to the rotors observed in laboratory models by Long (1955) as well as observations of air-flow over mountains.

Each of these features may be examined with the Doppler probe. However it should be emphasized that interpretation of the Doppler measurement must take account of the beam orientation. Flow transverse to the beam does not contribute to the Doppler signal. Thus the measurements made as the vessel advances over the upstream face of the sill, in which the vertical component of velocity must be positive (upwards), will tend to understate the resolved horizontal component, while the opposite will apply on the downstream face.

Upstream of the sill the Doppler probe shows a steadily increasing speed which has maximum values at about 10m depth, just over the sill crest. A rapidly increasing flow then evolves downstream of the crest in the form of a jet, bounded below by the sill and above by the rotor. Speeds of 1.2 ms^{-1} occur in the core of this jet.

The flow speed changes very quickly between the jet and the rotor. Within the rotor itself reverse flows of 0.2 - 0.25 ms^{-1} are encountered, thus providing confirmation of our interpretation of this portion of the acoustic image. This large rotating mass of water must provide for efficient vertical mixing. Analysis of the Doppler and echo-sounder returns shows that irregularities in the imaged flow occur in the deeper portion of the rotor in the boundary separating the downstream jet and the return flow. This observation supports the hypothesis that instabilities are generated in this highly sheared zone accompanied by entrainment into the rotor. Thus the rotor may not be a simple recirculating mass of fluid but rather a more complex and irregular structure that

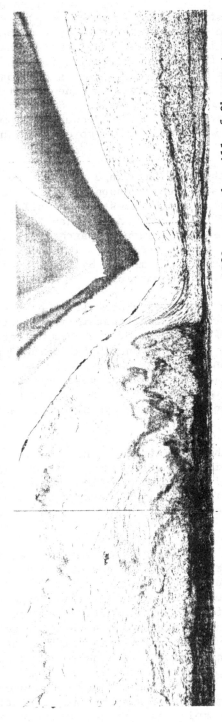

Fig.9a.   An echo-sounding image made during an ebb tide in June 1982, over the sill of Observatory Inlet, B.C. (see map, Fig.8a).   The flow is from left to right.   The sill itself appears as a relatively smooth line above a shaded band.   A shorter, steeper image of the sill beneath is due to multiple reflection from the sea-surface.   Movement of the water is apparent from coherent scattering from biota in the water column.   Individual particles remain within the beam, which is 5×10^{0}, for several seconds thus allowing the eye to infer actual shapes of the streamlines.   A large and turbulent lee wave has developed downstream of the crest, accompanied by instabilities in the flow.   Some scatterers on the lee face of the sill appear to dive into the sea-floor; this is an artifact of the echo-sounder presentation.   Since the transmitted beam is of finite width, the sill depth appears shallower than it really is when the bottom is steeply sloping.

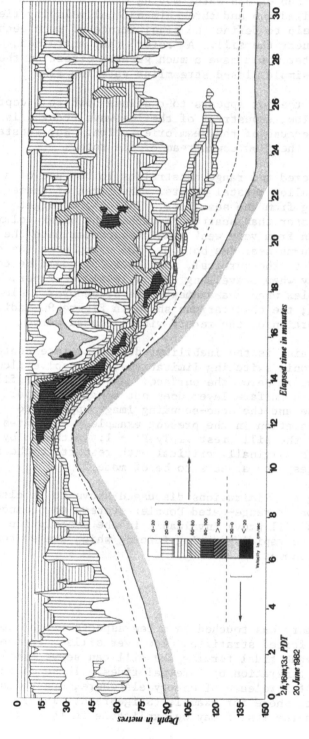

Fig.9b. Doppler flow speeds, resolved on to the horizontal component, obtained at the same time as the echo-sounding of Figure 9a. During the 30 minute traverse the ship moved a total of 1146m. The dashed curve just above the sill separates useful data from signals contaminated by side-lobe reflections from the bottom. The figure illustrates the acceleration as water moves up over the crest, the formation of a strong jet, just downstream of the crest and reverse flow (stippled) in the turbulent "rotor". Comparison with Figure 9a allows various smaller scale features to be interpreted.

entrains, mixes and then expels water moving over the crest. It is
clear that more detailed two and three-dimensional velocity field
measurements would help to define the significance of such mechanisms
to mixing processes near the sill. A 'rotor' that entrained, mixed
and then expelled water would have a much greater effect on the local
oceanography than a simple closed streamline rotating flow.

Below the rotor the jet appears to continue down to a depth of
120m. The upwards flow, downstream of the lee wave trough, is
less well resolved because of the beam orientation. But substantial
structure appears in the flow downstream of the rotor.

We have encountered two range constraints in the use of a
Doppler system in shallow coastal waters. The first of these
relates to scattering from the sea floor. This has a target
strength so much greater than scatterers within the water column
that reflections even from very weak sidelobes contaminate the
volume scattered returns near the bottom. For example in the case
of a beam tilted 30° to the vertical, the lower 13% could be contaminated in this way when travelling over a level sea-floor. The
problem is more complex when measurements are taken over a sloping
floor such as a sill; the observations shown in Figure 9b indicate
that contaminated portion of the record with a dashed curve.

A second constraint is the inability, due to surface contamina-
tion and also electronic switching limitations, to resolve flow
speeds shallower than 7m below the surface. However CTD profiles
suggest that the fresh surface layer does not always take part in
the lee wave response and the echo-sounding image appears to
confirm this interpretation in the present example. Estimates of
the Froude Number at the sill crest imply $F_1 \approx 1$. Although by this
criterion the flow is marginally critical with respect to the first
internal mode, the response appears to be of mode 2.

Notwithstanding the limitations discussed above it is clear
that the introduction of range-gated Doppler systems to oceanographic measurements will prove invaluable in the description of small
scale flows subject to rapidly varying forcing which are so common
in the coastal environment.

CONCLUDING REMARKS

This brief summary has touched on a few aspects of the local
dynamics of tidally forced stratified flow over sills. In the
case of relatively weak tidal forcing the sill can serve simply as
an obstacle for the generation of internal tides. These tides may
provide a mechanism for release of energy elsewhere, due to interaction with a sloping shore for example (Stigebrandt, 1979), but
the generation mechanism itself may be explicable in terms of

linear theory (c.f. Cushman-Roisin and Svendsen, 1982, this
volume). However this cannot generally be true and especially in
areas of large tides or where the constriction is small, or where
estuarine flow results in critical conditions over the sill, the
flow will be highly nonlinear and may be accompanied by strong
mixing.

The theoretical framework for analyzing such nonlinear flows
is insufficiently developed to allow realistic modelling of the
complex processes actually observed. There is a pressing need for
laboratory experiments carried out under carefully controlled
conditions to determine the mixing rates that occur for different
types of response. The laboratory modelling must be guided by
field observations; indeed many of the historical laboratory
modelling efforts have limited application to flows over large
sills, due to the very special nature of the stratification that
occurs in nature. The thin fresh surface layer that often occurs
in coastal waters subject to strong run-off tends to inhibit the
development of a first mode response, but enhances the opportunity
for higher mode effects with rotors, free shear layers and similar
phenomena. Linear or two-layer stratifications are inadequate for
modelling these processes in a way that permits ready comparison
with observations.

Measurement of sill flows indicates several similarities to
sattelite observations of flows over mountains (c.f. Gjevik, 1980).
Moreover the remote sensing approaches now being applied in the
atmosphere have their analogue in acoustic remote sensing in the
ocean. Figure 9 illustrates two such techniques now being used to
resolve such flows. Related acoustic methods designed to shed
light on the turbulent structure together with profiles of
temperature and velocity microstructure, will lead to a much better
and more detailed account of the flow. These observations will
present an increased challenge to fluid dynamicists to reconcile
the measurements with an adequate theoretical description.

REFERENCES

Armi, L.D., 1974, The internal hydraulics of two flowing layers of
        different densities. Ph.D. Thesis, University of
        California, 147pp.
Baines, P.G., 1977, Upstream influence and Long's model in
        stratified flows, J.Fluid Mech., 82(1):147-159.
Baines, P., 1981, On internal tide generation models, Deep-Sea
        Res., 29(3A):307-338.
Bell, T.H., 1975, Lee waves in stratified flows with simple harmonic
        time dependence. J. Fluid Mech., 67:705-722.
Blackford, B.L., 1978, On the generation of internal waves by tidal
        flow over a sill-possible nonlinear mechanism. J. Mar. Res.,
        36:529-549.

Brighton, P.W.M., 1977, Boundary layer and stratified flow over
        obstacles. Ph.D. Thesis, University of Cambridge, England,
        201pp.

Cushman-Roisin, B. and Svendsen H., 1983, Internal gravity waves in
        sill fjords: vertical modes, ray theory and comparison with
        observations, in: Proc. of Coastal Oceanog. Workshop, Os,
        Norway, June 6-11, 1982, Plenum Press, New York. (This vol.)

Dubreil-Jacotin, M.L., 1937, Sur les théorèmes d'existence relatifs
        aux ondes permanentes périodiques à deux dimensions dans les
        liquides hétérogènes. J. Math. Pures Appl., (9), 16, 43-67.

Farmer, D.M. and Smith, J.D., 1980a, Tidal interaction of stratified
        flow with a sill in Knight Inlet, Deep-Sea Res. 27A:239-254.

Farmer, D.M. and Smith, J.D., 1980b, Generation of lee waves over
        the sill in Knight Inlet, in: "Fjord Oceanography,"
        H.J. Freeland, D.M. Farmer and C.D. Levings, eds., Plenum
        Press, New York, 259-269.

Freeland, H.J. and Farmer, D.M., 1980, Circulation and energetics
        of a deep, strongly stratified inlet, Can. J. Fish. and
        Aquatic Sci., 37(9): 1398-1410.

Gade, H.G. and Edwards, A., 1980, Deep-water renewal in fjords,
        in: "Fjord Oceanography," H.J. Freeland, D.M. Farmer and
        C.D. Levings, eds., Plenum Press, New York, 715 pp.

Gardner, G.B. and Smith, J.D., 1980, Observations of time-dependent,
        stratified shear flow in a small salt-wedge estuary, in
        "Stratified Flows," T. Carstens, T. McClimans, eds., Tapir,
        Trondheim, Norway.

Geyer, W.R. and Cannon, G.A., 1982, Sill processes related to
        deep-water renewal in a fjord, J. Geophysical Res.,
        87:7985-7996.

Gjevik, B., 1980, Orographic effects revealed by sattelite
        pictures: mesoscale phenomena, in "Orographic effects in
        planetary flows," GARP Publ.23, World Meteorological Org.

Haury, L.R., Briscoe, M.G. and Orr, M.H., 1979, Tidally generated
        internal wave packets in Massachusetts Bay, U.S.A.;
        preliminary physical and biological results, Nature,
        278: 312-317.

Houghton, D.D. and Kasahara, A., 1968, Nonlinear shallow fluid
        flow over an isolated ridge, Comm. Pure and Appl. Maths.,
        21:1-23.

Hunt, J.C.R. and Snyder, W.H., 1980, Experiments on stably and
        neutrally stratified flow over a model three-dimensional
        hill, J. Fluid Mech., 96(4): 671-704.

Huppert, H.E., 1980, Topographic effects in stratified fluids,
        in: "Fjord Oceanography," H.J. Freeland, D.M. Farmer and
        C.D. Levings, eds., Plenum Press, New York, 117-140.

Janowitz, G.S., 1968, On wakes in stratified fluids, J. Fluid Mech.,
        33:417.

Janowitz, G.S., 1981, Stratified flow over a bounded obstacle in a
        channel of finite height. J. Fluid Mech., 110:161-170.

Lee, C.Y., 1972, Long nonlinear internal waves and quasi-steady
        lee waves. Ph.D. Thesis, Massachusetts Inst. of Technology
        and Woods Hole Institution.
Lee, J.D. and Su, C.H., 1977. A numerical method for stratified
        shear flows over a long obstacle. J. Geophys. Res.,
        82(3):420-426.
Long, R.R., 1953, Some aspects of the flow of stratified fluids.
        I. A theoretical investigation. Tellus, 5:42-58.
Long, R.R., 1954, Some aspects of the flow of stratified fluids.
        II. Experiments with a two-fluid system. Tellus, 6:97-115.
Long, R.R., 1955, Some aspects of the flow of stratified fluids.
        III. Continuous density gradients, Tellus, 7: 341-357.
Long, R.R., 1970, Blocking effects in flow over obstacles. Tellus,
        22:471-480.
Long, R.R., 1972, Finite-amplitude disturbances in the flow of
        inviscid rotating and stratified fluids over obstacles.
        An. Rev. Fluid Mech., 4:49-92.
Long, R.R., 1974, Some experimental observations of upstream
        disturbances in a two-fluid system, Tellus, 26:313-317.
Macagno, E.O. and Macagno, M.C., 1975, Mixing in interfacial
        hydraulic jumps. Pro. XVI, IAHR Congress, Sao Paulo,
        Vol/Ser.3:373-381.
Rummler, W.D., 1968, Introduction of a new estimator for velocity
        spectral parameters, Tech. Memo. MM-68-4141-5 Bell Telephone
        Laboratories, 24pp.
Sambuco, E. and Whitehead, J.A., 1976, Hydraulic control by a wide
        weir in a rotating fluid. J. Fluid Mech., 71:529-540.
Smith J.D. and Farmer, D.M., 1980, Mixing induced by internal
        hydraulic disturbances in the vicinity of sills, in:
        "Fjord Oceanography," H.J. Freeland, D.M. Farmer and
        C.D. Levings, eds., Plenum Press, New York.
Stigebrandt, A., 1977, On the effect of barotropic current
        fluctuations on the two-layer transport capacity of a
        constriction, J. Phys. Oceanogr., 7:118-122.
Stigebrandt, A., 1979, Observational evidence for vertical
        diffusion driven by internal waves of tidal origin in the
        Oslofjord, J. Phys. Oceanogr., 9:435-441.
Stigebrandt, A., 1980, Some aspects of tidal interaction with
        fjord constrictions, Estuarine and Coastal Mar. Sci.,
        11:151-166.
Stigebrandt, A., 1981, A mechanism governing the estuarine
        circulation in deep, strongly stratified fjords,
        Estuarine, Coastal and Shelf Sci., 13:197-211.
Stommel, H. and Farmer, H.G., 1952, on the nature of estuarine
        circulation, Parts I, II, III., Refs. 52-51, 52-88, 52-63.
Su, C.P., 1976, Hydraulic jumps in an incompressible stratified
        fluid, J. Fluid Mech., 73:33-47.
Sykes, R.I., 1978, Stratification effects in boundary layer flow
        over hills, Proc. Roy. Soc. 361 A, 225-243.

Tunnicliffe, V., 1983, Corals move boulders, an unusual mechanism
        for sediment transport, <u>Limnology and Oceanography</u>, *in
        press*.
Yih, C.-S., 1980, Stratified Flows," Academic Press, Inc., New
        York, 418pp.

SUBCRITICAL ROTATING CHANNEL FLOW ACROSS A RIDGE

Karin Borenäs

Department of Oceanography
University of Gothenburg
Gothenburg, Sweden

ABSTRACT

The flow of an inviscous rotating fluid over a ridge in a
channel of constant width is considered. The channel width is of
the same order or less than the Rossby radius of deformation and
the depth varies in the downstream direction. The flow is subcritical and the streamlines are developed from the concept of potential vorticity conservation. Flow separation and stagnant regions
may appear and some laboratory experiments were carried out, which
demonstrate these features.

INTRODUCTION

During the last ten years several papers dealing with the
hydraulics of rotating channel flow have appeared in the literature.
The papers have mainly been concerned with the problem of hydraulic
control and thus the models involve critical flow with respect to a
Froude number in some section of the channel, see for example
Whitehead, Leetmaa and Knox (1974) and Gill (1977). Subcritical
flows have been examined in the limit where the width of the ridge
is much smaller than the width of the channel (Boyer, 1971; Huppert
and Stern, 1974).

The present paper discusses how a subcritical flow responds
to a slowly varying topography in a narrow channel whose width is
of the same order or less than the Rossby radius of deformation.
This flow condition may perhaps occur in fjords when the bottom
water flows across a sill or in straits with a sloping bottom.

A simple analytic model of a frictionless forced flow is

presented. Some laboratory experiments were run in order to compare qualitatively the observed flow pattern with the theoretical predictions. Photos from the experiment are presented also here.

THEORY

A channel of constant width B rotates around a vertical axis with angular velocity f/2. The channel is aligned along the y-axis of a right-handed coordinate system. The bottom of the channel is flat except for a certain part where a transverse ridge with length 2L and slow variation in the y-direction is situated. There is a constant flow Q of homogeneous fluid through the channel, see Fig. 1.

The theory is based on the inviscid and time-independent shallow water equations. These can be used to derive an expression which states that the potential vorticity is conserved along streamlines (Ψ)

$$(1) \qquad \frac{f + \zeta}{H} = F (\Psi)$$

Here $\zeta = \partial v/\partial x - \partial u/\partial y$ and v and u are the longitudinal and transverse velocity components respectively. The depth H of the fluid is given by two parts

$$(2) \qquad H = h(y) + \eta (x,y)$$

The first part gives the water depth in the absence of motion and

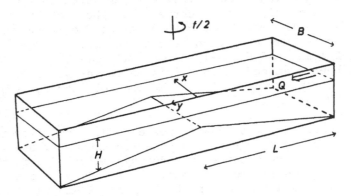

Fig. 1. Geometry of the channel.

the second, the variation of the free surface.
The form of $h(y)$ is

(3) $h(y) = H_o (1 - k(y))$

where H_o is the depth outside the ridge in the absence of motion
and $k(y)$ is a slowly varying function describing the bottom topography. Since the bottom varies slowly, we assume that the relative vorticity ζ is approximately equal to $\partial v/\partial x$.

    In the following derivation the tilt of the surface is
ignored in the expression for the water depth. To do this, we have
to compare the magnitude of the two terms in the denominator and
give a condition for ignoring the η-term. If we assume that the
velocity across the channel and its derivatives are small, the
balance for the x-component will be geostrophic. η will then be
scaled by $\eta = \dfrac{V_o f B}{g} \eta'$ where η' is of order one and $V_o = Q/BH_o$. The
condition needed is hence

(4) $\dfrac{R}{B} F^{-1} \gg 1$

where $R = (gH_o)^{\frac{1}{2}}/f$ is the Rossby radius of deformation and
$F = V_o/(gH_o)^{\frac{1}{2}}$ is the Froude number. If we choose to make R/B very
large, we approach the non-rotating case. So in order to keep the
effect of rotation we have instead to assume that $F \ll 1$.

    Let us now look at the special case where the relative
vorticity $\zeta = 0$ upstream of the ridge. The $F (\Psi) = f/H_o$ and the
vorticity equation is

(5) $\dfrac{f + \partial v/\partial x}{h} = \dfrac{f}{H_o}$

This equation is valid for a region occupied by stream-lines originating from the upstream side of the ridge. Eqn.(5) can be integrated
to obtain the velocity

(6) $v = f(h/H_o -1)x + C(y)$

From the requirement of constant transport through the channel

(7) $\displaystyle\int_o^{x'} vh \, dx = Q$

$C(y)$ can be determined and Eqn. (6) becomes

(8) $v = \dfrac{Q}{X'h} - \dfrac{fk(y)}{2} (2x - X')$

The velocity field has two components, one rotational and one irrotational and at this stage it is not evident what X' actually is. We can see from Eqn. (8) that the velocity may become negative and that separation from the right side-wall (looking downstream) may occur. Now, the flow can behave in two different ways after separation: it can either drive an eddy on its right side or have a stagnant pool there. In the first case the eddy drains energy from the main flow and since the flow then becomes dissipative the concept of constant potential vorticity is no longer valid. In the experiment, described later, no reversal was observed on the right side though there was separation. Since the kinetic energy flux is less in this unidirectional flow than in one containing eddies, it seems reasonable to assume that the system chooses a state where the kinetic energy flux is minimal. We therefore let X' be the width for which the kinetic energy flux E is minimized. If we assume that $v^2 \gg u^2$ because $k(y)$ varies slowly, E is given by

$$(9) \qquad E = \int_0^{X'} \frac{v^3 h}{2}\, dx$$

and we then choose that particular X' ($\leqslant$ B) for which $\partial E/\partial X' = 0$. This gives

$$(10) \qquad X' = B\left(k(1-k)B/\lambda \right)^{-\frac{1}{2}}$$

where $\lambda = 2V_0/f$ is twice the radius of inertia. Insertion of Eqn. (10) into Eqn. (8) shows that $v = 0$ at $x = X'$. The value of X' should thus be X' = B before separation and X' = Eqn.(10) afterwards.

Let us introduce a transport stream function and non-dimensional quantities defined by

$$u = -\frac{1}{h}\frac{\partial \Psi}{\partial y} \quad v = \frac{1}{h}\frac{\partial \Psi}{\partial x} \quad x = Bx^* \quad y = Ly^* \quad \Psi = Q\Psi^*$$

We can now write the non-dimensional expression for the stream function, which satisfies the boundary conditions as $u^* = 0$ at $x^* = 0$ as

$$(11) \qquad \Psi^* = x^* \frac{B}{X'} - \frac{B}{\lambda} k\,(1-k)x^* \left(x^* - \frac{X'}{B}\right)$$

The stream pattern depends greatly on the magnitude of B/λ. The streamlines separate from the right wall on the upstream side for $B/\lambda > 1/k_{max}\,(1-k_{max})$ when $k_{max} < 0.5$ and $B/\lambda > 4$ when $k_{max} \geqslant 0.5$. The results obtained here are also valid for the lower layer in a two-layer model if, in condition (4), R and F are replaced by R' (the internal radius of deformation) and F' (the internal Froude number). Here g', the reduced gravity and v', the velocity difference between the two layers are used instead of g and V_o.

AN EXAMPLE

In the experiment described below, the ridge profile is triangular, given by

$$h = H_o/2 \ (\ 1 - y^*) \qquad\qquad k = \tfrac{1}{2}(\ 1 + y^* \ ) \qquad\quad -1 \leqslant y^* \leqslant 0$$

$$h = H_o/2 \ (\ 1 + y^*) \qquad\qquad k = \tfrac{1}{2}(\ 1 - y^* \ ) \qquad\quad 0 \leqslant y \leqslant 1$$

and the expression for the stream function is

$$\psi^* = x^* \frac{B}{X'}, - \frac{B}{\lambda} \frac{1}{4} \ (\ 1 - y^* \ )(x^* - \frac{X'}{B})x^* \qquad\quad -1 \leqslant y^* \leqslant 1$$

In Fig. 2 contours of the dimensionless transport stream function are shown for different values of B/λ. Separation from the right hand side of the channel occurs for $B/\lambda > 4$. The width of the separated flow is in this case given by

$$X' = 2B \ (\ B/\lambda \ (\ 1 - y^{*2} \ ))^{-\tfrac{1}{2}}$$

THE EXPERIMENT

In the experiment a rectangular clear plastic tank, 100 cm long and 15 cm wide, was used. The ridge wad 40 cm long with maximum height of 2.5 cm. The tank was on a turnable which rotated with an angular velocity f/2. The flow Q was varied from 60 $cm^3 \ s^{-1}$ to 170 $cm^3 \ s^{-1}$ and f from s^{-1} to 1.8 s^{-1} in order to get different values of B/λ. The depth H_o was 5 cm. A pump recirculated the fluid and the outlets and inlets were well diffused. The flow was visualized by means of the pH–technique. A camera above the table took photos for different values of B/λ. Some of these results are shown in Fig. 3.

DISCUSSION OF THE EXPERIMENTAL RESULTS

If we compare the theoretical results shown in Fig. 2 with the experimental, we note that there is a qualitative agreement on the upstream side whereas the downstream side differs. When B/λ is increased sufficiently in the experiment we see that the flow is blocked and it separates. The separation occurs earlier and the stagnant blocked region increases when B/λ is increased, as predicted by the model.

At the crest of the ridge, the flow separates from the left channel wall and crosses the channel in a jet-like manner. The jet becomes unstable on its left flank where there is a large horizontal

current shear and, for large B/λ small scale eddies can clearly be
seen on this flank. The jet also drives an eddy over the left part
of the down-slope. The assumption of non-dissipative flow, where
the potential vorticity is conserved, apparently does not hold for
the downstream side.

To investigate the importance of friction in the experiment,

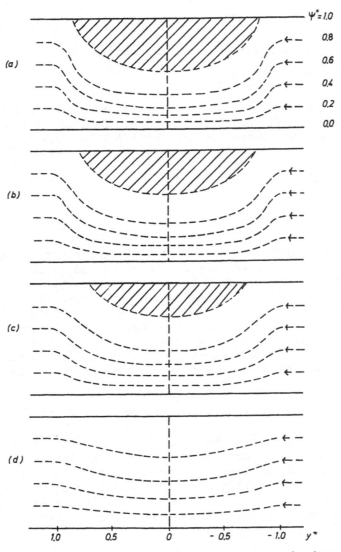

Fig. 2. a-d. Contours of the dimensionless transport stream function
in a channel, looking from above. The hatched areas are
stagnant. a) B/λ = 14 b) B/λ = 11 c) B/λ = 8 d) B/λ = 3.

Fig. 3. Experimental results for B/λ > 20. The flow is from right
to left. The top of the ridge is marked by the dark line.
The shadow in the middle is from the camera.

we compare the Rossby number ($\varepsilon = V/fL$) with the square-root of the
Ekman number ($E^{\frac{1}{2}} = (v/fH^2)^{\frac{1}{2}}$). If $\varepsilon \gg E^{\frac{1}{2}}$ we can neglect the effect
of viscosity. Though ε is greater than $E^{\frac{1}{2}}$, it is not sufficiently
large; hence the discrepancy between the experimental results and
the theoretical model may be due to the friction.

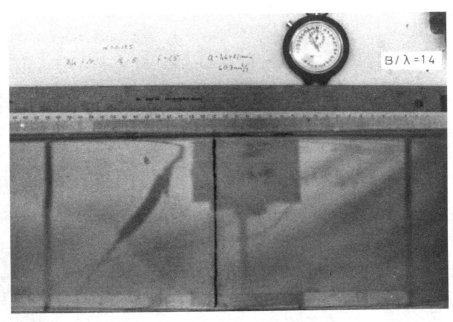

Fig. 3. (continued) Experimental results for $B/\lambda = 14$ (top) and $B/\lambda = 11$ (bottom).

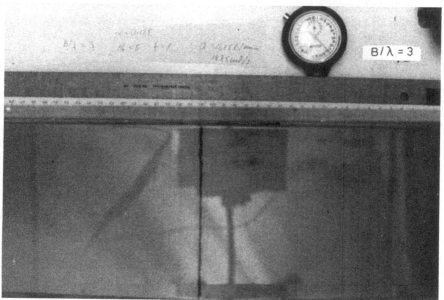

Fig. 3. (continued) Experimental results for B/λ = 8 (top)
         and B/λ = 3 (bottom).

ACKNOWLEDGEMENTS

    I would like to thank Dr. A. Stigebrandt for suggesting this
problem to me and for many helpful discussions. My thanks also go
to Dr. G. Walin for valuable comments.

REFERENCES

    Boyer, D.L., (1971), Rotating flow over long shallow ridges,
        Geophysical Fluid Dynamics, 3:165-184.
    Gill, A.E., (1977), The hydraulics of rotating channel flow,
        Journal of Fluid Mechanics, 80:641-671.
    Huppert, H.E., Stern, M.E., (1974), The effect of side walls
        on homogeneous rotating flow over two-dimensional
        obstacles, Journal of Fluid Mechanics, 62:417-436.
    Whitehead, J.A., Leetmaa, A., Knox, R.A., (1974), Rotating
        Hydraulics of strait and sill flows, Geophysical Fluid
        Dynamics, 6:101-125.

INTERNAL GRAVITY WAVES IN SILL FJORDS: VERTICAL MODES, RAY THEORY AND COMPARISON WITH OBSERVATIONS

Benoit Cushman-Roisin* and Harald Svendsen**

*Florida State University
Tallahassee, Florida, USA
**University of Bergen
Bergen, Norway

ABSTRACT

It is well known that marked topographic variations are an important feature by which surface gravity waves can generate internal gravity waves. Typical examples are the generation of internal tides on an abrupt continental shelf and in sill fjords. Two methods of description are available: vertical modes and ray tracing. Both have severe limitations. Decomposition into vertical modes is rigorously justified only if the bottom is horizontal, whereas ray tracing is asymptotically valid only if the wavelengths present are at most a small portion of the total depth. In view of these restrictions, neither method is strictly applicable to the study of internal waves in sill fjords. However, lacking any other applicable techniques, the methods have been applied to data from Skjomen, a sill fjord in northern Norway.

Qualitative conclusions can be clearly stated: (i) interactions of the surface tide and the sill topography is certainly the mechanism responsible for the existence of internal waves away from the sill, (ii) surface or bottom reflection can account for changes in the direction of vertical phase propagation, (iii) waves tend to be a combination of the first few modes only, and (iv) there is no evidence of standing waves or waves coming from the head of the fjord.

Finally, currents caused by various wind conditions, both outside the fjord (up or downwelling) or locally in the fjord, are likely to affect the horizontal propagation of the internal wave energy through the fjord. The influence of such currents on the direction and curvature of the rays is briefly discussed.

373

This is the first attempt to apply ray tracing to the study of internal gravity waves in fjords.

INTRODUCTION

As a result of numerous field investigations it is now well established that internal gravity waves constitute an important aspect of fjord dynamics. These waves are found in almost any fjord, below a surface layer where wind effects dominate. The importance of the waves resides, among other aspects, in their effect upon biological communities such as phytoplankton. How are these displaced horizontally, uplifted, dispersed or even stirred? What are the random fluctuations in sampling owing to wave displacements? These are questions of fundamental interest in fjord ecology. From a dynamical point of view, there is a need to understand the mechanisms of generation, propagation and absorption of internal gravity waves in fjords. The questions are then as follows: Where can these waves be expected? How much energy do they carry? How extensive are the horizontal and vertical displacements that they cause? Is breaking of internal waves likely to be an efficient mixing mechanism? The aim of the present work is limited to discussing by simple methods the mechanisms of generation and propagation of internal waves in the vicinity of the sill of a fjord and to predicting where and when internal gravity wave energy may be found.

Helland-Hansen and Nansen (1909) were the first to be aware of the tidal character of many of the observed internal waves in the sea, although they did not suggest any physical mechanism for their generation. A few years later, Zeilon (1912) demonstrated that an oscillating current in a two layer system can interact with the bottom topography to generate interfacial waves. Later, Proudman (1953), Rattray (1960), Mork (1968), Prinsenberg and Rattray (1975) and others clarified, recast the problem and studied the interaction of topography and stratification. The focus of these investigations was mostly the shelf-break topography, so as to explain the characteristics of internal tides on the continental shelf and in the deep ocean.

The first observations of internal waves of tidal periods in sill fjords were made by Zeilon (1914) in the Gulmar fjord in west Sweden. Fjeldstad (1933) advanced further by analyzing measurements from the Herdlafjord in west Norway. More recently, Buckley (1980) and Stigebrandt (1980) used vertical modes to determine the amplitude of internal waves for various fjord models. Independently, Prinsenberg and Rattray (1975) and Baines (1974 and 1982) applied ray theory to the internal wave field along the continental shelf.

The present study, in line with previous work, is of internal waves at a tidal period in a sill fjord. Data were gathered during a biological investigation of the productive fjord Skjomen, part of a west Norwegian fjord system. The current meter observations are therefore important per se but, beyond this, the work offers the opportunity to apply, for the first time, ray tracing to a fjord wave field and to compare this method with decomposition into vertical modes. Neither method is strictly rigourous in a sill fjord like Skjomen. On one hand the sill is too steep for decomposition into vertical modes to be well founded and on the other hand the observed internal waves are characterised by vertical wavelengths too large compared to the water depth for ray tracing to be applicable. However, lacking any other technique, these methods are applied to the present data and shown to supply qualitative information which sheds insight on and provides strong support for the conjecture that the topographic obstacle to the barotropic tide is the cause of internal gravity waves in sill fjords. Both techniques provide a coherent and complementary picture of the internal wave field and can account for most of the observed characteristics. The next step towards a more rigorous study is two or three dimensional numerical modelling of the sill area.

FJORD DESCRIPTION

Skjomen is a fjord near the head of a system formed by the Vestfjord and Ofotfjord in northern Norway (Fig. 1a). Skjomen is about 25 km long and extends inland between precipitous mountain walls. The width varies (Fig. 1b): it is narrow at the mouth (300 m), broad in the middle (∿3 km) and narrow towards the head of the fjord (∿1 km). Inward from the sill plateau in the mouth, the depth increases towards a deep basin from which the bottom slopes gently towards the head.

During the period from January 18, 1977, to November 30, 1979, currents were measured with Aanderaa current meters (Aanderaa, 1979) at ten depths under an anchored surface buoy in Skjomen. The buoy was about four kilometers in from the shallowest point of the sill (B1 in Fig. 2). The observation depths were 1.5, 5, 10, 15, 20, 40, 60, 100, 125 and 148 meters. The current meters recorded speed, direction and temperature every 10 minutes. Using a CTD-probe, temperature and salinity were sampled once a month at fixed positions in Skjomen during the period. The measurements were made at selected depths, every meter in the depth interval 0-25 meters and every 5 meters below.

In the present paper only a part of the data is used, periods of nine days of current measurements in each of the four seasons and four density profiles calculated from temperature and salinity data at the CTD-station (05) closest to the buoy station (Fig. 1b). Each profile is assumed to represent the density distribution for the corresponding 9-day period. Finally, surface currents are

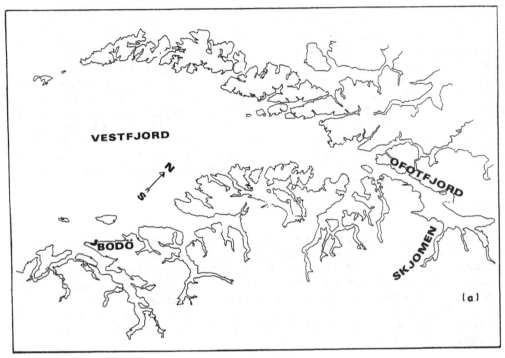

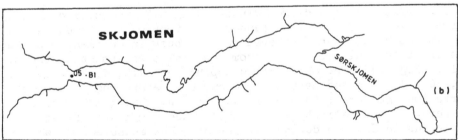

Fig. 1. (a) Map of the Vestfjord-Ofotfjord and the surrounding
        fjord system. (b) Map of the Skjomen.

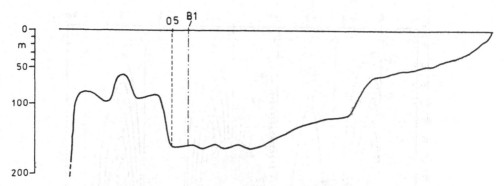

Fig. 2. Cross-section of Skjomen, showing the bottom topography,
including the sill at the mouth, the hydrographic station
(05) and buoystation (B1).

predominantly affected by surface winds and current measurements at
1.5 m were therefore ignored.

INTERNAL TIDES

        Along the coast of Norway, the tide is a progressive northward
wave of period 12.42 hours. As it passes a coastal fjord, the tide
forces a wave into the fjord with the same period. In the Skjomen
region, the amplitude of the barotropic tide is among the largest in
Norway, probably because of resonance effects. This barotropic signal, in turn, generates internal tides. Such waves have been observed
in many sill fjords and the data here provide additional support for
their existence. Progressive and standing waves can coexist but,
as pointed out by Gordon (1978) and Stigebrandt (1980), absorption
rather than reflection occurs at the head of the fjord so that progressive waves are for the most part dominant. Other tidal motions
and their harmonics are present in the oceanic wave field. These
too generate fjord waves at their frequencies but are expected to
behave much like the fundamental M2 tidal motions and are therefore
omitted from this analysis.

        The interaction between the barotropic tide and the bottom
topography in continuously stratified water is well described by
Baines (1982). In brief, the barotropic tide, passing over the
region of topography variations, induces a vertical velocity which
displaces the isopycnals in an oscillatory manner. This generates
a vertical buoyancy force which causes the internal tides, is proportional to

$$F = -\frac{zh'(x)}{h^2(x)}$$

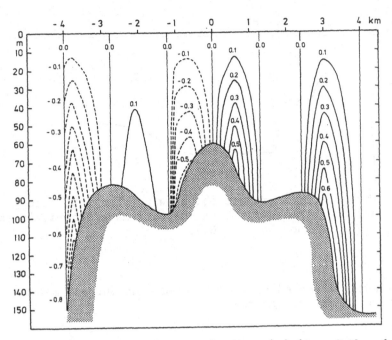

Fig. 3. Plot of the vertical force in the vicinity of the shallowest
point of the fjord sill. This vertical force is a result
of the surface tide and is the generating mechanism of the
internal tide. Arbitrary scale.

and oscillates at the tidal period; z is the vertical coordinate
(zero at the surface and negative downward), h(x) is the water depth
as a function of the in-fjord coordinate x and h'(x) is dh/dx, the
local topographic slope. The amplitude of the generating body force
has been plotted in the case of the mouth of Skjomen (Fig. 3). This
figure readily shows four regions of internal gravity wave generation on the steep flanks of the fjord sill. Scaling of the momentum
equations leads to a scale for the horizontal component of the internal tide velocity, u_i:

$$u_i \sim \frac{N}{\omega} \frac{h'}{h} (gh)^{\frac{1}{2}} \zeta \, ,$$

where N is the Brunt-Väisälä frequency, ω the tidal frequency, ζ the
barotropic tide surface elevation and g is gravity. For Skjomen
$N/\omega \sim 14$, $h \sim 150$ m, $h' \sim 0.05$, $\zeta \sim 3$ m, so that $u_i \sim 50$ cm s^{-1}. Observed current fluctuations are of the order of 20 cm s^{-1}, of the
same order of magnitude as u_i. It is therefore evident that the velocity signals recorded just inside the sill correspond to internal
gravity waves (internal tides) which originate at the sill and propagate into the fjord. Note that internal waves should also exist
outside the sill but these are not studied here, for there was no

mooring on that side. The next two sections study wave propagation
by means of vertical modes (section 4) and ray theory (section 5),
in order to explain the characteristics observed at the mooring
inside the sill in terms of barotropic tide interaction with bottom
topography.

A stratified body of water can support an infinite set of internal gravity waves. Because of bottom and surface boundaries to
the system, this set is a discrete ensemble. If only the first few
modes characterized by the lowest wavenumbers are present, it is
helpful to discuss the motion in terms of these modes. However,
decomposition into vertical modes is rigorously justified only if
the bottom is flat and horizontal. On the contrary, if many higher
modes characterized by high wavenumbers close to one another are
present, it is advantageous to discuss the observations in the light
of ray theory – "a relative simple and versatile theory" according
to Baines (1982). This theory is asymptotically valid only when the
wavenumbers present can be considered as a continuous range but it
is easily applicable to any bottom topography.

In what follows, only the first few modes dominate the wave
field in Skjomen; although the vertical mode approach is not rigorous
owing to the steep topography, and ray tracing is not strictly valid
because of lack of wavenumber continuity.

VERTICAL MODES

The method of vertical modes in continously stratified fluids
is a well known tool. It is important to recall and to stress that
the method of vertical modes is rigorously valid only if the bottom
is a flat, horizontal surface. After assuming that all variables
change with time like horizontal progressive waves, $u(x, z, t) =$
$U(z) \cos(kx - \omega t)$, $w(x, z, t) = W(z) \sin(kx - \omega t)$, where ω is the
known tidal angular frequency and k an unknown horizontal wavenumber,
(see Phillips, 1966) we obtain

$$\frac{d^2 W}{dz^2} + k^2 \left[\frac{N^2(z)}{\omega^2} - 1 \right] W = 0$$

For wanishing $W(z)$ at the surface ($z = 0$) and at the bottom ($z = -h$),
this constitutes an eigenvalue problem. To each eigenvalue k corresponds a vertical mode characterized by a vertical velocity $W(z)$,
from which can be derived a horizontal velocity:

$$U(z) = \frac{1}{k} \frac{dW}{dz}$$

The numerical values of the wavenumber eigenvalues as well as the

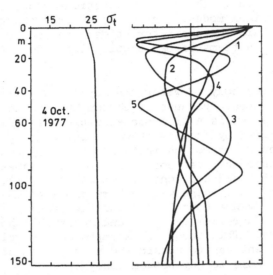

Fig. 4. Modal vertical profiles of the horizontal velocity corresponding to the density profile observed on October 4, 1977,
near the mooring location (Fall condition). The wavenumbers
are k = 0.29, 0.50, 0.74, 1.08 and 1.39 km^{-1}, and the respective wavelengths are λ = 21.5, 12.5, 8.51, 5.82 and 4.52
km.

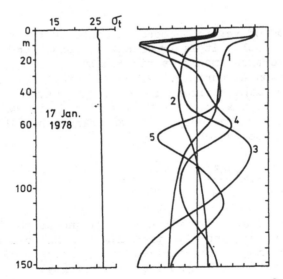

Fig. 5. Same as for Fig. 4, but for January 17, 1978 (Winter condition). The wavenumbers are k = 0.46, 0.74, 1.24, 1.80 and
2.19 km^{-1}, and the respective wavelengths are λ = 13.6,
8.52, 5.06, 3.48 and 2.87 km.

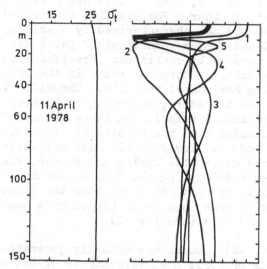

Fig. 6. Same as for Fig. 4, but for April 11, 1978 (Spring condition). The wavenumbers are k = 0.42, 0.79, 1.42 , 1.82 and 2.38 km^{-1}, and the respective wavelengths are λ = 14.9, 7.92, 4.44, 3.46 and 2.64 km.

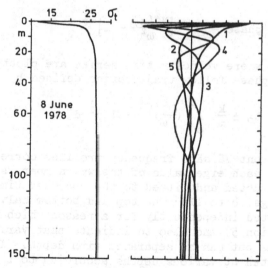

Fig. 7. Same as for Fig. 4, but for June 8, 1978 (Summer condition). The wavenumbers are k = 0.25, 0.40, 0.54, 0.79 and 0.98 km^{-1}, and the respective wavelengths are λ = 25.0, 15.8, 11.6, 7.98 and 6.41 km.

shapes of the vertical profiles strongly depend upon the Brunt-Väisälä frequency profile, N(z). There are noticeable differences from one season to another. For example, winter and spring conditions (Figs. 5 and 6) are characterized by a well mixed surface layer. The fall and summer conditions (Figs. 4 and 7), on the other hand, are characterized by continuous stratification from the surface down to the bottom. More important is the variation of the level of vanishing horizontal velocity. The sill rises to 60 m below the surface and therefore, according to Buckley (1980), any mode for which the horizontal velocity vanishes around 60 m deep is preferred among the modes that the barotropic tide could excite. As a rule, the first mode is likely in all seasons, while the second mode would be preferred during the spring and summer, and higher (third and fourth) order modes during the fall and winter. This is only an indication since the sill does not have the appearance of a knife edge, but rather of broad shoulders topped by a head, with depths equal to 82, 60 and 87 m (see Fig. 2).

To identify which modes are actually generated, it is best to study the records of horizontal velocity in or out of the fjord at the mooring station at several depths. These data are striking by their evidence of regular tidal oscillations whose phase varies with depth. The study of this phase shift identifies the modes present in the wave field. It can be shown that the phase propagates vertically with a depth dependent speed:

$$\frac{dz}{dt} \bigg|_{\text{phase}} = \pm \frac{\omega}{k} \left[\frac{N^2(z)}{\omega^2} - 1 \right]^{-\frac{1}{2}} .$$

In a z-t plane, where velocity time series are plotted versus depth, lines of equal phase follow trajectories defined by

$$t = t_0 \pm \frac{k}{\omega} \int_0^z \left(\frac{N^2(\xi)}{\omega^2} - 1 \right)^{\frac{1}{2}} d\xi .$$

For the known Brunt-Väisälä frequency profiles corresponding to each season, and for each eigenvalue of the wavenumber, such trajectories have been constructed and fitted to the recorded time series. Results are shown in Figs. 8 to 11. The top and bottom halves of the time series were fitted independently for a reason which will become clear later (see section 5) and also to indicate that various modes are present at one time but can be separated with depth. Indeed, different modes are detected at various depths according to their amplitude, although all modes are present at all depths.

Results can be summarized as follows:

- The fall data of 1977 (see Fig. 8) are composed of the first three

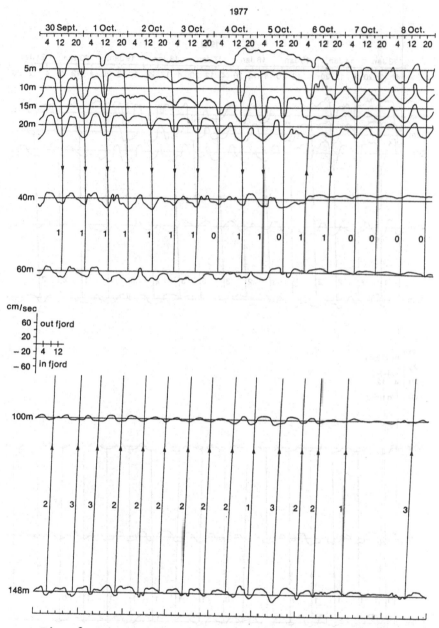

Fig. 8. Time series of the horizontal in-fjord velocity for various depths during nine days in
the Fall season of 1977. Curves joining the
minima represent fitted lines of vertical
phase propagation; the adjacent number
corresponds to the mode number.

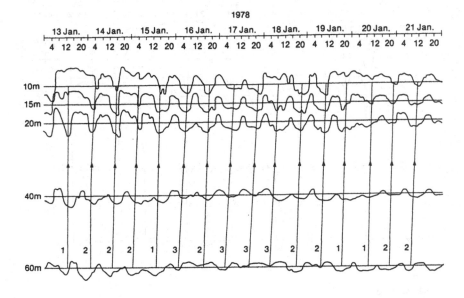

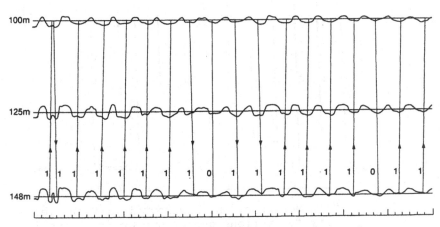

Fig. 9. Same as for Figure 8, but for nine
days during the Winter of 1978.

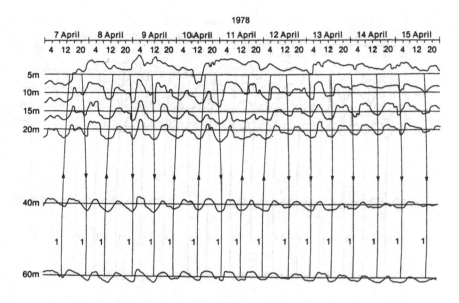

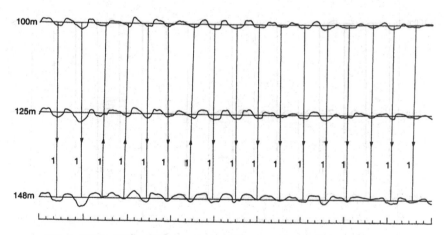

Fig. 10. Same as for Fig. 8, but for nine days
during the Spring of 1978

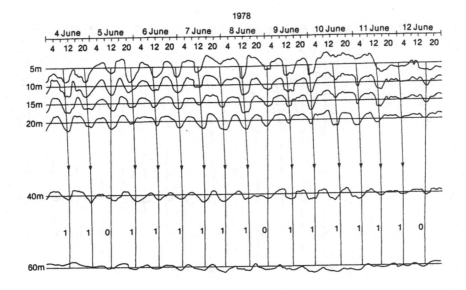

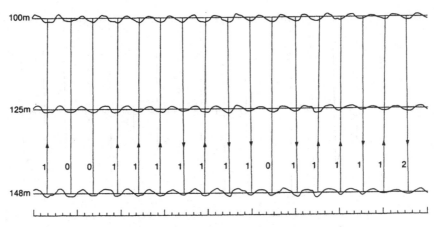

Fig. 11. Same as for Fig. 8 but for nine days
during the Summer of 1978.

modes: the first mode is detected in the upper half while mostly
the second and third modes are found in the bottom half. This
is due to the weak amplitude of the first mode with depth. Caution has to be exercized, for no data are present at 125 m during
this nine day period and the second and third modes have a small
amplitude around 100 m. Note that the wave phase propagates
downward in the upper half while it is propagating upward in the
lower half. This observation will be supported and explained by
ray tracing. Also, note the absence of tidal oscillation at 40,
60 and 100 m during 6, 7 and 8 October; vertical mode theory is
inadequate to explain such missing energy in a broad band.
- The winter data of 1978 (Fig. 9) show first, second and third modes.
  The first mode is well observed throughout the water column. This
  observation agrees with the fact that below the surface mixed
  layer and near the bottom the amplitude of the first mode is about
  the same, and differs from the Fall condition, during which the
  first mode had a small amplitude near the bottom and was therefore not detected in the bottom half of the water column. Moreover, the zero of the first mode is situated at 68 m, nearly at
  the level of the sill's highest point; this explains the presence
  of this mode. Finally, phase generally propagates upward throughout the depth, in contrast to all other conditions, which are
  characterized by upward phase propagation only in the lower half
  of the water column. This observation cannot be explained by
  vertical modes but will be interpreted by means of ray theory.
- The spring data of 1978 (see Fig. 10) are composed of the first
  mode only, which is detected throughout the water column. Although
  the phase propagates mostly downward near the surface and upward
  near the bottom, there is an indication that the reverse also
  occurs.
- The summer data of 1978 (see Fig. 11) show the first mode throughout the water column and the second mode in the bottom half only.
  These are the two modes with vanishing horizontal velocity near
  60 m, the top of the sill. Again, phase propagates downward near
  the surface and upward near the bottom, although there is an
  indication that the reverse also occurs near the bottom.

In summary, the observations can be explained by the first few
modes, the first mode being detected throughout the water column if
it is of sufficient amplitude, or otherwise only in the upper half.
In fact, all modes are present but are detected or unseen at various
depths according to their amplitude. There is also an indication
that modes with vanishing horizontal velocity near the top of the
sill (60 m) are more likely to be excited. In general, the wave
phase propagates downward near the surface and upward near the bottom.
Exceptions are the winter condition, for which the phase propagates
upward near the surface, and the winter, spring and summer conditions, for which the phase propagates both upward and downward near
the bottom. These observations may seem random in the context of
the vertical modes, but will be clearly interpreted by means of ray
theory.

RAY THEORY

One of our purposes is to determine how much a ray theory can account for the observations. It is helpful to recall the general lines of ray tracing, as established by Lighthill (1978, section 4.5 in particular). The fundamental assumption is two-fold: (i) waves are presumed to retain small amplitudes so that the equations of motion can be linearized, and (ii) waves ought to be so much dispersed that the wavenumber vector varies between them only gradually on a scale of wavelengths. Furthermore, for our purpose, attention will be restricted to a two dimensional system (no across fjord variations). In this case, each variable can be expressed in a locally sinusoidal form:

$$q = Q(x, z, t) \exp\left[i\alpha(x, z, t)\right] ,$$

where Q is the modulated amplitude. The latter varies much slower than the phase α; x and z are the in-fjord and upward coordinates respectively. The wavenumber components and the angular frequency are derived from the phase function

$$k = -\frac{\partial\alpha}{\partial x} , \quad m = -\frac{\partial\alpha}{\partial z} \quad \text{and} \quad \omega = +\frac{\partial\alpha}{\partial t} .$$

As the wave is locally sinusoidal, the governing linear equations yield the dispersion relation, which, in the case of gravity waves in a stratified fluid, takes the form

$$\omega^2 = N^2(z)\frac{k^2}{k^2+m^2} , \tag{1}$$

where $N(z)$ is the Brunt-Väisälä frequency, varying with depth only (assumption of horizontal homogeneity). Waves can exist (k^2, $m^2 > 0$) only in regions where N is greater than ω. In this specific case, $N(z)$ is generally much larger than ω. From the above expression, the group velocity components can be computed:

$$C_x = \frac{\partial\omega}{\partial k} = N \frac{m^2 \text{ sgn}k}{(k^2+m^2)^{3/2}} , \quad C_z = \frac{\partial\omega}{\partial m} = -N \frac{km \text{ sgn}k}{(k^2+m^2)^{3/2}} .$$

Note that the group velocity (C_x, C_z) is everywhere perpendicular to the wavenumber (k, m). This is a direct result of the fluid incompressibility. Wave energy propagates with the group velocity and thus follows trajectories satisfying

$$\frac{dx}{dt} = C_x \quad \text{and} \quad \frac{dz}{dt} = C_z ,$$

which are called rays. Energy therefore propagates at a right angle to the local direction of phase propagation. In Skjomen, N is typically 10 to 20 times greater than ω, and so is m compared to k (see dispersion relation (1)). Because $|m| >> |k|$, the phase propagates

almost vertically while the energy propagates almost horizontally, at
angles of about 3° to 5° from the vertical and horizontal respectively.
Consequently, while the wave energy travels horizontally from the
sill to the mooring line, it also travels vertically over a depth of
the order of the fjord depth.

Because the equations of motion are autonomous (time independent coefficients), the angular frequency ω must remain constant
along a ray (Lighthill, 1978). Moreover, in the fjord considered
here, dominant internal waves are caused by the M2 tide of period
12.42 hrs. ω is thus an absolute constant, the same for all rays, and
equal to $1.405 \times 10^{-4} s^{-1}$. From the differentiation of $\partial \alpha / \partial t =$
$\omega(z, k, m)$ with respect to x, it follows that

$$\frac{\partial k}{\partial t} + C_x \frac{\partial k}{\partial x} + C_z \frac{\partial k}{\partial z} = 0 \,,$$

showing that k is constant along each ray although it may differ from
ray to ray, unlike ω. In summary, for ω and k constant along a ray,
the dispersion relation (1) provides a means to determine m(z) from
N(z) along that ray:

$$m(z) = \pm k \left[\frac{N^2(z)}{\omega^2} - 1 \right]^{\frac{1}{2}} .$$

Rays can be traced most adequately in the x–z physical space.
Ray trajectories are defined by their slope

$$\frac{dz}{dx} = \frac{dz}{dt} \Big/ \frac{dx}{dt} = \frac{C_z}{C_x} = -\frac{k}{m} = \pm \left[\frac{N^2(z)}{\omega^2} - 1 \right]^{-\frac{1}{2}}$$

so that the rays passing through any given point (x_0, z_0) obey the
equation

$$x = x_0 \pm \int_{z_0}^{z} \left[\frac{N^2(\xi)}{\omega^2} - 1 \right]^{\frac{1}{2}} d\xi .$$

It is readily seen from this expression that the trajectories are
steady and can be computed once and for all from the knowledge of
the Brunt-Väisälä profile and the angular frequency. They are independent of k; the particular values of k and m along a ray prescribe
only the magnitude of the velocity at which energy propagates, not
its direction. Through every point pass two rays at angles
$\theta = \pm \tan^{-1} \left[\frac{N^2(z)}{\omega^2} - 1 \right]^{\frac{1}{2}}$ from the vertical and, along each ray, the
energy can propagate either upward or downward. There are there-
fore four directions of energy propagation at each point: upward
(m<0) or downward (m>0), and in-fjord (k>0) or out-fjord (k<0).

Finally, as a general rule, phase and energy propagate vertically in opposite directions: for positive m, phase propagates upward and energy downward, and vice versa for negative m.

Figures 12 to 15 show the various rays into the fjord which emanate from points on the sill. Because the shallow sill is the obstacle to the surface tidal waves and the generating body force is greatest at certain points of the sill (see Fig. 3), only the rays into the fjord emanating from these points have been constructed; it is reasonable to regard these as the only possible rays inside the sill. These rays are grouped in two major families: an upward beam carrying energy from the sill toward the surface and a downward beam originating from the inner flank of the sill and carrying energy towards the bottom. At the mooring line, the two beams are separated by a shadow zone, which lies approximately between 60 and 100 meters and the extent of which fluctuates seasonally with the density distribution. Caution has to be exercized in delimiting rays and shadow zones, for here only the first few modes are present (see section 4): it is indeed awkward to speak of spatially limited rays, when a combination of only the first few modes must lead to wave motions throughout the water column. Nevertheless, ray tracing is not limited by steep topography unlike the method of vertical modes and has proved in this case to be a valuable tool for interpreting the observations.

Reflection takes place at the surface, the bottom, and at levels where $N(z)$ drops to ω, beyond which internal gravity waves cannot be sustained. For the spring, summer and fall conditions, reflection occurs only at the surface and the bottom but, for the winter condition, the presence of a well mixed layer causes reflection about 5 meters below the surface. This, combined with a steeper ray slope somewhat below the mixed layer, contributes to a sub surface reflection between the sill and the mooring line (see Fig. 13), so that in the upper 50 meters of the mooring line there coexist upward and reflected downward propagating rays. For the spring and summer conditions, bottom reflection occurs between the sill and the mooring line (Figs. 14 and 15), so that there also coexist downward and upward propagating rays.

These considerations explain the upward and downward propagation of wave phase (see Figs. 8 to 11). The downward phase propagation in the upper half of the water column during the fall, spring and summer conditions is a result of upward propagation of wave energy from its source on the sill towards the surface (see Figs. 12, 14 and 15). The upward phase propagation in the upper half of the water column during the winter case corresponds to upward energy propagation from the sill to the base of the mixed layer, followed, after reflection, by downward energy propagation in the region of the mooring line (see Fig. 13). The upward phase propagation generally observed in the lower half of the water column

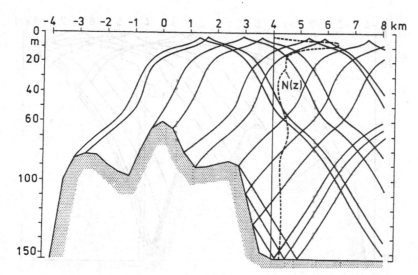

Fig. 12. Rays emanating from the sill during the case of October 4, 1977 (Fall condition). The Brunt-Väisälä frequency profile is plotted along the mooring line, on the inside of the sill (dashed line).

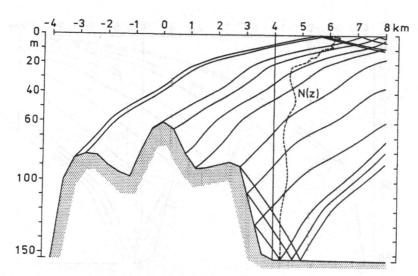

Fig. 13. Same as Figure 12, except for January 17, 1978 (Winter condition). Note the reflection at the base of the well mixed surface layer between the sill and the mooring line.

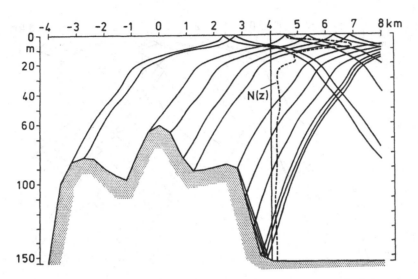

Fig. 14. Same as Figure 12, except for April 11, 1978 (Spring con-
        dition). Note the bottom reflection between the sill
        and the mooring line.

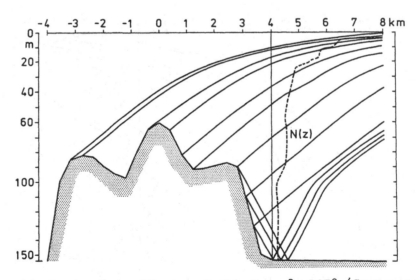

Fig. 15. Same as Figure 12, except for June 8, 1978 (Summer con-
        dition). Note the absence of reflection between the sill
        and the mooring line.

is a result of downward propagation of wave energy from the flank
of the sill. The spring and summer conditions exhibit some upward
propagation of energy in this region, owing to bottom reflection
between the sill foot and the mooring line. In summary, the perfect agreement between the directions of phase propagation as revealed from the current records and as deduced from ray theory
leaves no doubt that the sill is the actual source of internal gravity wave energy in the fjord.

Note that if the slope of the sill were less than the slope
of the ray, only upward propagation of energy away from the sill
would be possible. Data collected in fjords with gentle sills
(Svendsen, unpublished data) indeed show no significant internal
wave energy in the lower part of the water column in the proximity
of the sill. No ray could reach such a region.

Another observation revealed by the phase analysis of the
current records may be qualitatively explained by ray tracing.
During October 6, 7 and 8, 1977, no signal was detected at 40, 60
and 100 meters. Why? Probably, the Brunt-Väisälä frequency during
those days was such that it led to a well defined shadow zone at mid
depth between the upward propagating surface beam and the downward
propagating bottom beam.

SUMMARY AND DISCUSSION

Internal gravity waves can be analyzed in terms of vertical
modes or rays. However, each method has its limitations and is not
strictly valid for a shallow system with abrupt topography such as
a sill fjord. Indeed, the method of vertical modes is rigorous only
if the bottom is flat and horizontal or approximately so over one
horizontal wavelength. Obviously, this is not the case for a sill
fjord whose depth doubles within four kilometers in the sill area.
On the contrary, ray tracing is applicable without discrimination
to a system with any topography but is limited to wave packets
characterized by wavenumbers sufficiently close to be thought of as
varying continuously. This is not the case either, for data show
the presence of the few first modes whose wavenumbers are quite
separated. This results primarily from the relative shallowness
of a fjord (as opposed, for example, to the deep ocean next to the
shelf break).

In spite of these limitations and in the absence of any general yet convenient and simple method, wave data from Skjomen have
been studied by both methods, which have proved coherent and complementary. No quantitative inference has been drawn, but some
qualitative conclusions can be clearly stated: (i) the interaction of the surface tide and the sill topography is the mechanism

responsible for generation of internal waves, for at any time or depth the vertical phase propagation is always correctly predicted by the vertical phase propagation of a ray emanating from the sill, with or without intermediate reflection; (ii) surface or bottom reflection can account for changes in the direction of vertical phase propagation; (iii) waves tend to be a combination of the first few modes only, and (iv) there is no evidence of standing waves or waves coming from the head of the fjord. Moreover, ray theory provides a simple tool to predict shadow zones (regions not affected by internal waves) and, therefore, to predict where and when internal waves will be present and in which direction their phase and energy will propagate.

Advection by a mean current, an effect not so far discussed, may have some importance in transporting internal wave energy in or out of the fjord. According to Lighthill (1978), the energy is carried with a velocity equal to the vectorial sum of the group velocity at rest and the mean current velocity, in general a function of depth:

$$\frac{dx}{dt}\Big|\text{energy} = C_x(z) + \overline{u}(z) \ , \ \frac{dz}{dt}\Big|\text{energy} = C_z \ ,$$

in the absence of any knowledge of the mean vertical velocity. The horizontal position of a point along the ray emanating from a point at the depth z_0 ought to be corrected by

$$\Delta x = \int_{z_0}^{z} \frac{\overline{u}(z)}{C_z(z)} dz \ ,$$

which is, for a mean current of 5 to 10 cm/s, as much as 150 to 300 m. This is not large compared to the several kilometers that the ray travels horizontally. Another way to quantify the error introduced on the ray trajectory by neglecting the horizontal mean current is to form the ratio $\overline{u}/C_x$. For a Brunt-Väisälä frequency about ten times or more larger than the tidal frequency, C_x is approximately ω/k, and the ratio is

$$\frac{\overline{u}}{C_x} \sim \frac{k\overline{u}}{\omega} \sim 18\% \text{ to } 36\% \ .$$

Evidently the mean current, passing over a steep topography where internal tides are generated, produces a mean vertical velocity of the order of $\overline{u}\ dh/dx$, a relative correction to C_z

$$\frac{\overline{u}}{C_z} \frac{dh}{dx} \sim \frac{k\overline{u}}{\omega} \frac{N}{\omega} \frac{dh}{dx} \sim 13\% \text{ to } 25\% \ .$$

It is therefore advisable to include advection of energy by mean currents if they are close to 10 cm/s or larger. Qualitatively, rays emanating from the sill and propagating into the fjord are more horizontal if the mean current is into the fjord and more vertical if it is out of the fjord.

Other aspects of internal waves have been omitted. Absorption of energy by friction, dissipation or breaking cannot be important between the sill and the mooring line, because of their proximity, but absorption can be important along greater distances, owing to multiple reflections, and is presumed to be very important near the head of the fjord (Stigebrandt, 1978). This leads to the question of non-existence of standing waves, which at this stage has not been firmly established by observation. Finally, cross fjord variations and lateral reflections could also occur, especially in meandering fjords, and interactions between short branched fjords may play a role. These problems are still to be explored.

ACKNOWLEDGEMENTS

The authors wish to thank Drs. J. J. O'Brien and K. Takeuchi for support and constructive comments throughout this research. They also wish to acknowledge the helpful advice given by Mary Cushman-Roisin. The first author acknowledges support from the Office of Naval Research under grant No. N00014-80-C-0076. For the second author, support was provided by Statens Konsesjonsavgifts-fond (Norwegian Governmental Foundation) and by the Geophysical Institute, Div. A, of the University of Bergen. This constitutes contribution No.189 of the Geophysical Fluid Dynamics Institute of the Florida State University.

REFERENCES

Aanderaa Instruments, 1979, RCM 4/5. Recording current meter models 4&5. Technical description, No. 119.

Baines, P.G., 1974, The generation of internal tides over steep continental slopes, Phil.Trans.R.Soc.London, A277:27-58.

Baines, P.G., 1982, On internal tide generation models, Deep Sea Res., 29:307-338.

Buckley, J.R., 1980, A linear model of internal tides in fjords, in: "Fjord Oceanography", Freeland, Farmer and Levings, Eds., 165-171.

Fjeldstad, J.E., 1933, Interne Wellen, Geofysiske publika-sjoner, 10:No.6.

Gordon, R.L., 1978, Internal wave climate near the coast of northwest Africa during JOINT-1, Deep-Sea Res., 25:625-643.

Helland-Hansen, B., and Nansen, F., 1909, The Norwegian Sea,
        Rep. Norw. Fish. Mar. Invest. Christiania 1909.
Lighthill, J., 1978, Waves in Fluids, Cambridge University
        Press, 504 pp.
Mork, M., 1968, The response of a stratified sea to atmospheric
        forces. Report No. 11, Geophysical Institute, University
        of Bergen, Norway.
Phillips, O.M., 1966, The dynamics of the upper ocean,
        Cambridge University Press, Cambridge, 336 pp.
Prisenberg, S.J., and Rattray, M. Jr., 1975, Effects of conti-
        nental slope and Brunt-Väisälä frequency on the coastal
        generation of internal tides, Deep-Sea Res., 22:251-263.
Proudman, J., 1953, Dynamical Oceanography, Methuen & Co, Ltd.,
        London, 351 pp.
Rattray, M., Jr., 1960, On the coastal generation of internal
        tides, Tellus, 12:54-62.
Stigebrandt, A., 1980, Some aspects of tidal interaction with
        fjord constriction, Est. Coastal Mar. Sci.,11:151-166.
Zeilon, N., 1912, On the tidal boundary waves and related
        hydrodynamical problems. Kungl. Svenska Vetenskaps-
        akademiens Handlingar. Band 47, No.4, 46 pp.
Zeilon, N., 1914, On the seiches of the Gulmarfjord. Svenska
        Hydr.-Biol. Komm. Skrifter 5.

NUMERICAL SIMULATIONS OF INTERNAL WAVE GENERATION IN SILL FJORDS

Gunnar Mørk* and Bjorn Gjevik**

*Institute of Geophysics
**Department of Mechanics
University of Oslo
Oslo, Norway

ABSTRACT

A linear two-layer depth integrated model is used for simulating internal waves generated by fjord sills. A long narrow fjord is connected to a wide ocean region which enables the wave energy to be radiated away from the fjord mouth. On open boundaries the surface and internal wave modes are treated separately. The energy dissipation at the fjord head is simulated by means of a partially open boundary. An explicit forward-backward numerical scheme is employed. Wave generation by tidal motion in the ocean and seiches in the fjord are investigated. The dependence of the wave amplitude on the depth and location of the sill is discussed and some comparisons with analytical results are made. Internal waves of tidal period are found to propagate away from the sill as Kelvin waves. Short periodic internal waves propagate into the ocean mainly as Poincaré-waves. The damping of the barotropic Helmholtz seiche is briefly discussed.

INTRODUCTION

It is well known that internal waves of tidal frequency are generated at fjord sills and these waves have been described by several authors. Fjeldstad (1964) reported observations of internal tidal waves in Herdlefjorden west of Bergen, Norway. He found progressive waves propagating away from the shallow sill in the northwestern end of the fjord. The vertical displacement amplitudes associated with the wave motion were of the order of 2 m. Stigebrandt (1979) shows that internal tidal waves were generated at the shallow Drøbak sill in the Oslofjord, Norway. He found that the waves propagated away from the sill on both sides. Stigebrandt

argued that the progressive character of the waves inside the sill
was due to wavebreaking along the sloping bottom at the fjord head.
He used a two-dimensional channel model with a two-layered ocean to
estimate the amplitude of the internal tide. In Stigebrandt's analysis the sill completely blocks the lower layer.

The aim of this work is to develop a numerical model for simulation of internal tides in sill fjords. In particular we consider
the effect of rotation and radiation of energy into the ocean in
the three-dimensional models. We also examine effects of sill length,
sill height and sill location on the wave motion. In this study
progressive internal waves in the fjord are obtained by radiating
the waves out of the model at the fjord head boundary. This is a
simple way of simulating energy dissipation at the fjord head and
it may provide a realistic model for internal tides in long fjords.

MODEL AND BASIC EQUATIONS

We restrict the study to cases where the stratification in
the fjord and in the ocean can be approximated by a two-layer fluid
model with uniform density in each layer. We introduce a Cartesian
coordinate system x, y, z with the z-axis vertically upwards and
x- and y-axes in the horizontal undisturbed interface between the
layers. In the undisturbed state the depth of the upper layer h_1
is constant and the depth of the lower layer h_2 is a function of x
and y due to bottom topography. The displacement of the surface
and the interface are denoted by η and ζ respectively and these
are functions of x, y and time t. We use a hydrostatic, depth integrated, linear and frictionless model and we write the equations
of motion for the upper and lower layer respectively:

$$\frac{\partial U_1}{\partial t} - fV_1 = -gh_1 \frac{\partial \eta}{\partial x} \tag{1}$$

$$\frac{\partial V_1}{\partial t} + fU_1 = -gh_1 \frac{\partial \eta}{\partial y} \tag{2}$$

and

$$\frac{\partial U_2}{\partial t} - fV_2 = -gh_2 \left[(1-\varepsilon) \frac{\partial \eta}{\partial x} + \varepsilon \frac{\partial \zeta}{\partial x} \right] \tag{3}$$

$$\frac{\partial V_2}{\partial t} + fU_2 = -gh_2 \left[(1-\varepsilon) \frac{\partial \eta}{\partial y} + \varepsilon \frac{\partial \zeta}{\partial y} \right] \tag{4}$$

where U_i and V_i are the x- and y-component of the volume fluxes
per unit width in the upper (i = 1) and lower (i = 2) layer. The
density difference between the layers relative to the lower layer
density is denoted by ε and f is the Coriolis parameter. From the
continuity equations we obtain

$$\frac{\partial \eta}{\partial t} = -\frac{\partial}{\partial x}(U_1 + U_2) - \frac{\partial}{\partial y}(V_1 + V_2) \qquad (5)$$

and

$$\frac{\partial \zeta}{\partial t} = -\frac{\partial U_2}{\partial x} - \frac{\partial V_2}{\partial y} \qquad (6)$$

.The geometry of the ocean-fjord model is depicted schematically in Fig. 1, where ABCDA represents the fjord and EFGHE represents a section of the ocean.

The sill is considered as a rectangular block of length 2L and the total depth at the sill is $h_1 + h_s$. The lengths from the midpoint of the sill are l_1 to the fjord head and l_2 to the fjord mouth. The width of the fjord is b.

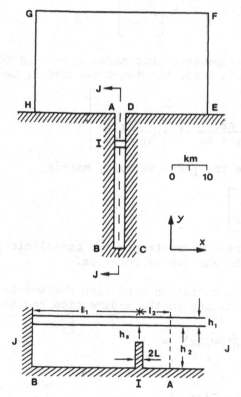

Fig. 1. Definition sketch and the geometry of the ocean-fjord model.

The tidal forcing is modelled by feeding a barotropic Kelvin wave into the model at the boundary GH. Waves reaching GH from inside will be reflected. All other boundaries except BC and EF are no-flux boundaries. The boundary EF is open with a condition that allows barotropic and baroclinic Kelvin waves that propagate along the coast to radiate out of the model. The boundary BC at the fjord head may be treated as a no-flux boundary or as a partially open boundary where baroclinic waves radiate out and barotropic waves are fully reflected.

On open boundaries the different wave modes can be treated separately, see for example Klemp and Lilly (1978). In our model this technique is easily accomplished since the waves reaching the boundaries are predominantly nondispersive Kelvin waves. The fluxes consist of barotropic and baroclinic contributions

$$U_i = U_{iBT} + U_{iBC} \quad , \quad i = 1,2$$

Introduce the matrix notation

$$U = \begin{bmatrix} U_1 \\ U_2 \end{bmatrix} \quad \text{and} \quad W = \begin{bmatrix} U_{1BT} \\ U_{1BC} \end{bmatrix}$$

The fluxes can be separated into modes according to $U = AW$, where the splitting matrix A for the two-layer model, to order ε, is

$$A = \begin{bmatrix} 1 & 1 \\ \dfrac{h_2}{h_1} - \dfrac{\varepsilon h_2}{h_1 + h_2} & -1 + \dfrac{\varepsilon h_2}{h_1 + h_2} \end{bmatrix}$$

Further, introduce the phase velocity matrix

$$C = \begin{bmatrix} c_s & 0 \\ 0 & c_i \end{bmatrix}$$

where c_s and c_i are the barotropic and baroclinic phase velocities respectively in the absence of rotation.

The Sommerfeld radiation condition for barotropic and baroclinic waves propagating in the x-direction can then be written

$$(\partial W/\partial t) + C(\partial W/\partial x) = 0$$

or

$$(\partial U/\partial t) + ACA^{-1}(\partial U/\partial x) = 0 \tag{7}$$

Eq.(7) is employed at the boundary EF. Numerical experiments show that it is essential to retain the terms of order ε in the matrix A in order to radiate both modes properly.

For a partially open boundary which reflects the barotropic but radiates the baroclinic mode we use the conditions

$$U_{\ell_{BT}} = 0$$

$$(\partial U_{1_{BC}}/\partial t) + c_i(\partial U_{1_{BC}}/\partial x) = 0$$

which give

$$(\partial U/\partial t) + c_i(\partial U/\partial x) = 0 \tag{8}$$

NUMERICAL METHOD

The numerical scheme employed is an adaption of the scheme used by Martinsen et al.(1979) in the study of barotropic storm surges. It is a nearly forward-backward scheme, nonstaggered in time and staggered in space ("C-grid", Mesinger and Arakawa (1976)). The boundary intersects only flux grid points. The time integration proceeds as follows: Eqs.(1) and (3) are integrated one time step forwards. Eqs. (2) and (4) are then integrated forwards, apart from the Coriolis terms, which are integrated backwards. Eqs. (5) and (6) are finally integrated backwards. The Coriolis terms are evaluated as mean values of four grid points.

The scheme is neutrally stable if

$$\sqrt{2} \, c_s \Delta t/\Delta s \leq 2,$$

where Δs is the grid size and Δt is the time step, provided that $f\Delta s/c_s \leq 1$.

The last inequality is always satisfied when we apply a grid size which resolves baroclinic Kelvin waves. The radiation conditions are integrated using an upstream space differencing and a forward time differencing.

ANALYTICAL SOLUTIONS FOR A 2-D CHANNEL

To our knowledge no analytical solution for this ocean-fjord model is available. It may however be useful to compare the numerical results with analytical results for a two-dimensional channel model of a sill fjord (Fig. 1).

For periodic two-dimensional motion without rotation the equations (1)-(6) are easily solved in the ocean, sill and fjord regions provided the depths are uniform in each region. By requiring that the volume fluxes and the displacement of the surface and the interface are continuous the solutions can be matched. We consider an incoming barotropic tidal wave with period T and surface amplitude a_s which is totally reflected at the fjord head. Hence the amplitude of the surface tidal elevation is $2a_s$. The tidal flow over the sill generates internal waves radiating away from the sill. We assume that the internal waves are progressive in the fjord and use a radiation condition (Eqn. (8)) at the fjord head.

If the sill is below the pycnocline ($h_s > 0$) we find that the ratio between the amplitude a_i of the internal wave and the amplitude $2a_s$ of the surface tide is

$$\frac{a_i}{2a_s} = 4\pi^2 \frac{h_1/h_2}{(1+h_1/h_2)^2} \frac{1-h_s/h_2}{h_s/h_2} \frac{\mathit{l}_1}{c_i T} \frac{L}{c_i T} \qquad (9)$$

This expression is valid provided the wave-length of the internal wave over the sill is much larger than the length of the sill ($L \ll c_i T$) and the wave-length of the barotropic tidal wave is much larger than the length of the fjord ($\mathit{l}_1 \ll c_s T$). It can also be shown that the fluxes in the upper and lower layer at the sill are in phase and that the displacement of the interface corresponding to the internal wave is phase-shifted by an angle π across the sill. We note that in this case the amplitude of the internal wave depends on the length of the sill.

If the top of the sill is at or above the pycnocline ($h_s \leq 0$) the lower layer is completely blocked and the approximation leading to (9) is invalid. This case needs separate treatment. We find

$$\frac{a_i}{2a_s} = 2\pi \frac{1}{1+h_1/h_2} \frac{\mathit{l}_1}{c_i T} + 0 \left(\frac{L}{c_i T}\right) \qquad (10)$$

In this case, too, the displacement of the pycnocline is phase-shifted an angle π across the sill but the displacement inside the sill is phase-shifted an angle $\pi/2$ relative to the non-blocking case. The expression (10) for the internal wave amplitude (with $h_s = 0$ and $L \to 0$) corresponds to what Stigebrandt (1979) used. This case with blocking of the lower layer was originally studied by Proudman (1953).

By evaluating the ratio $a_i/2a_s$ as a function of h_s/h_1 for some typical values of the other parameters we see from Eqns.(9) and (10) that the amplitude of the internal tide become significant only for very shallow sills with blocking of the lower layer.

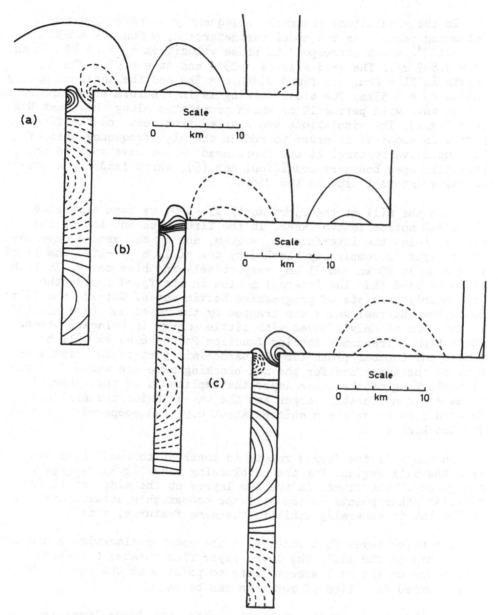

Fig. 2. Contour plot of displacement of the interface due to the
internal mode, $\zeta - ((h_2/(h_1 + h_2))\eta$.
Contour interval is 0.1m. Positive and negative values
are indicated by full lines and dashing respectively.
(a) t = 14hr. (b) = 17hr. (c) = 20hr.

RESULTS OF SIMULATIONS

In the simulations reported subsequently we have used the
following values for the model parameters: $h_1 = 20m$, $h_2 = 230m$,
$\varepsilon = 4.10^{-3}$ which correspond to phase velocities $c_s = 49.49$ m/s and
$c_i = 0.849$ m/s. The grid size $\Delta s = 250m$ and $\Delta t = 6.92$ s. The sill
length is $2L = 2km$, the fjord width $b = 3km$ and the internal Rossby
radius $R_i = 6.53km$. The tidal forcing is in the form of a barotropic
Kelvin wave with period 12 hr which propagates along the coast H E
(see Fig.1). The simulations are started from rest and the first
period is smoothed in order to reduce the high-frequency part of
the impulsive forcing. At the fjord head BC we have applied the
partially open boundary condition, Eqn.(8), which leads to a progressive internal tide in the fjord.

With the sill at the fjord mouth $l_2 - L = 0$, we have simulated
the tidal motion in two cases. In the first case the top of the
sill is below the interface, $h_s = 20km$, and in the second case the
lower layer is completely blocked by the sill, $h_s = -10m$. The fjord
length l_1 is 29 km and 31 km respectively in these cases. In both
cases we find that the internal motion in the fjord inside the
sill mainly consists of progressive Kelvin waves. Outside the fjord
mouth the internal waves are trapped by the coast DE (see Fig.1)
in the form of Kelvin waves with little energy in Poincaré waves.
For a higher frequency forcing function Poincaré waves will be
generated. Contour plots for the baroclinic part of the displacement of the interface for the non-blocking case are shown in Fig.2.
With blocking of the lower layer the amplitudes of the internal
waves are considerably larger and the waves inside the sill are
delayed 1.84 hr or phase shifted about 0.96 rad compared to the
non-blocking case.

On the sill the fluxes rotate in contrast to conditions away
from the sill region. For the nonblocking case Fig.3a depicts a
hodograph of the fluxes in the two layers at the midpoint of the
sill. At other points on the sill the hodograph's details may
differ but it generally exhibits the same features, viz:

The upper layer flux rotates in the counter-clockwise sense at
all points on the sill. The lower layer flux rotates clockwise
everywhere on the sill except close to point D at the mouth, where
no preferred direction of rotation can be seen.

The major axis of the flux ellipse for the lower layer is
aligned more along the fjord than for the upper layer and the
major axes for both layers are aligned more along the fjord at
points closer to the shore.

The largest flux values reached on the sill for the lower and
upper layer are $6.4m^2/s^{-1}$ and $2.2m^2/s^{-1}$ respectively, corresponding

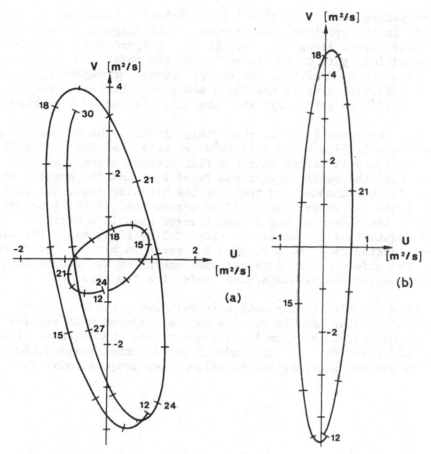

Fig. 3. Flux hodographs at midpoint of sill. Hours after start
       of simulation are marked.
       (a) For upper and lower layer in the non-blocking case.
           The small ellipse represents the upper layer.
       (b) For upper layer with blocking of lower layer.

to velocities of $0.32m/s^{-1}$ and $0.11m/s^{-1}$. With $c_i = 0.85m/s^{-1}$, the
densimetric Froude number is thus well below one, which justifies
the neglect of the non-linear terms in the equations of motion.

   The largest values of the gradient of the interface displacement
are found at the sill. The gradient at the sill rotates in the
counter-clockwise sense.

   In the second case, with complete blocking of the lower layer,
the flux vector at the midpoint of the sill (Fig.4) shows a clockwise rotation which is also found at all points on the sill. We
note that in this case the cross fjord flux component on the sill

is everywhere less than $1m^2/s^{-1}$ (mean velocity less than $0.1m/s^{-1}$) whereas in the previous case it could be as large as $3.6m^2/s^{-1}$ in the lower layer. Along the side AB of the fjord the amplitude of the baroclinic part of the interface displacement is 6.21m and at the ocean coast DE 2.24m. The energy fluxes corresponding to the internal Kelvin waves in the fjord and along the coast are $3.03 \cdot 10^4$W and $3.74 \cdot 10^3$W respectively, when the sill lies below the pycnocline.

When the lower layer is completely blocked, the corresponding figures are $1.39 \cdot 10^6$W and $2.73 \cdot 10^5$W. We note that the energy flux into the fjord is larger than the flux trapped along the ocean coast. Also the ratio between the fjord flux and the ocean flux is larger for the nonblocking than for the blocking case; the ratios are 8.1 and 5.1 respectively. If we assume that 5% (Stigebrandt, 1976) of the energy of the internal waves goes into mixing of the fjord stratification, an energy flux of $1.39 \cdot 10^6$W, as in the simulated blocking case, will require 3.7 years to mix the fjord completely. The effect of the internal wave energy on the stratification is thus negligible within a time scale of a few days.

Since the effect of rotation is relatively small in these numerical simulations the results may be compared with results for the channel model (section 4). The analytical channel model (Eqns. 9 and 10) gives for the amplitude of the internal waves 0.66m and 4.5m in the non-blocking and blocking cases respectively. The

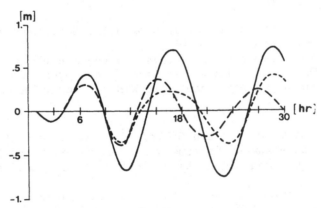

Fig. 4. Cross-fjord mean values of baroclinic interface displacement, $\zeta - ((h_2/(h_1 + h_2))\eta$, 750 m inside the sill for three different sill locations. The fjord lengths inside the sill are identical. $(\text{———})\ l_2 - L = 0$
                                                                            $(\text{— —})\ l_2 - L = \frac{1}{4}c_i T$
                                                                            $(\text{-----})\ l_2 - L = \frac{1}{2}c_i T$

corresponding cross-fjord mean values from the numerical model are
0.73m and 5.0m. The phase shift between blocking/nonblocking situation of displacement inside the sill is 0.96 rad in the fjord-ocean
model compared with $\pi/2$ in the channel model. This shows that when
the sill is situated at the fjord mouth the amplitudes of the generated internal tide can be estimated reasonably well by the simplified channel model, as anticipated by Stigebrandt (1979). Numerical
simulations further confirm the analytical result that the amplitudes
increase linearly with increasing fjord surface area.

In order to study the effect of the location of the sill on
the internal tide we have made a few simulations with the sill at
different distances from the fjord mouth. Some results of these
simulations are displayed in Fig.5. The internal waves that radiate
from the sill towards the ocean are partially reflected by the fjord
mouth and the reflected waves are partially transmitted over the
sill into the fjord, thereby modifying the amplitude and phase of
the progressive waves inside the sill. In addition, we note that
the results for different sill locations also differ before any
reflected waves have reached the sill. Fig.4 shows that the amplitudes of the generated waves are smaller when the sill is not
situated at the fjord mouth. This phenomenon can most likely be
attributed to the differences in lateral geometry in the proximity
of the sill. Between the sill and the fjord mouth the wave pattern
resembles that of a closed basin as the internal waves in this
region are not purely progressive.

In the above simulations the first period of the forcing
function was smoothed in order to avoid excitation of possible

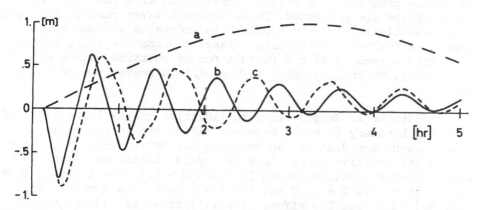

Fig. 5. (a): Surface elevation outside fjord mouth when forcing
function is not smoothed. (b) and (c): 10 times the
difference between surface elevation at fjord head and
outside fjord mouth, with no sill (b) and with sill (c).

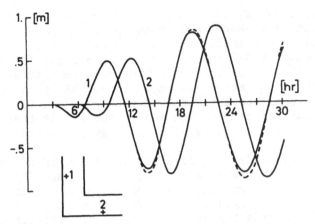

Fig. 6. Baroclinic interface displacement in two points situated
at each side of a fjord bend. The numbering shows corresponding points and displacements. The dashed curve is
the result for a straight fjord in a point corresponding
to point 1.

resonances with period shorter than the tidal forcing. Fig.6 displays some results when the tidal forcing is applied impulsively
and an open boundary condition is applied along the outer ocean
boundary GF. The model parameters are otherwise as in the non-
blocking case discussed above. The high frequency part of the forcing function excites a Helmholtz mode barotropic seiche in the
fjord. No baroclinic effects are observed when the depth is uniform
but in the presence of a sill internal waves with the barotropic
seiche period are generated. These internal waves radiate into the
ocean essentially as cylindrical Poincaré-waves without being
trapped by the coast. The period observed without a sill, 2630 s
is in good agreement with a formula for no rotation (Miles and
Munk (1961)) which gives 2689 s. The sill increases the period to
3190 s.

The seiche is damped owing to radiative energy losses. Use of
a no-flux boundary GF would severely impede the energy radiation
to the ocean. The damping can be expressed by the Q-factor, where
$2\pi/Q$ is the relative energy loss per cycle. Estimates of the Q-
values are for a uniform depth (Fig.6b) Q = 8.3, for a homogeneous
fluid with a sill Q = 10.7 and for the layered fluid with a sill
(Fig. 6c) Q = 9.6. The effect of a sill in a layered ocean will
thus be to decrease the barotropic and increase the baroclinic
energy loss. Based on the Q-values we estimate the ratio between
the energy losses caused by the baroclinic and barotropic waves to
be of the order of 0.1 for Fig. 6c.

We have also made one simulation with an L-shaped fjord which has the same width, surface area and parameters as in the previously mentioned non-blocking case, except that halfway between mouth and head there is a 90° bend. For the internal Kelvin wave then, $R_i/b=$ 2.18 and $c_iT/b=12.2$. With these parameters the bend has but small effect on the propagation of the internal Kelvin wave, see Fig.7.

SUMMARY

The technique of splitting of modes in the open boundary condition performs equally well for fully open and partially open boundaries when the waves are nondispersive. A simplified analytical channel model gives reasonable estimates of baroclinic wave amplitude in the fjord when the sill is located at the mouth and the effect of rotation is small. Significant amplitudes occur only for shallow sills. At the sill the fluxes rotate and significant transverse flow may take place. The internal tides appear as Kelvin waves in the fjord and along the ocean coast. A smaller forcing period creates more Poincaré waves and decreases the wave trapping along the coast. A channel between the sill and the ocean reduces the wave amplitudes and modifies the wave pattern. With the parameters used, a bend in the fjord has negligible effect on the propagation of the internal tides. High frequency internal waves may be generated by a barotropic Helmholtz seiche. The energy loss arising from the internal wave generation has small effect on the damping characteristics of the Helmholtz mode.

ACKNOWLEDGEMENT

This work has been partially funded by The Royal Norwegian Council for Scientific and Industrial Research (NTNF) through project grant 1583.11009.

REFERENCES

Fjeldstad, J.E., 1964, Internal waves of tidal origin. Part I. Theory and analysis of observations, Geof.Publ., Vol. XXV, No 5.

Klemp, J.B., and Lilly, D.K., 1978, Numerical simulation of hydrostatic mountain waves, J.Atmos.Sci., 35:78.

Martinsen, E.A., Gjevik, B., and Røed, L.P., 1979, A numerical model for long barotropic waves and storm surges along the western coast of Norway, J.Phys.Oceanogr., 9:1126.

Mesinger, F., and Arakawa, A., 1976, Numerical methods used in atmospheric models. GARP Publ.Ser., No.17.

Miles, J., and Munk, W., 1961, Harbor paradox, Proc. ASCE, Journ.Waterways Harbors Div., 87:111.

Proudman, J., 1953, "Dynamical oceanography", Methuen & Co,

Stigebrandt, A., 1976, Vertical diffusion driven by internal waves in a sill fjord, J.Phys.Oceanogr., 6:486.

Stigebrandt, A., 1979, Observational evidence for vertical
        diffusion driven by internal waves of tidal origin in
        the Oslofjord, J.Phys.Oceanogr., 9:435.

TWO-YEAR OBSERVATIONS OF COASTAL-FJORD INTERACTIONS

IN THE STRAIT OF JUAN DE FUCA

James R. Holbrook, Glenn A. Cannon and David G. Kachel

Pacific Marine Environmental Laboratory, NOAA
Seattle, Washington
U.S.A.

ABSTRACT

During the two-year period between March 1979 and March 1981, continuous measurements of currents, water temperature and salinity, winds, sea level and water pressure were made in the Strait of Juan de Fuca, a fjord-like estuary between Washington State, U.S.A. and Vancouver Island, Canada. These observations along with satellite imagery and CTD surveys show that surface intrusions of less-dense coastal water into the Strait are a major, recurring feature of its general circulation. These intrusions, which are accompanied by reversals in the normally vigorous, two-layer estuarine flow, are correlated with southwesterly coastal winds (conducive to coastal downwelling) and rising sea level at the entrance. These data suggest that a wind-induced, shoreward Ekman flux along the coast pushes less-dense coastal water into the Strait where it feeds a complex, 3-dimensional density driven flow which has been observed to travel as far as 135 km up-strait. The relationship between these baroclinic events and the along-strait and across-strait sea surface slopes and pressure gradients will be discussed.

INTRODUCTION

The role played by coastal winds in forcing a circulation response in adjacent estuaries, embayments and straits has received growing interest and observational support over the last five years (e.g. Svendsen, 1977; Holbrook and Halpern, 1977; Wang and Elliott, 1978; Smith, 1978; Holbrook, et al. 1980a; Cannon and Holbrook, 1981; and Holbrook and Halpern, 1982). Most studies have found that coastal winds are coherent with axial estuarine currents or sea level fluctuations or both and consequently a mechanism incorporating coastal Ekman dynamics has often been suggested to explain the phenomena. Indeed, numerical modeling efforts by Klinck et al. (1981) and Niebauer (1980) have clearly shown that a wind-driven coastal regime can induce a response qualitatively similar to many of the observations.

The Strait of Juan de Fuca is a relatively deep ($\sim$ 200 m), U-shaped channel connecting the Pacific Ocean with the inland waters of the Pacific Northwest (Fig. 1). Redfield (1950) described the net circulation in the Strait as two-layered and related this circulation to the region by means of a composite temperature-salinity diagram (Fig. 2). Recent field observations (e.g. Cannon, 1978; Holbrook et al., 1980a) show a seaward-flowing upper layer with typical speeds of 30 cm/s above a landward-flowing deep layer with speeds of 10 cm/s, confirming the net circulation inferred by Redfield (1950). However, large current fluctuations about this mean occur, with time scales of 5-20 days and amplitudes larger than the mean, that dominate the regional, subtidal circulation (Holbrook, et al., 1980). These subtidal fluctuations cause major reversals in the estuarine circulation (i.e. inflow near-surface and outflow at depth) that have been observed up to 135 km from the entrance (Frisch, et al., 1981). Holbrook and Halpern (1982) have shown that current reversals are associated with surface intrusions of less dense coastal water, southerly coastal winds and rising sea level at the entrance. They suggest that the mechanism responsible for these intrusions involves the combination of a shoreward Ekman flux (driven by southerly coastal winds) that, in turn, moves less-dense coastal water to the entrance of the strait. This lighter water then feeds a density-driven flow which propagates up-strait as a gravity current similar to Benjamin's (1968) model.

In this report we present time series measurements of currents, water properties, winds and sea level obtained over a two-year period that further delineate the seasonal and spatial variability of coastally-forced intrusions in the strait. These observations along with satellite imagery and CTD surveys provide additional support for the Ekman flux-density current mechanism proposed by Holbrook and Halpern (1982).

DATA

Time series of currents, water properties, winds and sea level were obtained between March, 1979 and March, 1981, in the western Strait of Juan de Fuca (Fig. 1). EG&G (formerly AMF) Vector-Averaging Current Meters (VACM's) suspended below a taut-line surface mooring measured near-surface (4 and 20 m) currents and temperature at ST-14 and ST-15, and deep (150 m) currents and temperature at ST-15. Since a continuous two-year record was obtained at ST-15 and only intermittent (2-4 month) records were obtained at ST-14, data from ST-15 will be used in this report to indicate the

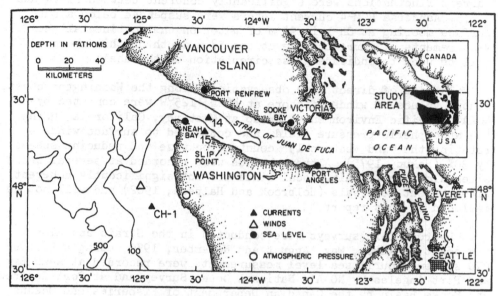

Fig. 1. The Strait of Juan de Fuca showing positions of moorings and measurement sites.

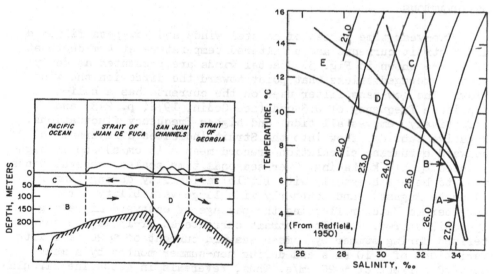

Fig. 2. Schematic diagram showing net circulation and temperature-
salinity diagram for the Strait of Juan de Fuca.
(Adapted from Redfield, 1950).

occurrence of current reversals and coastal intrusions. Subtidal current fluctuations were significantly coherent between ST-14 and ST-15. Aanderaa RCM-4 current meters were suspended below a sub-surface mooring at CH-1 in 90 m of water on the continental shelf. Measurements of currents made at the 58 m depth on this mooring will be used to indicate gross circulation on the shelf.

In lieu of direct wind observations along the Washington coast, 6-hour values of winds offshore at 48°N, 125°W were computed by NOAA's Pacific Environmental Group at Monterey, California, using the atmospheric pressure field, and converted to surface winds by rotating the wind vectors 15° counterclockwise and reducing speeds by 30% (Bakun, 1975). Although winds were recorded at several shore sites along the Strait, local winds are not significantly coherent with current reversals (Holbrook and Halpern, 1982) and will not be included in this report.

Hydrographic surveys were conducted in the Strait and along the coast in March, May, August and November, 1979, using a Plessey CTD system. Hourly sea level measurements were recorded at Neah Bay and Port Angeles by NOAA's National Ocean Survey and at Port Renfrew and Sooke Basin by the Canadian Department of Fisheries and Oceans' Institute of Ocean Sciences. Selected sea-surface-temperature-enhanced infrared satellite images were furnished by NOAA's National Environmental Satellite Service (Breaker, personal communication).

OBSERVATIONS

Two-year time series of coastal winds and low-pass filtered along-strait currents and unfiltered temperature at 4-m depth at ST-15 are shown in Fig. 3. Coastal winds are presented as daily vector-averaged sticks that point toward the direction the wind blows. The low-pass filter used on the currents has a half-amplitude frequency of 0.3 cpd (see Godin, 1972, p. 243) and effectively removes all tidal and higher frequency fluctuations. Negative subtidal flow into the Strait indicates a reversal in seaward estuarine circulation (record mean = 15 cm/s) and is high-lighted by black shading. Over seasonal time scales, coastal winds tend to be northwesterly with little variability during summer (June to August), and southerly with large variability during November to Match reflecting the passage of eastward migrating cyclonic storms. The along-strait currents at ST-15 may be charac-terized during summer by a mean seaward current of ∿ 20 cm/s with variations of ± 10 cm/s and during non-summer months by a mean seaward current of ∿ 20 cm/s. Thus, reversals in estuarine circula-tion seldom occur during summer and frequently occur during the remainder of the year, with typically 2-3 reversals lasting from 2-7 days each month. Temperature also experiences a strong seasonal cycle: both temperature values and their variations have maxima in late summer and minima in late winter. Temperature maxima are

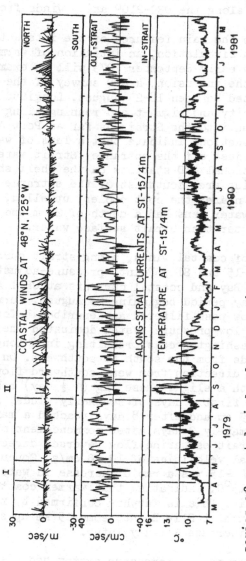

Fig. 3. Time series of coastal winds, low-pass filtered, along-strait currents at ST 15 (4 m depth), and unfiltered temperature at ST-15(4 m depth).

preceded by a rapid rise in temperature and a reversal in subtidal currents. The current reversals are, in turn, generally preceded by periods of southwesterly coastal winds. Broad frequency bands of significant (at 95% level) coherence exist across the entire subtidal spectrum between coastal winds and along-strait currents; a maximum of 35% of the subtidal current variance is explained by wind fluctuations along the 030-210° axis, winds from the southwest.

To illustrate the main features of the interaction between coastal and Strait circulation, an intrusion of warm coastal water into the Strait in early September 1979 will be examined in more detail. Prior to the intrusion, a CTD survey of the shelf and Strait region was conducted between 18-24 August. Longitudinal sections of temperature, salinity and sigma-t and running along the shelf, up Juan de Fuca Canyon into the Strait as far as Port Angeles, are shown in Fig. 4. These sections indicate that a lens of warm (T>15°C) water was located just off the entrance; Strait water had temperatures <11°C. Additional CTD stations on the shelf show a warm water mass covering a wide area southwest of the entrance ∿ 10 km seaward of the coastline; nearer the cost cooler, upwelled, water was found. The coastal warm water lens had a depth of 25 m and was ∿ 0.6 sigma-t units lighter than Strait surface water.

Time series of coastal winds, along-strait currents at ST-15, temperature at ST-15 and Slip Point, pressure-adjusted (1 mb = 1 cm) sea level at Neah Bay and coastal currents at CH-1 are shown in Fig. 5 for a 30-day period beginning 25 August. Currents and sea level have been low-pass filtered as described before. Between 1-8 September a large low-pressure system dominated the mesoscale wind pattern off the Washington coast resulting in a long 9-day period of persistent winds from the south to southwest. On the shelf, currents reversed direction from weak southward flow to strong northward flow with a slight phase lag (∿ ½ day) behind the coastal winds. During the first 3 days of southerly winds, adjusted sea level rose by ∿ 15 cm and after 8 days reached a maximum increase of 21 cm. Approximately 2 days after commencement of southerly winds, the out-strait estuarine flow reversed direction to up-strait with typical velocities of 10-30 cm/s. Concurrent with the flow reversal at ST-15 was a rapid increase in water temperature from ∿ 10°C to ∿ 15°C, indicating that warmer coastal water was intruding up-strait. This is further confirmed by the temperature record at 20 m depth at Slip Point, some 15 km up-strait, which showed a similar increase ∿ 1 day later.

On 9 September the low-pressure system had dissipated off the coast and winds shifted to northerly. At about the same time, currents in the Strait returned to out-strait and, by 11 September, currents on the shelf had shifted to southward. Even though estuarine circulation was reestablished quickly, it took several days for the intruded coastal water to move seaward past Slip Point and

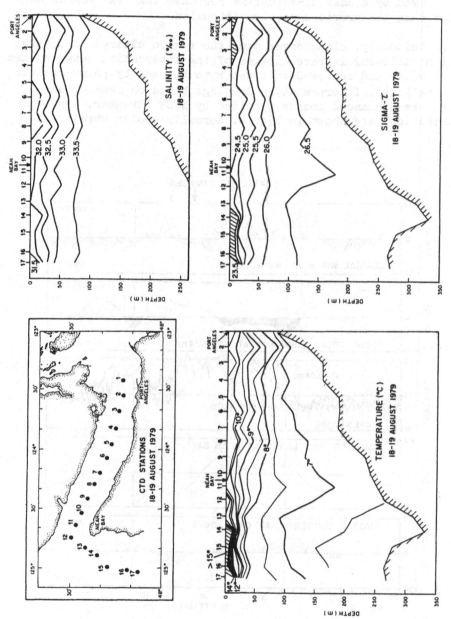

Fig. 4. Longitudinal sections of temperature, salinity and sigma-t obtained in the Strait of Juan de Fuca between 18-19 August, 1979.

ST-15. Adjusted sea level responded to the change in coastal winds by dropping to its pre-storm level by 13 September. By 14 September another low pressure system had developed with southerly winds that was followed by a coastal-estuarine response that was weaker and shorter than the previous massive intrusion.

Fortuitously, cloud cover over the region cleared on several occasions allowing infrared images of the interaction between warm coastal water and cooler Strait water to be made by polar-orbiting NOAA satellites. Interpretative drawings based on sea-surface-temperature- enhanced images provided by NESS (Breaker, personal communication) are shown in Fig. 6. Normally cooler Strait water

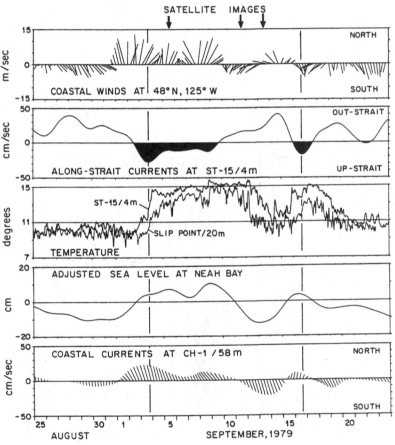

Fig. 5. Time series of coastal winds, along-strait currents at ST-15, temperature at ST-15 (4 m depth) and Slip Point (20 m depth), adjusted sea level at Neah Bay and coastal currents at CH-1 (58 m depth).

(light stippling) flows northwest onto the shelf, hugging the west coast of Vancouver Island. Also during summer months cooler water is often found within 10-20 km of the Washington Coast as a result of coastal upwelling (not shown in figure). On 5 September, five days after the start of southerly coastal winds (see Fig. 5), warm coastal water (heavy stippling) had intruded ∿ 75 km along the southern half of the Strait. By 11 September, one day after coastal winds had shifted to northerly and surface currents had begun to flow seaward, the warm-water intrusion appears to be very near its maximum extent into the Strait, past Port Angeles some 135 km from the entrance. Near the mouth, coastal water has spread across the entire Strait. However, further east a filament of cooler Strait water still exists on the north side. Three days after the southerly winds had stopped and one day after cooler water was measured at ST-15; the distribution of warm intruded water is more complex, with a warm-water pool seemingly isolated in the north-central portion of the Strait and masses of slightly cooler water intruding onto the shelf. These satellite images graphically illustrate that coastal intrusions with cross-channel scales of ∿ 10 km can penetrate well into the Strait along the southern shoreline.

DISCUSSION

The observations presented in this and other reports (Cannon, 1978; Holbrook et al., 1980b; Cannon and Holbrook, 1981; and Holbrook and Halpern, 1982) demonstrate that surface intrusions of coastal water into the Strait are a major and recurring feature of its general circulation. Intrusions, which are accompanied by reversals in both the surface and deep currents of the normally vigorous two-layer estuarine flow, occur following southwesterly coastal winds and rising sea level. The characteristics of the intruding water mass, which is invariably less-dense than ambient Strait water, change with season, being warmer during summer owing to thermal heating and less-saline during winter owing to the north-ward movement of the Columbia River plume (Barnes et al., 1972). The effects of seasonal heating and freshwater discharge into the inland waters of the Pacific Northwest are greatly reduced owing to vertical mixing of surface and deep water by strong tidal currents in the channels and passages leading into the Strait.

We feel the critical factors responsible for generating intru-sions of coastal water are the combined effects of wind-forced circulation along the coast and the presence of a less-dense coastal surface water mass. On the Washington shelf, our observations are consistent with other investigators' conclusions: that coastal wind stress fluctuations generate, through Ekman dynamics, a current and sea level response at "event" time scales of days (see the compre-hensive review of the California Current System by Hickey, 1979). Southerly winds (conducive to downwelling) force northward currents and rising sea level, whereas northerly winds (conducive to up-

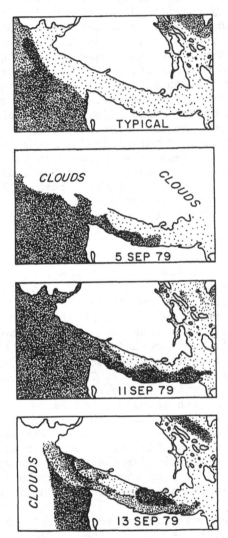

Fig. 6. Schematic diagram based on sea-surface-temperature enhanced
satellite images showing progression of warm water (heavy
stippling) intrusion into the Strait of Juan de Fuca.

welling) force southward currents and falling sea level. Klinck et al. (1981) have demonstrated, through the use of a linear two-layer numerical model, the dynamical interaction that can occur between a narrow fjord and a wind-driven coastal regime. For a wide variety of wind conditions, bottom topography and model parameters, the wind-forced coastal circulation with its geostrophic alongshore currents has a strong effect on the circulation within a fjord. However, Klinck et al. (1981) give results for model runs simulating the Strait of Juan de Fuca that indicate only a limited region of the Strait (perhaps the outer 10-30 km) is affected by a disturbance at the coast, even after five days of steady forcing. Observations in this and other papers indicate that coastal intrusions

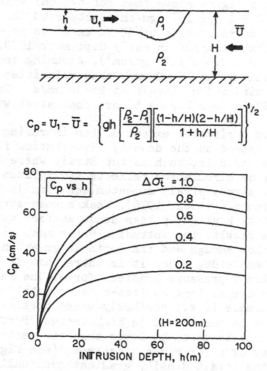

$$C_p = U_1 - \overline{U} = \left\{ gh \left[\frac{\rho_2 - \rho_1}{\rho_2} \right] \left[\frac{(1-h/H)(2-h/H)}{1+h/H} \right] \right\}^{1/2}$$

Fig. 7. Two-layer gravity current model showing contours of relative current speed C_p for different intrusion depths (h) and density differences ($\Delta\sigma_t = (\rho_2 - \rho_1) * 10^3$) in the Strait of Juan de Fuca ($H \sim 200$ m).

are major features of circulation up to 135 km into the Strait.

    Holbrook and Halpern (1982) suggest that the less-dense coastal water mass plays a critical role in the intrusion process by supplying a source of light water that, in turn, feeds a gravity-driven current that advances up-strait. According to Benjamin's (1968) hydraulic model of an intrusive gravity current, in total depth H, the relative velocity (C_p) of a lens of lighter fluid of density P_1 and depth h flowing over fluid of density ρ_2 ($>\rho_1$) is given by Fig. 7).

$$C_p = \left[g\left(\frac{\rho_2 - \rho_1}{\rho_2}\right)\left(\frac{h(1-h/H)\ (2-h/H)}{1 + h/H}\right) \right]^{\frac{1}{2}} \tag{1}$$

This simple model, adapted to the Strait, implies that if the intruding water mass is deep enough and the density difference is great enough, then the intrusion can advance up-strait against a seaward-flowing estuarine current. Shown in Fig. 7 are contours of C_p vs h for different values of $\Delta\rho = \rho_2 - \rho_1$ for the Strait (H = 200 m). This figure shows that for typical values of h (20-50 m) and $\Delta\rho$ (0.2-1.0 x 10^{-3} gm/cm^3) observed in the Strait, relative intrusion velocities of 30-70 cm/s can be expected. For the intrusion that occurred in early September 1979, $C_p \sim$ 45 cm/s (h = 25 m and $\Delta\rho$ = 0.6 x 10^{-3} gm/cm^3). Assuming the mean estuarine current to be seaward at 15-20 cm/s then velocities in that intrusion should be directed up-strait at 25-30 cm/s. The observed velocities at ST-15 (see Figure 5) are consistent with this range.

    The maximum horizontal extent of the intruding water mass will, in part, depend on the density distribution found in the Strait. In wide channels, such as the Strait where the width $\sim$ 20 km, net estuarine circulation reflects the influence of rotational effects and the across-strait momentum equation is in approximate geostrophic balance (Fissel, 1976). Weak across-strait density gradients exist with slightly less-dense surface water found to the north. As a result, the intruding water mass will advance (if its density is low enough and its depth is great enough) first along the southern side, since it is there that it "feels" the strongest up-strait pressure gradient force. The intrusion will continue to advance as long as less-dense coastal water is available at the entrance (i.e.: southerly winds persist) and density of surface water in the Strait is high enough. Surface density in the eastern Strait tends to be lower than the western Strait owing to the proximity of major river discharges (see Fig. 2). As a result, the along-strait density gradient eventually becomes too small to allow the gravity current to propagate further east against the seaward estuarine flow and the intrusion stops.

    The horizontal scale associated with coastal intrusions can be estimated by the internal Rossby radius of deformation defined by $\lambda = (g'h)^{\frac{1}{2}}/f$ where g' is reduced gravity, h is the depth of the intruding water mass and f is the coriolis parameter. Typically λ has a value $\sim$ 7 km ($\Delta\rho = 10^{-3}$ gm/cm^3; h = 50 m; f = 10^{-4}s^{-1}). Such a characteristic horizontal scale is in reasonable agreement with the across-strait intrusion scale ($\sim$ 10 km) evident in satellite images (e.g. Fig. 6).

The sea level response in the Strait to coastal forcing has two time scales: a barotropic response that propagates rapidly through the Strait at $(gH)^{\frac{1}{2}} \sim$ 44 m/s and a baroclinic response that moves up-strait with the intrusion at speeds of 20-30 cm/s. The barotropic response is reflected in the sea level records (at Neah Bay, Port Renfrew, Port Angeles and Sooke Basin) by extremely high coherence across the entire subtidal frequency band with most coherence-squared estimates > 0.90 and phase lags not significantly different (at 95% level) from zero. The baroclinic sea level signal is primarily due to the isostatic head associated with less-dense intruding coastal water. CTD observations indicate that sea level elevations (0/25 db) off the coast are typically greater than in the Strait by 10-15 dyn cm. Higher sea level elevations occur on the southern side of the Strait near the entrance at Neah Bay as the intrusion begins, resulting in along- and across-strait sea level difference fluctuations that are significantly coherent with coastal winds, along-strait currents and temperature fluctuations. The along-strait sea level difference along the southern side (Neah Bay minus Port Angeles) and across the entrance (Port Renfrew minus Neah Bay) account for 27% and 33%, respectively, of the subtidal along-strait current variance at ST-15. No significant coherence estimates were found between the along-strait sea level diffences on the north side and along-strait currents at ST-15 or the along-strait sea level difference on the south side, indicating that intrusions generally favour the south side and do not tend to occupy the entire width of the Strait. High coherence was also found between across-strait sea level difference fluctuations at the entrance (Port Renfrew minus Neah Bay) and 90 km up-strait (Sooke Basin minus Port Angeles), Indicating that intrusions frequently reach the Port Angeles region. Time series of along-strait currents at ST-15 and along-strait and across-strait sea level fluctuations during the September 1979 intrusions are shown in Fig. 8. The time series have been demeaned using the two-year average. Negative sea level differences during reversals (up-strait flow) in near-surface currents indicate higher sea levels occur on the south side and near the entrance.

The annual cycle of intrusions into the Strait can be explained in terms of seasonal variations of coastal winds and coastal water properties. During summer, northerly winds predominate, resulting in upwelling of cooler and higher salinity water along the coast and an offshore flux of less-dense surface water. Consequently, conditions are not conducive to intrusions. Occasionally

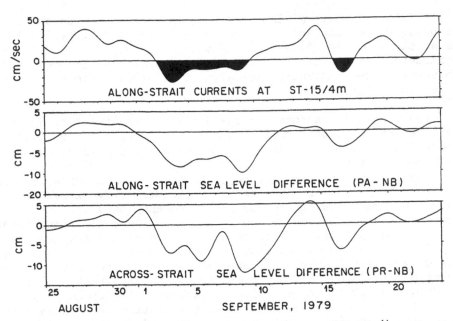

Fig. 8. Time series of along-strait currents at ST-15 (4 m depth),
and along-strait (Port Angeles minus Neah Bay) and across-
strait (Port Renfrew minus Neah Bay) sea level difference.

during summer, an atmospheric low pressure system will sit off the
coast for a few days and winds will shift to southerly, causing an
Ekman convergence near the coast and the appearance of warm and
less-dense water near the entrance to the Strait. Brief intrusions
will occur during such periods if the coastal surface water is warm
enough to establish a critical up-strait density gradient. During
September and October, water off the coast reaches its maximum temperature and autumn storms and southerly coastal winds become more
frequent. These conditions generally result in major intrusion of
warm water well into the Strait (as illustrated in Fig. 6). As
winter advances, coastal water cools and southerly winds push the
Columbia River plume northward along the Washington Coast and,
consequently, coastal surface water is less-dense than Strait surface water owing to Columbia River discharge. (Recall that fluctuations in Strait surface water salinity are not as great as off the
coast because of extensive vertical mixing of surface and deep
water by strong tidal currents in the channels to the east). Thus,
between November and May, frequent cyclonic storms cause low salinity surface water to move near the entrance of the Strait, establishing a source of less-dense water that can flow into the Strait
as a density current.

SUMMARY

    Two-year measurements of currents, water properties, winds and
sea level near the entrance to the Strait of Juan de Fuca have been
briefly reviewed. These observations along with satellite imagery
and CTD surveys show that surface intrusions of less-dense coastal
water into the Strait are a major and recurring feature of its general circulation throughout the year. These intrusions, which are
accompanied by reversals in the normally vigorous two-layer estuarine flow, are coherent with southwesterly coastal winds (conducive
to coastal downwelling) and rising sea level at the entrance. The
data suggest that a wind-induced shoreward Ekman flux along the
coast "pushes" less-dense coastal water into the Strait where it
feeds a complex 3-dimensional density driven flow which has been
observed to travel as far as 135 km up-strait. The isostatic head
associated with the intruding coastal water mass establishes along-
strait and across-strait sea surface slope fluctuations that are
significantly coherent with reversals in current and coastal winds.

ACKNOWLEDGEMENTS

    We wish to thank Mr. Larry Breaker of NOAA's National Environ-
mental Satellite Service for providing sea-surface-temperature-
enhanced images of the warm water intrusions and Dr. W. Crawford
of Canada's Institute of Ocean Sciences for furnishing sea data
along Vancouver Island. This work was jointly sponsored by NOAA's
Puget Sound Project Office of the Marine Ecosystems Analysis (MESA)
Program and the Pacific Marine Environmental Laboratory (PMEL).
Contribution No. 629 from PMEL/NOAA.

REFERENCES

    Bakun, A., 1975, Daily and weekly upwelling indices, west coast
        of North America, 1967-73, NOAA Tech. Rept. NMFS SSRF-693.
        United States Dept. of Commerce, 114 pp.
    Barnes, C.A., Duxbury, A.C., and Morse, B., 1972, Circulation
        and selected properties of the Columbia River effluent
        at Sea, in: "The Columbia River Estuary and Adjacent
        Ocean Waters," A.T. Pruter and D.L. Alverson, eds.,-
        University of Washington Press, Seattle, 41-80.
    Benjamin, T.B., 1968, Gravity currents and related phenomena,
        J.Fluid.Mech., 31(2):209-248.
    Cannon, G.A., editor, 1978, Circulation in the Strait of Juan
        de Fuca; some recent oceanographic observations, NOAA
        Tech. Rept. ERL-PMEL 29, Supt. of Documents, U.S. Govt.
        Printing Office, Washington, D.C. 20402, 49pp.
    Cannon, G.A., and Holbrook, J.R., 1981, Wind-induced seasonal
        interactions between coastal and fjord circulation, in:
        "The Norwegian Coastal Current," R. Sætre and M. Mork,
        eds., University of Bergen, Bergen, Norway, 131-151.

Fissel, D.E., 1976, Pressure differences as a measure of
        currents in Juan de Fuca Strait, Pacific Marine Science
        Rept. 76-17, Institute of Ocean Sciences, Patricia Bay,
        Victoria, B.C., 63 pp.

Frisch, A.S., Holbrook, J.R., and Ages, A.B., 1981, Observa-
        tions of a summertime reversal in circulation in the
        Strait of Juan de Fuca, J. Geophys. Res., 86(C3):2044-
        2048.

Godin, G., 1972, "The analysis of tides," University of Toronto
        Press, Toronto, Canada.

Hickey, B.M., 1979, The California Current System- hypotheses
        and facts, Progress in Oceanography, 8:191-280.

Holbrook, J.R., and Halpern, D., 1977, Observations of near-
        surface currents, winds and temperature in the Strait of
        Juan de Fuca during November 1976-February 1977, Trans.
        of the American Geophysical Union, 58(12):1158.

Holbrook, J.R., Muench, R.D., and Cannon, G.A., 1980a, Seasonal
        observations of low-frequency atmospheric forcing in the
        Strait of Juan de Fuca, in: "Fjord Oceanography,"
        H.J. Freeland, D.M. Farmer and C.D. Levings, eds., Plenum
        Publishing Corp. New York, 305-317.

Holbrook, J.R., Muench, R.D., Kachel, D.G., and Wright, C.,
        1980b, Circulation in the Strait of Juan de Fuca: recent
        oceanographic observations in the eastern basin, NOAA
        Tech. Rept. ERL 412-PMEL 33, Supt. of Documents, U.S.
        Govt. Printing Office, Washington, D.C. 20402, 42 pp.

Holbrook, J.R., and Halpern, D., 1982, Wintertime near-surface
        currents in the Strait of Juan de Fuca, Atmospher-Ocean,
        20(4): (In press).

Klinck, J.M., O'Brien, J.J., and Svendsen, H., 1981, A simple
        model of fjord and coastal circulation interaction,
        J. Phys. Oceanogr., 11:1612-1626.

Niebauer, H.J., 1980, A numerical model of circulation in a
        continental shelf-silled fjord coupled system, Estuarine
        and Coastal Marine Science, 10:507-521.

Redfield, A.C., 1950, Note on the circulation of a deep
        estuary-the Juan de Fuca-Georgia Straits, Proc. Colloquim
        Flushing of Estuarines, Woods Hole Oceanog. Inst.,
        Woods Hole, Mass., 175-177.

Smith, N.P., 1978, Long-period, estuarine-shelf exchanges in
        response to meteorolical forcing, in: "Hydrodynamics of
        Estuaries and Fjords," J.C.J. Nihoul, ed., Elsevier
        Scientific Publishing Company, Amsterdam, 147-159.

Svendsen, H., 1977, A study of the circulation in a sill fjord
        on the west coast of Norway, Mar.Sci.Comm., 3(2):151-209.

Wang, D.P., and Elliot, A.J., 1978, Non-tidal variability in
        the Chesapeake Bay and Potomac River: Evidence for non-
        local forcing, J. Phys. Oceanogr., 8:225-232.

WATER EXCHANGE BETWEEN THE SEA AND COMPLICATED FJORDS WITH SPECIAL REFERENCE TO THE BALTIC WATER EXCHANGE

Anders Stigebrandt

Department of Oceanography
University of Gothenburg
Gothenburg, Sweden

ABSTRACT

Two-layer exchange between fjords of different kinds and the sea is discussed. The complicated exchange between the Skagerrak and the Baltic is discussed at some length and a model is briefly presented. The model is forced by the water level variations in the Kattegat, the freshwater supply to the Baltic and the regional wind over the Baltic entrance area. Salinities and pycnocline depths are calculated for the Kattegat and the Belt Sea for a period one and a half year in length.

INTRODUCTION

As the Baltic entrance model and the prototype both arc rather complex it was felt that an outline of the model would benefit from a preceding broad description of the estuarine circulation in more simple systems. The complex system under interest is in this manner hopefully approached in a logical and comfortable way.

STEADY ESTUARINE CIRCULATION IN FJORDS

The simplest kind of estuarine circulation in fjords occurs in so called N-fjords (see Stigebrandt, 1981, for the classification scheme) where there is a thin brackish layer overlying a thick layer of sea water, see Fig. 1a. The stratification in N-fjords is assumed to be dynamically controlled by the mouth where the flow is supposed to be critical with respect to a densimetric Froude number. In the paper mentioned a simple analytical model for the stratification in N-fjords was presented. The model makes use of a dynamical control condition at the mouth ($F_{d1}^2 = u_i^2/g'h_1 = 1$,

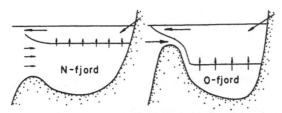

Fig. 1. Some important features of N- and O-fjords are shown in a)
and b) respectively. The arrows through the pycnoclines
indicate flow of dense water caused by the mixing
processes.

where $u_1(h_1)$ is the velocity (thickness) of the upper layer in the
mouth and g' is the buoyancy parameter), a generally accepted formula
for winddriven entrainment flows in two-layer stratification (Kato
& Phillips formula) and a simple model for the flow from the interior of the fjord to the mouth. The theory is believed to be especially good for wide fjords (wide compared to the width of the
mouth).

In fjords with shallow mouths the depth of the upper layer in
some cases is not small compared to the sill depth. Then also the
flow in the lower layer in the mouth is dynamically important and
the control condition is here the wellknown Stommel & Farmer condi-
tion ($F_{d1}^2 + F_{d2}^2 = 1$, where $F_{di}^2 = u_i^2/g'h_i$ and $u_i(h_i)$ is the velocity (thickness) of the i:th layer in the mouth, $i = 1.2$). The
consequences of this condition for steady estuarine circulation
were originally investigated by Stommel & Farmer and later for
example the present author (Stigebrandt, 1981). It can be shown
that there is a maximum for the transport of sea water into the
fjord (the so-called overmixed state. Fjords in this state may be
termed O-fjords). The physical interpretation of this state is that
the transport capacity of the mouth with respect to two-layer
exchange is fully utilized, see Stigebrandt (1977, 1981).

INFLUENCE OF BAROTROPIC CURRENTS IN THE FJORD MOUTH UPON THE
ESTUARINE CIRCULATION

Barotropic currents in the mouth, caused by sea level variations outside the fjord, may disturb the steady circulation sketched

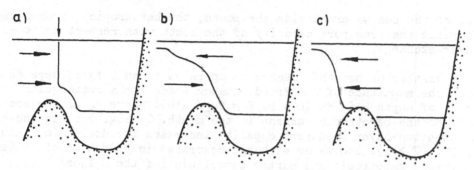

Fig. 2. The hydrographic situation in the fjord mouth for three
different phases of the barotropic current: a) maximum
inwards (note the so called tidal intrusion front, marked
by an arrow, inside the mouth), b) zero and c) maximum
outwards.

above. Serious disturbances will occur when the barotropic currents
have amplitudes larger than the baroclinic velocities calculated
for the steady estuarine circulation. The dynamic control in the
mouth is then eliminated for shorter or longer periods by the
fluctuating barotropic current and during these periods there is
essentially only one water mass present in the mouth, see Fig.2.
This process was described in detail by Stigebrandt (1977). For
cases where the pycnocline in the fjord is situated below the sill
level and where the brackish water forms only a very thin layer on

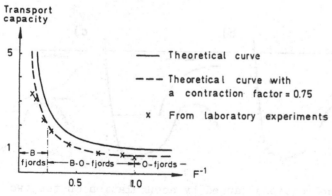

Fig. 3. The relationship between the normalized transport capacity
and F^{-1}. (From Stigebrandt, 1977).

top of the sea water outside the mouth, the barotropic currents may increase the transport capacity of the mouth with respect to two-layer exchange.

Whether or not this mechanism operates in an O-fjord depends upon the magnitude of the Froude number which for a rectangular mouth of depth H is defined by $F = 2u_b/(g'H)^{\frac{1}{2}}$ where u_b is the amplitude of the barotropic current in the mouth. In Fig. 3 it is shown how the two-layer transport capacity increases for decreasing values of F^{-1} (F > 1). The water exchange described in this kind of system is partly barotropic and partly baroclinic (of the O-fjord type). Obviously the incomplete dynamical classification scheme suggested by the present author may, by the use of F, be extended to include B-fjords (fjords with F > 2 where the two-layer exchange thus is executed mainly by the barotropic flow). This regime is indicated in Fig.3.

COMPLICATIONS BECAUSE OF ACCUMULATION OF BRACKISH WATER OUTSIDE THE MOUTH AND SEAWATER INSIDE THE MOUTH

One complication that may occur in B-fjords is that the out-flowing brackish water creates a permanent, not thin upper layer outside the mouth. During barotropic inflows there will then be a transport of brackish water back into the fjord, see Fig.4. Some of the brackish water thus recirculates and the transport capacity of the mouth with respect to two-layer exchange thereby decreases. In order to determine the water exchange for this case one apparently has to know the thickness of the upper brackish layer outside the mouth. Hence, for this case there is a need to model the circulation of the brackish water in the area outside the mouth. If the area of importance is rather large, further complications such as

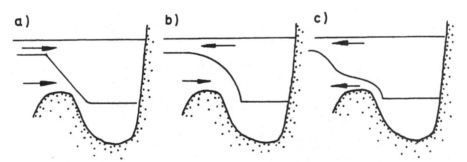

Fig. 4. Complications caused by accumulation on the two sides of the sill illustrated for three different phases of u_b: a) maximum inwards, b) zero and c) maximum outwards.

windmixing outside the mouth, raising the salinity in the brackish
layer there, may occur.

     The water exchange between the Baltic and Skagerrak is of the
kind depicted here. The area outside the mouth, which is covered by
a brackish layer of substantial thickness, coincides approximately
with the area of the Belt Sea and the Kattegat, see the map in Fig.6.

     A similar complication in B-fjords may occur if the bottom of
the fjord has only a small slope from the sill and inwards. This
means that some of the dense water brought into the fjord across
the sill may be swept back again by a strong barotropic current
directed outwards. This complication also occurs in the Baltic where
the bottom over a large area slopes very little from the sill (the
Darss Sill) and inwards.

ROTATIONAL EFFECTS

     Up to now we have not discussed possible effects of the rota-
tion of the earth upon the flow in a fjord mouth. If the width of
the mouth is approximately equal to (or larger than) the internal
radius of deformation, r_{Roi}, (calculated from the stratification in
the mouth) rotational effects can not be neglected. Experience shows
that currents tend to be confined to the right hand bank (in the
northern hemisphere seen in the direction of the current). The width
of the current is often comparable to r_{Roi}. For sufficiently wide

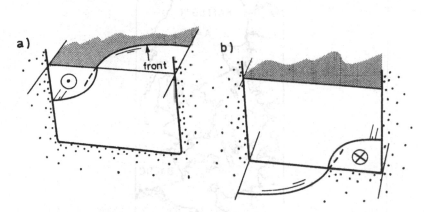

Fig. 5. Flow in the mouth of rotating fjords (as seen into the
        fjord): a) outflow from a fjord with a shallow pycnocline
        and b) flow into a fjord with a deep pycnocline.

mouths fjords with shallow pycnoclines will have outflow into a
coastal current at the right hand bank whereas fjords with a deep
halocline have an inflow which also is confined to the right hand
bank like a submerged dense coastal current, see Fig. 5.

For a two-layer stratification the geostrophic transport in a
coastal current is easily obtained from the stratification by integration of Margules' formula. In the "lucky" cases where the ambient
fluid may be considered as motionless we arrive at a unique relation
between the transport and the upstream stratification. This transport
is the rotational two-layer transport capacity of the mouth. This
concept was utilized by the present author in a model for the
Arctic Ocean upper layer (Stigebrandt, 1981b) and it is also used
in the present model for the exchange between the Baltic and the
sea.

A MODEL FOR THE WATER EXCHANGE BETWEEN THE BALTIC AND THE SKAGERRAK

For many of the cases described in the introduction there
should exist one section containing a dynamical control of the outflow through the mouth. However, the accumulation of brackish water
outside the mouth and seawater inside the mouth are indications of
downstream controls of the brackish layer and the sea water respec-

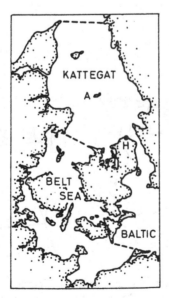

Fig. 6. Map of the Baltic entrance area. A = Anholt; H = Hornbaek.

tively. Thus, the control section for these cases is split into two
horizontally separated control sections. In the Baltic system, for
instance, the control of the outflowing brackish layer is found in
the northern Kattegat, where a rotational-baroclinic control is
assumed to exist. The control of the inflowing denser water is to
be found in the transition area between the Baltic and the Belt Sea.
This large-scale view of the Baltic water exchange constitutes the
basis for the model below. The model is presented in detail in
Stigebrandt (1982).

SHORT DESCRIPTION OF THE SYSTEM

The Baltic receives as a long-term mean 15.000 m^3/s of freshwater from rivers. On an annual time-scale evaporation annihilates
precipitation over the sea surface. The Baltic proper has a vertically homohaline surface layer (salinity about 7-8 o/oo about 50 m
in thickness. Below this there is an approximately 10 m thick halocline overlying the deep water (about 11-13 o/oo). The sill depth
between the Baltic and the Belt Sea is 18 m (the Darss Sill). There
is one direct connection between the Baltic and the Kattegat through
the Öresund. However, the sill depth here is only 8 m and the vertical cross-sectional area is a factor 0.25 times those encountered
in the narrows in the Belt Sea. The mean depths of the Belt Sea and
the Kattegat are 13 and 26 m respectively. As a long-term mean the
upper layer in the Belt Sea (Kattegat) is approximately 10 (15) m
thick with a salinity of the order of 15 (25) o/oo. Below the
pycnocline the typical Belt Sea (Kattegat) salinity is 18 (33) o/oo.
However, the time variations of the stratification in the two seas
are very large. Several fronts are usually found at the sea surface.
One of these is the northern Kattegat front (separating the Kattegat
surface water from Skagerrak water). Other fronts, of the tidal
intrusion type, are found in the narrows of the Belt Sea and inside
the Darrs Sill during barotropic inflows to the Baltic.

OUTLINE OF THE MODEL

The instantaneous transports in the Danish Sounds may by an
order of magnitude exceed those caused by the mean freshwater supply.
From a frictional model, driven by the sea level fluctuation in the
Kattegat (Hornbaek) and the freshwater supply to the Baltic, the
present author (Stigebrandt, 1980) obtained a good description of
these barotropic flows. As the local value of the Froude number F
is appreciably larger than one, both in the narrows of the Belt Sea
and on the Darrs Sill, the water exchange between the Belt Sea and
the Kattegat and the Baltic respectively is largely forced by the
barotropic flow.

In order to calculate the barotropically forced flow of salt,
when knowing the transport, one has to know the vertical stratifi-

cation. Therefore it is necessary to construct models for the vertical stratification in the Belt Sea and the Kattegat. In this first
version the stratification was assumed to be of the two-layer type.
The two seas are treated as horizontally homogeneous with respect
to the vertical stratification. Thus, as the boundaries between the
different seas there are salinity discontinuities that quite well
coincide with the known mean frontal positions in the system.

The complex model for the exchange of water and salt thus
consists of three sub-models: (i) the frictional model for the
barotropic exchange mentioned above, (ii) a model for the stratification in the Kattegat and (iii) a model for the stratification in
the Belt Sea. The Baltic is regarded as a reservoir with homogeneous
salinity as the pycnocline there is situated well below the sill.
Also the Skagerrak is considered as a homogeneous reservoir with
respect to salinity.

The lower layer in the Kattegat model is considered to be a
passive reservoir of Skagerrak water. The upper layer of the Kattegat model has some important dynamical properties. The northern
Kattegat front is assumed to be stationary and oriented essentially
in the east-west direction. However, near the Swedish coast the
front becomes parallel to the coast and between the coast and the
front a coastal current runs northward in the Skagerrak, draining
the surface layer of the Kattegat. The transport of this current
is given by the rotational-baroclinic transport capacity as explained in Introduction. In the model the wind over the Kattegat creates
entrainment flows from the lower to the upper layer, thus raising
the salinity of the Kattegat surface layer. Possible downward
entrainment flows caused by bottom currents below the pycnocline
are neglected in the Kattegat.

In the Belt Sea model the lower layer consists of essentially
surface water from the Kattegat (forced into the Belt Sea by barotropic currents) and water entrained from the upper layer in the
Belt Sea (in the model this entrainment is driven by the bottom
turbulence in the lower layer generated by the barotropic flow).
When the barotropic current switches from one direction to the other,
the salinity of the water crossing the boundary between the Kattegat
and the Belt Sea only gradually attains the value expected from the
vertical stratification in the Kattegat and the Belt Sea respectively.
This feature is accounted for in the model where a buffer volume is
inserted between the Kattegat and the Belt Sea. (A similar buffer
volume is also inserted between the Belt Sea upper layer and the
Baltic). Water from the dense lower layer in the Belt Sea flows
over the Darrs Sill. This flow is assumed to be in geostrophic
balance. Of course the wind may, by entrainment of water from the
lower layer in the Belt Sea, create a relatively salt upper layer
in the Belt Sea. A sketch of the complete model is given in Fig. 7.

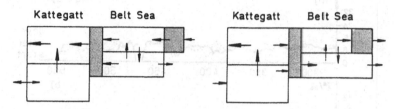

Fig. 7. A sketch of the model. The shaded boxes are the buffer
        volumes. a) barotropic flow out of and b) into the Baltic.

SOME RESULTS

        The Kattegat and the Belt Sea models were coupled and the
resulting model was integrated for a period of a year and a half
in length from a given set of initial conditions (starting 1 July,
1975). The model is driven by the observed wind (from Anholt) and
calculated barotropic flows (from the frictional barotropic model
driven by the water level variations in the Kattegat and the freshwater supply to the Baltic). The computed depths of the pycnoclines
and salinities for the Kattegat and the Belt Sea for the second
half of the period are shown in Fig. 8. The computed quantities can
be compared to measured ones from the light vessels Anholt Knob
(in the Kattegat) and Fehmarnbelt (in the Belt Sea). These light
vessels measure salinity at standard depths once a day. As there is
much high-frequency internal waves in the Baltic entrance area the
observed depths of the pycnoclines were smoothed by applying a 7
days moving average.

        As can be seen from Fig. 8 the model picks the right level of
salinities and pycnocline depths in the system. This could not have
been achieved in a model where the parameterization of the flows
was very strong. Also much of the low-frequency activity is well
modelled. Note that the model is integrated for a period which is
10-20 times longer than the residence time for water parcels in the
layers without "loosing the track". The output from a horizontally
integrated model should ideally be compared to horizontally averaged field data. Therefore some of the discrepancy between the results
of the model and the field data shown in Fig. 8 can be said to be
caused by incomplete sets of field data.

        In Stigebrandt (1982) a sensitivity analysis of the system was
performed. The effects of short term (1-19 years) changes in the
external parameters (barotropic flow, freshwater supply to the

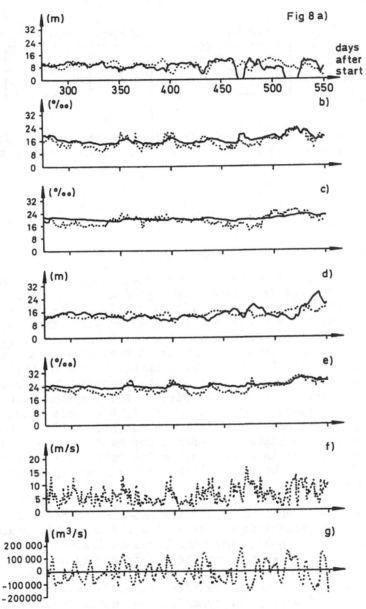

Fig. 8. This figure shows computed (——) and measured quantities:
a) thickness of the upper layer in the Belt Sea, b) salinity
of the upper layer in the Belt Sea, c) salinity of the lower
layer in the Belt Sea, d) thickness of the upper layer in the
Kattegat, e) salinity of the upper layer in the Kattegat, f)
wind speed (mixing wind) computed barotropic flow into/out
of the Baltic (positive inwards).

Baltic, regional winds and salinity of the Skagerrak water) upon
the stratification in the model area were investigated. Also long-
term changes in the external parameters upon the surface salinity
of the Baltic were investigated. The results of the sensitivity
analysis seem to be quite realistic which further supports the
conclusion that the model contains the basic physics of the system
and, besides, that the physical processes are adequately parame-
rized.

ACKNOWLEDGEMENT

        This work has been partially founded by the Swedish Natural
Science Research Council (NFR).

REFERENCES

    Stigebrandt, A., 1977, On the effect of barotropic current
        fluctuations on the two-layer transport capacity of a
        constriction, J. Phys. Oceanogr., 7:118-122.
    Stigebrandt, A., 1980, Barotropic and baroclinic response of
        a semi-enclosed basin to barotropic forcing from the
        sea, in: "Fjord Oceanography", H.J. Freeland, D.M. Farmer
        and C.D. Levings (eds.), Plenum Publishing Corp.,
        New York.
    Stigebrandt, A., 1981, A mechanism governing the estuarine
        circulation in deep, strongly stratified fjords,
        Est. Coast. and Shelf Sci., 13:197-211.
    Stigebrandt, A., 1981b, A model for the thickness and salinity
        of the upper layer in the Arctic Ocean and the relation-
        ship between the ice-thickness and some external para-
        meters, J. Phys. Oceanogr., 11:1407-1422.
    Stigebrandt, A., 1982, A model for the exchange of water and
        salt between the Baltic and the Skagerrak, J. Phys.
        Oceanogr.,13(3):
    Stommel, H., and Farmer, H.G., 1953, Control of salinity in an
        estuary by a transition, J. Mar. Res., 12:13-20.

SHELF-FJORD EXCHANGE ON THE WEST COAST OF VANCOUVER ISLAND

Dario J. Stucchi

Institute of Ocean Sciences
P.O. Box 6000
Sidney, B.C., V8L 4B2
Canada

INTRODUCTION

This paper deals with the circulation in a deep silled fjord that is forced by processes occurring in the adjacent coastal sea. The importance of the offshore forcing to the adjacent fjords was recognized early on, but it is only within the last 12 years that this interaction has received increased attention. For the Norwegian fjords, Gade (1976), Svendsen (1977, 1980) and Helle (1978) to name a few, have indicated and reported on the coupling between the offshore density fluctuations and water exchanges in the adjacent fjords, as have Muench and Heggie (1978) for the Alaskan subarctic fjords and Holbrook et al., (1980) for the Strait of Juan de Fuca. In particular, it is the above sill level or intermediate water exchanges that are the subject of this contribution. The investigation of the interaction between a fjord and the adjacent coastal sea was made possible because of a concurrent study of the coastal ocean dynamics on the continental shelf off the west coast of Vancouver Island during the period from January 1979 to June 1981.

The particular fjord is Alberni Inlet, a long (70 km), narrow (2 to 3 km) fjord located on the southwestern coast of Vancouver Island, B.C. (see Figure 1). The fjord comprises two main basins, a 15 km long inner basin of maximum depth 130 m which is separated from the outer basin by a sill at 40 m depth. The outer basin is considerably larger, about 45 km in extent with maximum depths of 350 m. The outer basin is connected to the Pacific Ocean over a short, shallow 30 m sill at the seaward end of Trevor

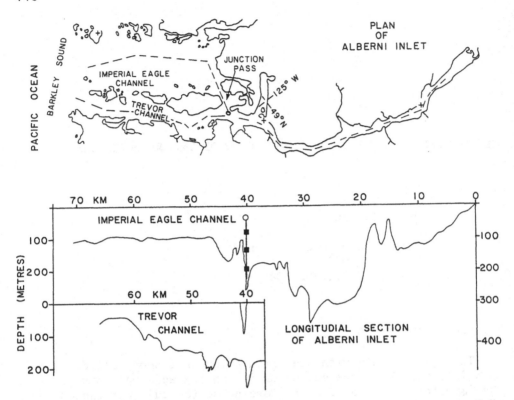

Figure 1. Plan and longitudinal profile of Alberni Inlet. The "+"
          indicates locations of bottom mounted pressure gauges
          and "o" indicates the location of the current meter
          mooring.

Channel, but the main deep passage (a constant 100 m) is via
Junction Passage and Imperial Eagle Channel, a distance of 30 km to
the Pacific Ocean.

        This short contribution will begin with the presentation and
description of some of the observations of the exchanges between
the fjord and adjacent coastal sea. Next, the magnitude and
relative importance of these exchanges in the context of the overall fjord circulation will be examined. An analysis of the relationship between the coastal wind driven regime and these exchanges will be presented and discussed. Finally, a simple numerical
model of these interactions, developed by Klinck et al.(1981)
will be briefly described and compared to the observations.

OBSERVATIONS

        The data presented here are Aanderaa current meter data from a mooring located on the inner slope of the sill, just beyond the eastern end of Junction Passage (see Fig.1). This sub-surface mooring had three current meters attached, one above sill level at 80 m and the other two below sill level at 140 m and 200 m. After October 1980 this mooring was moved into Junction Passage, with instruments placed at 35 m, 65 m and 95 m - all above sill level. The data from this last mooring, though not displayed here, will be included in the discussion. Density data were also obtained from all the current meters, but these data are not presented because of the poor long term stability of conductivity measurements. However, short term fluctuations in density are trustworthy and these will be briefly discussed. In addition to these moorings, three bottom mounted pressure gauges were deployed, one in the inner basin of Alberni Inlet, another in the outer basin about near Junction Passage and the last one at the mouth of the fjord in Barkley Sound, see Fig. 1 for locations.

The current data presented in Fig. 2 have been low- pass filtered with a cosine-Lanczos filter (half power point at 40 hours) and then transformed to flow directed into or out of the fjord. All the data from the Junction Passage mooring presented in Fig. 2 are biased towards inflow conditions. In the case of

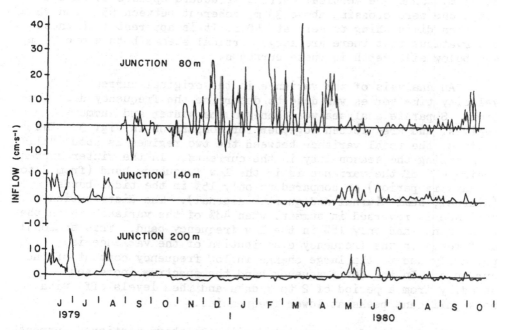

Figure 2. Plot of low passed inflow/outflow currents measured from 80 m, 140 m and 200 m depth in Alberni Inlet, just east of Junction Passage.

the instruments at 140 m and 200 m this can be explained in terms
of a density current inflowing during deep water renewal, which
occurs in the period from May to August. The large currents
observed above sill depth, at 80 m, during the winter months are
biased because of the location of the mooring in relation to
Junction Passage. The mooring was positioned just outside the
eastern end of Junction Passage, where inflows currents are larger
than outflows. Later measurements obtained in Junction Passage
show no such bias, with flow being equally large in both inflow
and outflow conditions.

There are several noteworthy features in the data from above
sill level. One is the marked change in magnitudes between the
summer and winter current regimes. During winter months inflow
speeds as high as 40 cm s^{-1} have been recorded, with typical
maximums during inflows of 15 to 25 cm s^{-1}. By contrast, the
currents are weaker in summer, with speed generally less than
5 cm s^{-1}. The vertical structure of these energetic winter
currents is also noteworthy. The strong currents do not extend
down to 140 m, as can be seen from the February to April, 1980
period. Current measurements made after October 1980, in Junction
Passage at 35 m, 65 m and 95 m show that currents are essentially
identical at 65 m and 95 m but weaker and sometimes out of phase
at 35 m. Thus the vertical current structure appears to have at
least one zero crossing above 35 m, coherent between 65 m and 95 m
and then diminishing to zero at 140 m. It is apparent from these
observations that there are large vertical shears both above 35 m
and below sill depth in these currents.

An analysis of the variance in the original current
velocity time series was done to determine the frequency distribution. Separate analyses were done on the winter and summer
regimes, and are plotted together for comparison in Fig. 3. The
ratio of the total variance between the two regimes is about 8:1,
emphasizing the seasonality in the currents. In the winter current
regime 67% of the variance is in the low frequency band (from 2 to
40 days in period) as compared to only 15% in the tidal bands
(diurnal, semi-diurnal and quarter-diurnal). The distribution of
variance is reversed in summer, when 48% of the variance is in the
tidal bands and only 18% in the low frequency band. This seasonal
difference in the frequency distribution of the variance is
primarily due to the large change in low frequency content of the
currents. In the low frequency band the spectrum increases
steadily from a period of 2 to 7 days and then levels off, with
no significant peaks at lower frequencies.

The density data associated with the above mentioned current
data exhibit fluctuations that are coherent with the fluctuating
current. The relationship is simple, the density at 80 m

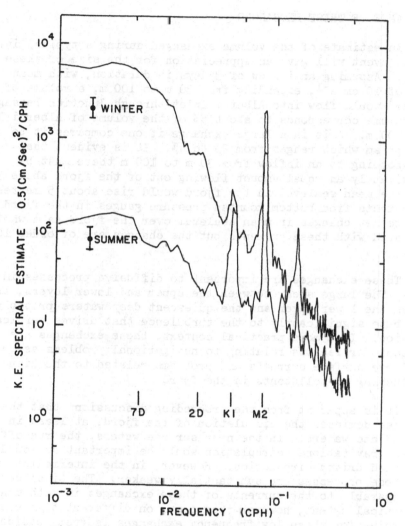

Figure 3. Kinetic energy spectra of the winter and summer
current regimes measured at 80 m depth near Junction
Passage. Confidence intervals at the 95% level are
indicated.

increases during inflows and decreases during outflow. The relative magnitude of the density fluctuations varies with stratification but the simple relationship mentioned above still holds.

IMPORTANCE OF THESE EXCHANGES

        An estimate of the volume exchanged during a typical inflow/
outflow event will give an appreciation for the size of these
events. Assuming an inflow of 3 days in duration, with mean
speed of 10 cm s^{-1}, extending from 30 m to 100 m, a volume of
0.75 km^3 would flow into Alberni Inlet through Junction Passage.
This volume corresponds to about 5% of the volume of Alberni Inlet
above 100 m. This is a large exchange if one compares it to the
tidal prism which ranges from 2% to 4%. It is evident that
corresponding to an inflow from 30 m to 100 m there must be
approximately an equal amount flowing out of the fjord above 30 m,
otherwise mean sealevel in the fjord would rise about 5 metres.
Measurements from bottom mounted pressure gauges in the fjord do
indicate net changes in mean sealevel over the fjord as a whole
associated with these currents but the changes are of order 10 or
20 cm.

        These exchanges are important to diffusive processes in the
fjord. The large shear between the upper and lower layers, and
between the lower layer and the quiescent deep waters probably
contribute significantly to the turbulence that drives vertical
diffusion. In a more practical context, these exchanges are
important for reasons relating to navigational problems associated
with large surface currents and problems related to the dispersion
and flushing of pollutants in the fjord.

        It is apparent from the proceeding discussions that these
exchanges dominate the circulation of the fjord, at least in the
intermediate waters. In the near surface waters, the run-off
induced gravitational circulation would be important as would the
local wind driven circulation. However, in the intermediate waters
both these processes are substantially weaker. The tidal currents
are comparable to the currents of these exchanges in both magnitude
and vertical extent, however they occur on different time scales.
The forcing for these low frequency exchanges is from outside the
fjord as it is difficult to conceive of any local forcing that
could produce such an intense response with the observed seasonal
variability.

FORCING OF SHELF-FJORD EXCHANGE

        Fluctuations in the stratification of the continental shelf
at the mouth of Alberni Inlet would force the fjord to adjust its

density field to match conditions at its mouth. The observed
changes in the density and current at Junction Passage support
this. The question that arises is what coastal ocean dynamics are
responsible for these fluctuations? It is evident from the
observations that the forcing must have a distinct seasonal
variation, and fluctuate with the time scale of the meteorology.
A number of processes such as wind driven upwelling, mixing over
shallow banks, and coastal currents could alter the stratification
in the coastal sea.

     Coastal winds were examined since they exhibit a marked
seasonal variation and fluctuate on the time scale of passing
weather systems. Coastal winds measured at Estevan Point, 120 km
up the coast from Alberni Inlet were used in the analysis.
Coastal winds were available from other closer locations but they
were biased by local topography. Estevan Point protrudes out from
the coast about 8 km on a low lying peninsula.

     Simple Ekman dynamics of wind induced upwelling were used to
examine the relationship between the coastal winds and the
exchanges in the fjord. The intensity of the wind induced
upwelling was computed in a manner similar to that used by Bakun
(1973). The hourly Estevan winds were used to compute the stress,
τ, at the sea surface according to

$$\overline{\tau} = \rho_a C_D |\overline{U}|\overline{U}$$

where ρ_a is the density of air, C_D is an empirically determined
drag coefficient and $\overline{U}$ is the wind velocity. Next, the mass
transport or Ekman transport, $\overline{M}$, for the surface layer is computed
from the following equation

$$\overline{M} = \frac{1}{f}\,\overline{\tau} \times \hat{k}$$

where f is the Coriolis parameter and k is the unit vector directed
vertically upward. Bakun (1973) called the component of $\overline{M}$ perpendicular to the coast the upwelling index. Negative upwelling
indices (downwelling) indicate a coastal convergence, an onshore
Ekman transport which tends to pile up surface waters at the coast.
During upwelling (positive indices) surface waters are transported
offshore bringing denser water to the coast and depressing sea
levels.

     Hourly upwelling indices computed from the Estevan winds were
averaged over a 24 hour period. For the purpose of comparison
these indices are plotted in Fig. 4 together with the current
from Junction Passage. Some visual correlation is discernible,
during downwelling periods the currents flow out of the fjord and
when downwelling relaxes or turns to upwelling inflow conditions
occur. The correlation is better when the currents lag the upwelling index by 1 or 2 days.

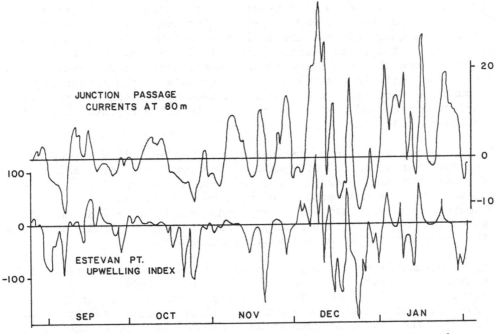

Figure 4. Plot of low passed currents measured near Junction
          Passage and daily averaged upwelling index from
          Estevan Point.

        A cross-spectral analysis of the two time series was under-
taken for the winter regime only. The computed coherence squared
(hereafter coherence) spectrum and phase spectrum are plotted in
Fig. 5. The coherence is plotted on an arctanh scale so that
confidence intervals are constant across the spectrum. The dashed
line in the coherence plot corresponds to the 95% level for the
null hypothesis test or 95% noise level as given by Julian (1975).
Confidence intervals for phase are not included because they
depend upon the coherence and number of degrees of freedom, such
that at zero coherence the confidence intervals are ± 180°.

        The coherence is generally above the 95% noise level,
indicating that there is a significant similarity between these
two time series. However, because of the large 95% confidence
intervals, most of the features or structure in the coherence
spectrum are not statistically significant. The phase spectrum
shows a marked linear trend, with relatively little scatter in
regions of low coherence estimates. There are features in the
phase but these are not significant. The negative phase angles
indicate that the currents lag the upwelling indices. The linear

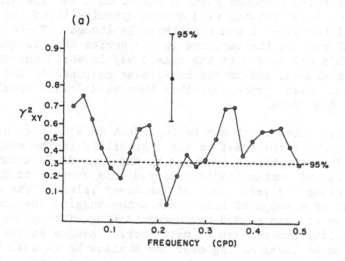

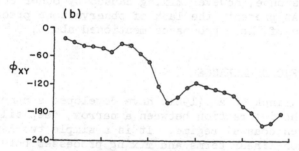

Figure 5. Coherence squared spectrum a) and phase spectrum b) between Estevan Point upwelling index (Series X) and Junction Passage currents (Series Y) for the winter regime.

trend in the phase spectrum reveals that there is a constant lag
of about one day across the spectrum.

The above analysis confirms that there is a correlation between
the coastal winds and the low frequency currents at Junction
Passage, although the coherence is far from perfect and features are
not well resolved. The phase relationship is in the correct sense,
in that the fjord responds about a day or so after the wind events.
The time scale for coastal wind driven dynamics is of the order of
an inertial period - 16 hours at these latitudes. There is still a
substantial part of the variance in the series that is uncorrelated.
In part, this may be due to the biased way in which the fjord response was measured, and to the non-linear response of the fjord to
its forcing. Also, processes other than wind driven upwelling could
be forcing the fjord.

One mechanism which may be important in altering the
stratification at the coast is the variability in the coastal
current. This current is produced by the Juan de Fuca outflow,
and Freeland and Denman (1982) reported this current to flow
northward along the west coast of Vancouver Island. The current
is observed as a wedge of less dense water hugging the coast.
Observations are scarce and our understanding of this current and
its variability are correspondingly poor. Changes in the stratification due to tidal mixing over the shallow banks near the fjord
are discounted since no clear fortnightly variability is observed
in the fjord response, however mixing caused by other currents could
be important. At present, the lack of observations precludes the
analysis of some of these processes mentioned above.

MODEL OF SHELF-FJORD EXCHANGE

Recently Klinck et al.(1981) have developed a numerical model
that examines the interaction between a narrow, deep silled sjord
and a wind driven coastal regime. It is a simple two-layer model
that neglects non-linear terms and mixing processes both in the
fjord and in the coastal regime. In the fjord, rotational effects
are ignored, as is the circulation forced by run-off. Coastal
upwelling dynamics are used to model the coastal field. The
important result of this model is that the alongshore geostrophic
currents control the circulation in the fjord. The displacement
of the free surface and pycnocline at the mouth is controlled by
the geostrophic coastal current, and the resulting pressure gradients drive the circulation in the fjord. An alongshore wind stress
produces either a filling or emptying of the fjord due to the
Ekman transport. Also the flow is mainly baroclinic since the slope
of the free surface mirrors the displacement of the pycnocline.

The model of Klinck et al.(1981) is applicable to Alberni

Inlet as the observations are in general consistent with the results of the model. The correlation of the fjord currents with the upwelling index is in a manner consistent with the model predictions. Observations of the mean sea level in the fjord show a net filling (emptying) during downwelling (upwelling) conditions. The changes in the fjord density may be explained by the deepening or shallowing of the pycnocline in response to the wind driven coastal regime. Furthermore, currents observed in Junction Passage (at 35 m, 65 m and 95 m) show that the flow is baroclinic. There are at least two layers, with a zero crossing just above 35 m, and a lower layer extending from 35 m to about 100 m.

The applicability of the model to other fjords on the west coast of Vancouver Island will depend upon the importance of non-linear, mixing and geometric effects in the fjord. In the case of Alberni Inlet, its interaction with the coastal regime dominates the circulation of the intermediate water. Investigations of the circulation of fjords adjoining coastal seas should not overlook the potentially significant interactions with the coastal circulation.

REFERENCES

Bakun, A., 1973, Coastal upwelling indices, west coast of North America, 1946-71. U.S. Dept. Commer., NOAA Tech. Rep. NMFS SSRF - 671, 103 p.

Freeland, H.J., and Denman, K.L., 1982, A topographically controlled upwelling centre off southern Vancouver Island, J. Mar. Res. (40, 4).

Gade, H.G., 1976, Transport mechanisms in fjords, in:"Fresh Water On The Sea", Skreslet et al., eds., Ass. Norwegian Oceanographers, Oslo 51-56.

Helle, H.B., 1978, Summer replacement of deep water in By-fjord, Western Norway: Mass exchange across the sill induced by coastal upwelling, in: "Hydrodynamics of Estuaries and Fjords", J.C.J. Nihoul, ed., Elsevier, Amsterdam, 441-464.

Holbrook, J.R., Muench, R.D., and Cannon, G.A., 1980, Seasonal observations of low-frequency atmospheric forcing in the Strait of Juan de Fuca, in: "Fjord Oceanography", Freeland et al., eds., NATO Conf. Ser., Ser. IV: Marine Sciences, Plenum, 305-317.

Julian, P.R., 1975, Comments on the determination of significance levels of the coherence statistic, J.Atmos. Sci., 32:836-837.

Klinck, J.K., O'Brien, J.J., and Svendsen, H., 1981, A simple model of fjord and coastal circulation interaction, J. Phys. Oceanog., 11:1612-1626.

Muench, R.D., and Heggie, D.T., 1978, Deep water exchange in
        Alaskan subarctic fjords, in:"Estuarine Transport Pro-
        cessea", B. Kjerfve, ed., Univ. of So. Carolina Press,
        Columbia, S.C., 239-267.
Svendsen, H., 1977, A study of the circulation in a sill fjord
        on the west coast of Norway, Marine Science Communi-
        cations, 3:151-209.
Svendsen, H., 1980, Exchange processes above sill level between
        fjords and coastal water, in: "Fjord Oceanography",
        Freeland et al., eds., NATO Conf. Ser., Ser. IV: Marine
        Sciences, Plenum, 355-361.

CONSIDERATIONS OF COASTALLY FORCED FLOW IN A BRANCHED FJORD

John M. Klinck, Benoit Cushman-Roisin* and James J. O'Brien*

Texas A&M University
Department of Oceanography
College Station, TX 77843 U.S.A.
*Florida State University, Tallahassee, Florida 32306, U.S.A.

ABSTRACT

Two techniques are combined to consider the dynamics of a narrow, stratified fjord system with two channels.

The first technique removes the barotropic mode from a two layer model but retains the influence of bottom topography. The second technique allows development of a numerical model of two narrow, connected channels. The resulting branched fjord model is forced by coastal wind stress through the mechanism of Ekman flux. The major question addressed with this model is, what determines the exchange between the two channels in a time dependent situation.

Various simulations are performed with different widths, depths, and lengths for the channels and locations for the junction. From the simulations three main conclusions are obtained:

1) The presence of a side channel increases greatly the variability of the forced flow in the main channel.

2) Geometric constrictions (sills and narrows) have only local effect on the flow if the flow remains subcritical.

3) The relative lengths of the two channels have the strongest effect on the variability of the resulting flow.

INTRODUCTION

The Ryfylkefjord system, comprising the Boknafjord, Sandsfjord, Hylsfjord, Saudafjord and others (Fig. 1), is a geometrically complicated interconnection of deep narrow channels and is typical of many fjord systems. Such a dynamical system can be analyzed with a three dimensional numerical model, but such models are expensive to develop and to run. This paper presents a model that approximates such geometric complexity in a way that is inexpensive enough for many situations to be examined.

A basic method for connecting channels depends on the use of a spatially-staggered numerical grid, with transport and thickness variables at alternating grid points. All connections between two channels are chosen to occur at thickness grid points. This procedure was first suggested by Dr. D. P. Wang (see Elliot, 1976) and was used in a model of Chesapeake Bay (Wang and Elliot, 1978).

Another procedure has been proposed by Narayanan (1979) in a study of the barotropic tidal response of the Douglas Channel along the coast of British Columbia. The linear wave propagation problem is solved for each tidal frequency, by following a non-staggered path of integration and introducing a certain number (say N) of arbitrary transport and depth values where necessary (junctions and open boundaries). The integration is performed N times with different values assigned to those constants. The solution is then sought as a linear combination, such that all flow constraints (also N in number) are met. In a second stage, the linear tidal response is used to initialize a non-linear model, using an explicit scheme. The basic limitations of Narayanan's procedure are two-fold: (i), for each frequency, the algorithm sweeps the system N+1 times and solves an N by N full linear system; (ii) the method of unknown constants cannot be generalized to non-linear models, while the initialization by a linear solution may fail in the study of baroclinic motions (particle velocity closer to wave speed).

The present model is based on Wang's method, to keep the number of passes over the system to a minimum of one Gaussian elimination per time step (or per frequency, if a modal decomposition were applied), and to open the way to direct and accurate extension to non-linear dynamics.

We chose to investigate narrow, deep silled fjords with a linear, two-layer, nonrotating model. The model is externally forced by coastal wind stress through Ekman flux. We ignore diffusively driven flow, such as that due to freshwater addition from rivers, and the effects of wind stress in the fjord itself. The simplified dynamics allow us to focus on the interaction of flow in several channels. Diffusion and local forcing can be considered as the understanding of the multibranched fjord system increases.

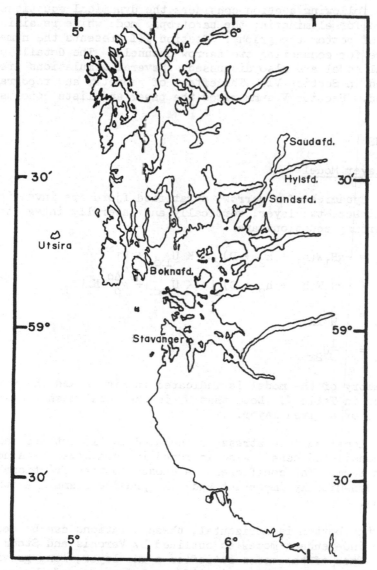

Fig. 1. Map of the Ryfylkefjords on the western coast of Norway.

The following questions are considered in this study. How do long gravity waves behave in a system with two channels? What controls the exchanges between the two channels? How do sills and narrows influence the exchange between the two channels?

The following section considers the dynamical equations and the procedure for eliminating the barotropic mode while retaining the effects of bottom topography. Section III presents the numerical technique for connecting two narrow channels. The details of the numerical model are also discussed. Several simulations are discussed in Section VI. The effect of geometry and topography are presented. Section V summarizes the paper and lists conclusions.

EQUATIONS

a) Two-layer Model

The dynamics of a narrow, stratified fjord are investigated with a linear, two-layer, vertically and laterally integrated model. The governing equations are

$$U_{1t} = -gH_1 W(h_1 + h_2 + D)_x + K\, U_{1xx} + W\tau^x$$

$$U_{2t} = -gH_2 W(h_1 + h_2 + D)_x + K\, U_{2xx} + g\frac{\Delta\rho}{\rho} H_2 W h_{1x} \qquad (1)$$

$$h_{1t} = -\frac{1}{W} U_{1x}$$

$$h_{2t} = -\frac{1}{W} U_{2x}$$

The geometry of the model is indicated in Fig. 2 and the variables are given in Table I. Note that U_i is the total mass transport (cm^3/sec) of a given layer.

The local surface stress is included in (1) but its influence is not considered here. Also included in each layer is a horizontal friction term. The coefficient is chosen so that frictional effects are not dynamically important and only provide a small amount of smoothing.

If the bottom is horizontal, these equations can be decoupled into two independent modes as outlined by Veronis and Stommel (1956). Each of the modes satisfies a one-layer, long gravity wave equation but the wave speed is different for each of the modes. To be more explicit about the modal separations, a linear combination of the variables is defined as

$$U = U_1 + \lambda U_2 \quad \text{and} \quad h = h_1 + \lambda h_2 \qquad (2)$$

Table I – Notation

A: area of junction region

c: local baroclinic wave speed

D: depth anomaly

g: acceleration of gravity

H_i: undisturbed thickness for layer i

H: total volume of junction region

h_i: thickness of layer i (main channel)

$\hat{h}_i$: thickness of layer i (side channel)

K: horizontal diffusion coefficient

$\hat{L}$: number of grid points in the main channel

t: time

U_i: volume transport for layer i (main channel)

$\hat{V}_i$: volume transport for layer i (side channel)

W: width of main channel

$\hat{W}$: width of side channel

x: position along main channel (positive toward head)

y: positive along side channel (zero at junction, positive towards head)

$\alpha\pm$: squared speed of barotropic (+) and baroclinic (–) modes

ρ_i: density of layer i ($\rho_1 < \rho_2 = \rho$)

$\Delta\rho$: density difference between layers ($\rho_2 - \rho_1$)

$\lambda\pm$: structure coefficient for ± mode

η: average elevation of junction region

μ: a function of order unity

τ^x, τ^y: surface wind stress

and the modal variables are required to satisfy an equation of the form

$$U_t = -\alpha W h_x + K \, U_{xx} - \alpha \lambda W D_x + W \tau^x \qquad (3)$$

$$h_t = - \frac{1}{W} U_x$$

The parameters α and λ must satisfy algebraic consistency relations, which reduce to a quadratic equation for α (or λ). The character of these parameters is easily shown in the limit of small $\Delta\rho/\rho$, which is $O(10^{-3})$ in most oceanic situations.

The approximate parameters, expanding in $\Delta\rho/\rho$, are

$$\lambda_+ = 1 \qquad \alpha_+ = g(H_1 + H_2)$$

and

$$\lambda_- = - \frac{H_1}{H_2} \qquad \alpha_- = g \frac{\Delta\rho}{\rho} \frac{H_1 H_2}{H_1 + H_2} \qquad (4)$$

The parameters denoted by the plus subscript correspond to the barotropic mode while the negative subscripts denote the baroclinic mode. Each alpha is the square of the celerity for the surface and internal gravity waves.

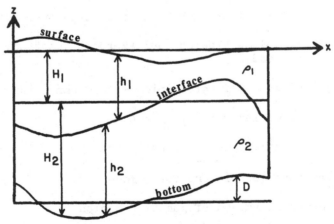

Fig. 2.   Model geometry and variables.

If topography effects are present (H_2 function of x), the above method of modal decoupling fails because of the interaction between the fast and slow motions: Passing a topography feature, a barotropic wave generates a baroclinic component and vice versa (Proudman, 1953). And, the more abrupt is the topography, the greater is that interaction. Therefore, only approximate methods can be developed to decouple the barotropic and baroclinic modes in the presence of bottom topography, and their accuracy depends on the strength of the coupling. To consider the barotropic mode in isolation, one simply sets H_2, h_2, U_2 to zero and lets H_1 be the total undisturbed depth. This is equivalent to neglecting $\Delta\rho$ in Equation (1). The accuracy is thus only $\Delta\rho/\rho$.

If the baroclinic mode is to be studied in isolation, the elimination of the barotropic mode is not so simple. The usual procedure involves an infinitely deep, passive, lower layer (the reduced-gravity model). This procedure also eliminates the effect of bottom topography. Its accuracy is H_1/H_2.

b) Baroclinic Model

In order to study the baroclinic mode in isolation while retaining the influence of bottom topography, a new procedure has been developed. It uses the fact that the baroclinic mode has almost no vertically integrated transport to eliminate the barotropic mode from the two-layer model. That is, from (4), $U_2 = -U_1$ to lowest order in $\Delta\rho/\rho$.

First, define h to be the displacement of the free surface from its reference position or $h = h_1 - H_1 + h_2 - H_2$. The two-layer equations (1) are rewritten (with K, $\tau^x = 0$) as

$$U_{1t} = -gH_1 W h_x$$

$$U_{2t} = -gH_2(x)Wh_x + \frac{\Delta\rho}{\rho}gH_2(x)Wh_{1x} \qquad (5)$$

$$h_{1t} = - \quad U_{1x}\frac{1}{W} \quad h_{2t} = - U_{2x}\frac{1}{W}$$

since $D_x = -H_{2x}$.

Assume that

$$U_2 = -\left[1 + \frac{\Delta\rho}{\rho}\mu(x,t)\right]U_1 \qquad (6)$$

where μ is a function (to be determined) of order unity. The system of equations (5) can be reduced to the single equation:

$$(\mu\ U_1)tt\ =\ g(H_1\ +\ H_2)W\frac{\partial}{\partial x}\left[\frac{1}{W}(\mu U_{1x})\right]\ +\ gH_2W\frac{\partial}{\partial x}(\frac{1}{W}U_{1x}). \qquad (7)$$

The time derivative term is of order $\Delta\rho/\rho$ compared to the other two terms in this equation. Indeed, it will be shown a posteriori that the baroclinic mode satisfies a wave equation which allows replacement of two time derivatives with the phase speed times two space derivatives. Since the baroclinic phase speed is approximately $\alpha^{-1/2}$, replace $\partial^2/\partial t^2$ with $\alpha-\ \partial^2/\partial x^2$, where $\alpha-$ is defined in (4). The first term is clearly of order $\Delta\rho/\rho$ compared to the other two terms.

If the function μ varies slowly in x or

$$\mu_x{<}{<}U_x \qquad\qquad\qquad\qquad\qquad\qquad\qquad\qquad (8)$$

then the solution for μ is

$$\mu(x)\ =\ -H_2(x)/(H_1\ +\ H_2) \qquad\qquad\qquad\qquad (9)$$

Condition (8) requires that the bottom topography vary slowly over a wavelength of the internal wave.

Now that $\mu(x)$ is known, the equations for the baroclinic mode are

$$U_{1t}\ =\ -c^2(x)Wh_{1x}$$
$$\qquad\qquad\qquad\qquad\qquad\qquad\qquad\qquad\qquad (10)$$
$$h_{1t}\ =\ -\ \frac{1}{W}U_{1x}$$

where $c^2(x) = \frac{\Delta\rho}{\rho}g\frac{H_1H_2}{H_1+H_2}|1\ +\ \frac{\Delta\rho}{\rho}\ \frac{H_1H_2}{H_1+H_2}|$, which is the square of the "local" baroclinic phase speed. Once U_1 and h_1 are calculated, U_2 is obtained from (6), and h_2 from an identical relationship.

The approximations have eliminated the coupling between the baroclinic and barotropic modes. Such coupling occurs through nonlinear interaction (hydraulic processes) or through creation of internal waves by barotropic flow over sharp bottom topography. It is therefore concluded that the coupling is weak and negligible (on the order of $\Delta\rho/\rho)^2$) as long as velocities are much smaller than the phase speeds and the topography smooth over one wavelength.

c) External Forcing

The effects of an along shore coastal wind on flow in a fjord are investigated with a coastal upwelling model and a fjord model which join at the coastline (Klinck, et al., 1981). The results of this study are paraphrased here: a more complete discussion is contained in the reference cited.

The first conclusion from the coupled model is that the coastal wind stress forces flow in the fjord through Ekman transport. The wind pumps water either in or out the upper layer at the ocean mouth of the fjord. The changes at the mouth induced by the forcing travel to the rest of the fjord as long gravity waves.

The second conclusion is that the barotropic disturbance created by changes in the wind is very quickly in balance (in a few hours for a 50 km long fjord) and thus the dominant response in the fjord is baroclinic. It is for this reason that we choose to focus on the baroclinic mode and its interaction with width and depth variations in the fjord.

The results of this previous study show that a coastal boundary condition can be imposed on the baroclinic wavefield. However, specifying the baroclinic transport at the ocean boundary is improper because it does not allow waves to leave the fjord and generates spurious reflections of waves. Fortunately, these reflected waves do not affect the interaction of the two channels. Therefore, for the present work, the model is forced by an imposed baroclinic flow at the ocean boundary, simulating the baroclinic flow.

DETAILS OF NUMERICAL MODEL

a) Branching Technique

The procedure for connecting two narrow channels uses the modal equations (3). As such, the technique can be used for any multi-layer model for which linear, nonrotating dynamics are appropriate.

The basic method for connecting a side channel to a main channel depends on the use of a spatially staggered numerical grid with transport and thickness variables at alternating grid points. The connection between the two channels occurs at a thickness grid point (see Fig. 3). That is, the two channels share a common grid point (called the junction). This procedure was first suggested by Dr. D. P. Wang (see Elliot, 1976) and was used in a model of Chesapeake Bay (Wang and Elliot, 1978).

The junction region is defined by the dashed lines in Fig. 3. The conditions at the junction are

1) continuity of thickness and,

2) conservation of mass.

The first condition is satisfied automatically since the two channels share a common thickness point.

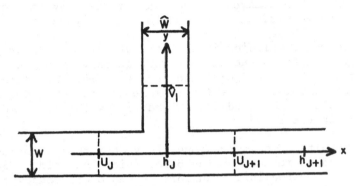

Fig. 3.   Geometry of the junction region and staggered grid.

The second condition is obtained by integrating the continuity equation over the junction region,

$$H_t = U_j - U_j+1 - \hat{V}_1 \tag{11}$$

For notational clarity, the main channel is along the x axis while the side channel is along the y axis. Variables with a caret refer to the side channel. H refers to the total volume of the junction region. This expression cannot be used directly in (3), but must be modified somewhat. Let A be the surface area of the junction region then the average interface elevation is η = H/A. Therefore, the continuity condition becomes

$$\eta_t = (U_j - U_j+1 - \hat{V}_1)/A$$

This development simply yields a modified continuity condition at the single grid point at the junction.

The branching condition is strictly kinematic, so inertia (dynamic) effects at the junction are ignored. This choice requires that the flow in the junction be slow enough that such inertia effects are negligible. The precise condition is $u^2 \ll c^2$ where c is the gravity wave speed for the mode under consideration. This condition is obtained by scaling the full momentum equation. The condition that $u^2 \ll c^2$ is also required for the flow in the channel

to be subcritical – a choice which is already made by the use of
linear dynamics.

Since inertia effects are unimportant in the junction region,
the angle between the two channels is not important. This
assumption may not be strictly valid for some fjord situations where
the junction is near a narrow (or shallow) region of the fjord.

The mathematical problem is now specified and can be solved by
a number of numerical techniques.

b) Numerical Model

The simplest numerical procedure is an explicit time integration of the equation in the two channels. Although this method
is the simplest, it can prove quite costly because the time step is
limited by the CFL stability condition,

$$\Delta t < \frac{\Delta x}{\sqrt{2} \, C_{max}}$$

where C_{max} is the maximum wave speed allowed by the dynamics. In
this case, the maximum speed is the barotropic wave speed. The CFL
condition can be quite restrictive for deep fjords.

The use of the baroclinic equation allows a large enough time
step for the calculation to be practical. This fact provided the
impetus to derive the baroclinic equations in the first place.

Since part of this work will compare the two-layer model with
the baroclinic model, another approach is taken. The equations are
integrated with a semi-implicit technique which is unconditionally
stable. This technique uses implicit time differences on the terms
in the equations which give rise to the fastest waves. Phase errors
are introduced for the barotropic waves but as the waves do not
participate actively in the overall circulation of the fjord, these
errors are not important (see Grotjahn and O'Brien, 1976, for a
discussion of the phase errors). The unconditional stability of
this integration procedure allows the choice of any time step
consistent with a reasonable truncation error.

However, this ability to chose a large time step has a price –
a set of linear equations must be solved at each time step. This
linear system must be amenable to a direct and fast solution
algorithm for the semi-implicit method to be computationally more
efficient than an explicit method.

For a one channel model, semi-implicit integration gives rise
to a linear system with a tridiagonal matrix which can be solved
with an "up-down" algorithm that is fast and direct. When the

semi-implicit method is used on the equations for the branched
system, the coefficient matrix of the resulting linear system is
"almost" tridiagonal. The "up-down" algorithm can be modified to
account for the off-tridiagonal terms yielding a direct solution
scheme. The appendix presents the details of the linear system and
of the solution procedure.

Both the two-layer model and the baroclinic model were
constructed using the semi-implicit integration scheme. Several
test calculations were made to compare the two models. For a
variety of depth and width variations, the two models give identical
results. The remainder of this paper considers simulations from the
baroclinic model alone.

SIMULATION AND DYNAMICS

The effects of width and depth changes in a fjord are now
considered with the baroclinic, branched model that is forced by
coastal Ekman flux. Particular emphasis is placed on the factors
which determine how the flow divides at the junction, and how the
two channels interact in a time-dependent situation.

The parameters for the simulation (Table II) are chosen to
correspond to fjords like the Ryfylkefjords. Data for these fjords
are provided by Svendsen (1981). The wind forcing is chosen to
simulate the reversals of coastal wind which are observed offshore
of the Ryfylkefjords (Svendsen, 1981). To simplify interpretation
of the model simulations, the wind is taken to be a sinusoid with a
5 day period. The magnitude of the wind is 2 dynes/cm^2 which gives
an average velocity at the fjord mouth of 10 cm/sec.

a) Case I

The first case presented has a main channel length of 60 km and
a side channel 40 km long. The junction is 20 km from the mouth of
the fjord. This geometry yields identical distances from the
junction to the end wall in both channels. Since gravity waves
travel at the same speed in each channel, the waves will split at

Table II

$H = 20$ m	$U(x=0) = 10. \times H_1 \times W \times \sin(2\pi t/60hr)$
$H = 480$ m	$\Delta\rho/\rho = .002$
$W = \hat{W} = 1$ km	$\Delta x = 2$ km
$K_x = K_y = 10^5$ cm^2/s	$\Delta t = 1$ h

the junction and return in phase after a reflection from the end
wall.

Fig. 4 is a phase plot of the pycnocline anomaly for the
situation just described. It is evident that the pycnocline
disturbance in the branch is the same as that in the main channel
between the junction and the end wall. This fact shows that the
junction region does not introduce any phase shifts or affect the
flow in an unrealistic manner. Also included on the figure is one
characteristic to show that the semi-implicit integration scheme
does not affect the phase of the gravity waves in the simulation.

To address the question of exchange between the two channels,
velocity is shown as a function of time (Fig. 5) for the first
velocity grid point in the side branch and for the first velocity
point toward the head of the fjord from the junction. The speeds
into each channel have the same magnitude and the same variation.
This fact is expected since both channels have the same cross
sectional area, thus splitting the flow equally between the two
channels. The velocity does not show a sinusoidal disturbance, from
the sinusoidal forcing, because waves reflect from the end wall and
mix with the waves produced by the direct forcing. There is also

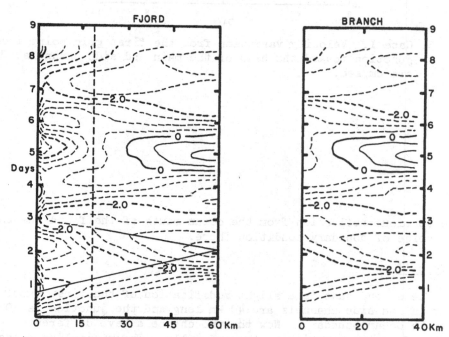

Fig. 4. Pycnocline anomaly for Case I. Ocean forcing has a 5 day
        period. Solid line denotes a characteristic. Contour inter-
        val is .5 m. The junction is denoted by the vertical dashed
        line.

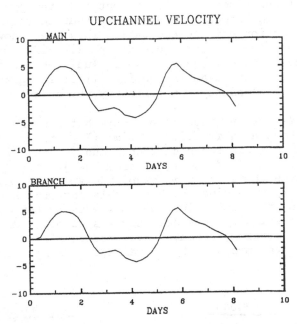

Fig. 5. Case I. Velocity variation from the first grid point from
        junction toward the head of the main and side channels. Units
        are cm/sec.

some spurious reflection from the ocean mouth of the fjord due to
the choice of boundary condition there.

b) Case II

    The second case is a slight modification of the first case.
The main and side channels are 40 km long and the junction is 10 km
from the ocean boundary. Now the two channels have different
lengths from the junction to the end walls. The pycnocline anomaly
(Fig. 6) is not the same in the two channels, but is greater in the
longer branch. To see the variation in the flow more easily,

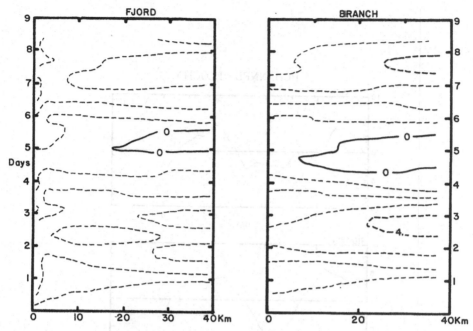

Fig. 6. Pycnocline anomaly for Case II. Contour interval is 1 m.

consider the time plot of the velocity at the entrance to each
branch (Fig. 7). Note that after the first two days, the velocity
is quite different in the two channels. This difference is due to
the phase of the waves when they return to the junction. This phase
difference can be seen easily through ray tracing arguments (not
presented here). From Fig. 7 it appears that there has been a net
increase (over the first eight days) in the volume of water in the
main channel downstream of the junction while the side channel shows
no increase in volume. These differences in transport are due
strictly to length differences of the two channels since both
channels have the same width and depth.

c) Case III

Case III considers the effect of channel width on the branched
system. The side channel has a constant width of .5 km while the
main channel is 1 km wide. The geometry of Case I is retained so
the two channels are the same length from the junction to the end
wall.

The pycnocline anomaly looks the same as Fig. 4, so only the
velocity time history is shown (Fig. 8). The velocity shows the

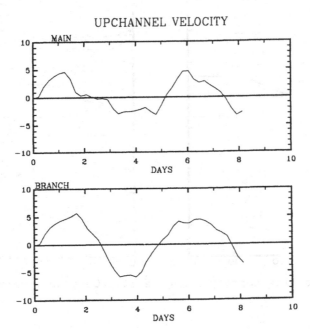

Fig. 7. Case II velocity variation. Units are cm/sec.

same structure as Fig. 5 but the amplitude is larger in both bran-
ches. The increase in speed over case I (15%) is proportional to
the decrease in total volume of the fjord system due to the narrower
side channel.

This simulation shows that the transport divides at the
junction in proportion to the cross sectional area of the two
"downstream" channels, where "downstream" depends on the direction
towards which the wave is moving. This division of the transport by
cross sectional area produces the same velocity for each of the
downstream channels even though the transport into each channel may
be quite different.

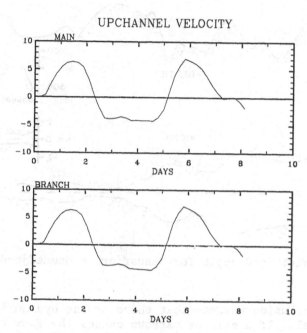

Fig. 8. Case III velocity variation. Units are cm/sec.

d) Sills and Narrows

Several simulations were calculated to consider the effect of sills and narrows on the exchange between two channels. These simulations had sills that were 200 m high or narrows to .5 km in one branch just inside the junction. None of these simulations are shown here because these constricting effects had only a local effect on the flow: the flow was greater in proportion to the decrease in cross sectional area. The presence of a sill narrows in one channel had no effect on the exchange at the junction. Therefore, for the dynamics included in this model, only the local geometry of the junction region and the relative lengths of the two channels have any influence on the exchange between the two channels in a branched fjord.

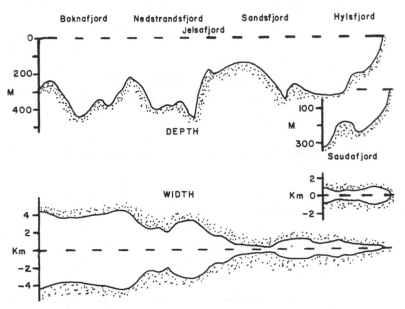

Fig. 9. Width and depth for Sandsfjord – Saudafjord system.

     This conclusion assumes that there are no hydraulic effects in
either channel. If a sill or narrows causes the flow to speed up
sufficiently to be supercritical, then this region becomes a control
section and must influence flow both up and downstream of that region.

     For deep-silled fjords which do not have hydraulic controls,
the baroclinic model should simulate the dynamics adequately, and
the geometry of a given branched fjord determines the exchange
between the two channels.

e) Sandsfjord System

     The ultimate aim of this research is to analyze an actual fjord
having two channels. Towards that end, one simulation is included
which considers a part of the Ryfylkefjord system. Because of the 2
km resolution of the model, the width and depth must be smoothed
somewhat to match the model resolution. Fig. 9 displays the
smoothed width and depth profiles for the main channel (Boknafjord
to Hylsfjord) and the side channel (Saudafjora).

     One simulation with this complicated geometry is performed with
forcing by a five-day period Ekman flux. The pycnocline anomaly is
presented in Fig. 10. The most notable feature of this simulation
is that the largest flows, indicated by the largest pycnocline
anomaly, appear at constrictions in the channel. It is difficult to

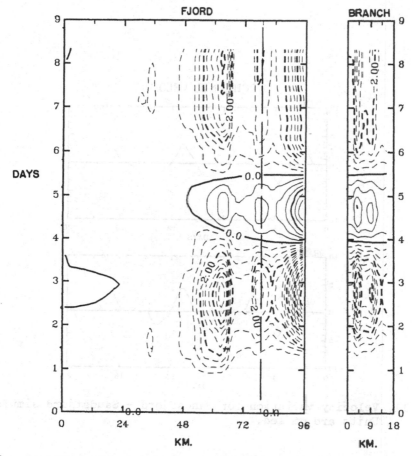

Fig. 10. Pycnocline anomaly for the Sandsfjord – Saudafjord
        geometry. Contour interval is .5 m.

detect much of the disturbance between the constrictions.

        The exchange between the two channels is considered in Fig. 11.
The velocities are not very different over the 20 day span of this
simulation, but the length of Hylsfjord and Saudafjord differ by only
2 km (Fig. 9), and their topographies are quite close. Therefore,
the waves return to the junction at about the same time. Notice
that even though the forcing has a 5 day period, the presence of a
side channel allows extra freedom to the waves in the system and
the disturbance at a given point does not reflect the forcing period
at all. The geometry of the Sandfjord system does not provide a

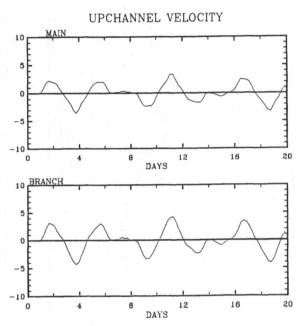

Fig. 11. Velocity variation for Sandsfjord - Saudafjord simulation.
        Units are cm/sec.

sensitive test of the exchange processes because the two channels are
so close in length and topography. A better place to consider exchanges might be the Jøsenfjord branch off the main channel.

    Since constrictions in a narrow channel amplify the flow so
markedly, as illustrated by Fig. 10, one speculates that these
regions would show the strongest response to coastal wind events. It
may also be that for a strong coastal wind storm, the flow in some
constriction may become critical and develop a hydraulic jump.
As the storm slackens, such a jump would be released as an internal
bore or solitary wave. It would be interesting to see if coastal
storms produce such jumps or enhanced mixing in constriction far
removed from the direct effect of the wind.

SUMMARY

Two techniques are combined to consider dynamics of a narrow, stratified fjord system with two channels.

The first technique removes the barotropic mode from a two-layer model but retains the influence of bottom topography. Earlier baroclinic models removed the barotropic mode by having a passive, infinitely deep lower layer. Such an assumption is not appropriate for fjords.

The second technique allows a numerical model to be constructed which connects two narrow channels. Only kinematic conditions of conservation of mass and continuity of the interface are specified. Therefore, there are no inertia effects at the junction and the angle between the two channels is not important.

The branched-fjord model is forced by coastal wind stress through the mechanism of Ekman flux. The major question addressed here is what determines the relative exchange of the two channels in under time dependent circumstances.

Various simulations are calculated with different widths, depths and lengths for the two channels and different choices for the location of the junction of the two channels. From the simulations, three main conclusions are obtained:

1) The presence of a side channel increases greatly the variability of the forced flow in a narrow channel.

2) Geometric constriction (sills and narrows) have only a local effect on the flow if the flow remains subcritical.

3) The relative lengths of the two channels have the strongest effect on the variability of the resulting flow.

ACKNOWLEDGMENTS

This work was supported by the National Science Foundation grant OCE-7925351. The authors are indebted to Dr. David Farmer for suggesting additional references.

REFERENCES

Elliot, A.J., 1976, A numerical model of the internal circulation in a branching tidal estuary, Chesapeake Bay Institute, The Johns Hopkins University, Special Report 54, Ref. 76-7.

Grotjahn, R., O'Brien, J.J., 1976, Some Inaccuracies in Finite
    Differencing Hyperbolic Equations, <u>Mon. Wea. Rev.</u>,
    104:180.
Klinck, J.M., O'Brien, J.J., Svendsen, H., 1981, A simple model
    of Fjord and Coastal Circulation Interaction, <u>J. Phys.
    Oceanogr.</u>, 11:1612.
Narayanan, S., 1979, Kitimat physical oceanography study 1977-
    1978. Tidal circulation model, Dobrocky SEATECH Ltd.
    <u>Report</u>, Victoria, British Columbia.
Proudman, J., 1953: Dynamical Oceanography, Methuen & Co. Ltd.,
    London, pp. 349-351.
Svendsen, H., 1981: A Study of the Circulation and Exchange
    Processes in the Ryfylke fjords, Vol. I and II, Univer-
    sity of Bergen, Report 55.
Veronis, G., Stommel, H., 1956, The Action of Variable Wind
    Stresses on a Stratified Ocean, <u>J. Mar. Res.</u>, 15:43.
Wang, D.-P., Elliot, A.J., 1978, Non-Tidal Variability in the
    Chesapeake Bay and Potomac River: Evidence for non-local
    forcing, <u>J. Phys. Oceanogr.</u>, 8:225.

APPENDIX

The linear algebraic system of equations to be solved at each
step of the semi-implicit integration is denoted

$$M \; \overline{u} = \overline{f} \tag{A1}$$

Where M is the coefficient matrix, $\overline{u}$ is a vector of the transport at
the new time and $\overline{f}$ is the forcing obtained in terms of variables at
the old time steps. The vector $\overline{u}$, is obtained by concatenating
the main channel variables and the side variables, $\overline{u} = (U_1, U_2, \ldots
\ldots, U_L, \hat{V}_1, \ldots, \hat{V}_L)$. The almost tridiagonal matrix is shown in Fig.
A1a where the asterisks indicate non-zero matrix elements and the
solid circles indicate the off tridiagonal elements that result from
the branching condition.

The "up-down" algorithm is modified in the following way to
form an "insweep-outsweep" procedure. The insweep step involves
Gaussian elimination of the matrix elements below the diagonal from
the ocean to the junction. A similar procedure eliminates the
elements above the diagonal starting at the end of each channel and
moving towards the junction. Fig. A1b shows the form of the matrix
at the end of the insweep part of the solution.

The matrix elements denoted by asterisks correspond to U_J, $U_J +1$
and $\hat{V}_1$, which are the three transport variables defining the
junction region (Fig. 3). These matrix elements compose an
independent, fully dense 3 x 3 linear system to be solved for the
three transport variables, analogous to the coefficient matrix

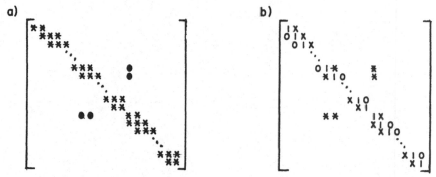

Fig. Al. Coefficient matrix for the branched model. a) asterisks
denote non-zero matrix elements. Solid circles denote
the "off tridiagonal" elements due to branching. b) After
insweep. X's denote changed elements. Asterisks define
a 3 x 3 matrix for the three velocity point defining the
junction.

obtained by Narayanan (1979). This 3 x 3 can be solved directly by
Gaussian elimination.

Once these three values are known, the "outsweep" part of the
procedure finds the values of the adjacent transport variables until
the complete solution is known.

This modified tridiagonal solution scheme requires the
computation of an ordinary "up-down" solution plus that for the
solution of a 3 x 3 linear system. This procedure gives a fast and
direct solution to the linear system and justifies the use of a
semi-implicit integration scheme over an explicit calculation (for
the two-layer model).

SOME ASPECTS OF CIRCULATION ALONG THE ALASKAN BEAUFORT SEA COAST

J.B. Matthews

Geophysical Institute
University of Alaska
Fairbanks, Alaska, U.S.A.

INTRODUCTION

Before the discovery of oil at Prudhoe Bay, Alaska in 1968 very little work had been reported on the coastal circulation along Alaska's Beaufort Sea Coast.The Prudhoe Bay oil discovery led to the development of that and other oilfields and subsequently to demands for oil leasing offshore. In 1975 the Outer Continental Shelf Environmental Assessment Program (OCSEAP) began to investigate the processes occurring on the Beaufort Sea Shelf in order to assess the potential impact of oil development. This paper attempts to bring together the findings of the physical oceanographers in order to give an overview of present knowledge of circulation along the Shelf.

PHYSICAL ENVIRONMENT

The Alaskan Beaufort Sea shelf extends for about 600 km from 156°30 to 141° W longitude at 70-71° N latitude (Fig. 1). It is typically 86 km wide with the western part being wider than the eastern part. The inner shelf is shallow; 20 km offshore the depth is typically only 10 m. Several long shallow lagoons protected by barrier islands and several large shallow open embayments typify the coastline. The tundra coastal plain is underlain by permanently frozen ground (permafrost) (Lewellen, 1974), which is slowly melting under the sea covered portion (Harrison and Oster-kamp, 1978). Osterkamp and Payne (1981) report permafrost 400 m thick off the Beaufort Sea coast thinning seaward associated with an unstable shoreline. Melting of coastal permafrost and subsequent impact of summer storms result in an annual erosion rate of 1.1 m (Naidu, 1982) along the coast and in some cases as much

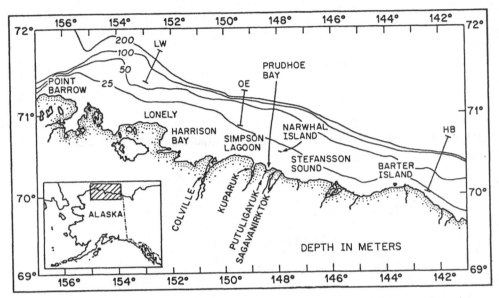

Fig. 1. Location map showing shelf transect sections and major
       features.

as 3–4 m in a single year (Short et al., 1974).

     The sea is ice covered for all but the three summer months
July – September. Seasonal ice growth is about 2 m and has a
salinity of .3–14 $^{o}/oo$ (Kovacs and Mellor, 1974) while multi–year
ice floes in the offshore region vary in thickness from 2.1 to 4.5 m
(Bushuev, 1964) and have salinities 0–6 $^{o}/oo$ (Kovacs and Mellor,
1974). The lagoons are generally ice free from July 15 to September 15 (Matthews and Stringer, 1982). The region from the coastline to the 20 m isobath is largely ice free from August 1 – September 15 although occasionally the region may not be ice free for
the summer (Stringer et al., 1981).

     Winds are reported to be bimodal ENE to WSW with the ENE winds
prevailing at mean speeds of 6 m sec^{-1} based on observations at
coastal stations (Searby and Hunter, 1971). During the summer
when the landmass is snow-free, Kozo (1979) showed that a strong
sea breeze regime modifies the prevailing winds. Details of the
weather over the shelf and the polar sea have begun to emerge from
the data reports of an arctic buoy program (Thorndike and Colony,
1980, 1981; Thorndike et al., 1982). Data from these buoys include
ice motion, atmospheric pressure and temperature. However, no
long-term meteorological data from the shelf region are available.

     There are no permanent tide gauges on the Alaskan Beaufort
Sea coast. The mean spring tidal range at Barrow was reported to

be 13.6 cm (Matthews, 1971), and 15.2 cm in Stefansson Sound
(Matthews, 1981a). The M2 tide along the Beaufort Sea coast is
reported to be in the range 5-8 cm (Kowalik and Matthews, 1982a).
The small tidal ranges result in small tidal currents with an M2
tide of 0.3 cm sec^{-1} reported for Stefansson Sound (Matthews,1981a).
Sea level changes due to storm surges can be an order of magnitude
larger than astronomic tides (Schafer, 1966; Hume and Schalk, 1967;
Reimnitz et al., 1972; Reimnitz and Maurer, 1979; Matthews, 1981a).
The largest positive surges appear to be associated with storms in
late summer and fall while large negative surges mainly occur in
winter (Henry and Heaps, 1976; Matthews, 1981a).

There are only a few small rivers draining into the Alaskan
Beaufort Sea. The largest of these is the Colville River and it is
one of the smallest draining into the Arctic Ocean (Walker, 1974).
The Colville has not been gauged unlike smaller rivers, the Saga-
vanirktok, Kuparuk and Putuligayuk near Prudhoe Bay (Carlson and
Kane, 1975). Walker (1973) estimates the total annual discharge
of the Colville to be 10^{10} m^3. Carlson and Kane (1975) indicate
that the arctic rivers flow only between June and September. For
the Kuparuk River typically 80% of the mean annual flow can occur
during the first 10 days of spring breakup floods (Matthews and
Stringer, 1982). The peak discharge for the Kuparuk river is 3340
m^3 sec^{-1} for the period since gauging began in 1971 (USGS, 1978).

The wave field in the arctic is unique in the world's oceans
in that, because of the ice cover, there is virtually no swell
(Hunkins, 1962). Hunkins (1962) reported that the spectral peak
associated with sea and swell is totally absent and that there is
a monotonic increase in displacement amplitude as the periods range
from 0.1 sec to 100 minutes over the continental shelf. The amplitudes (under an ice sheet) were about 0.5 mm at 30 sec. Wiseman
et al. (1974) reported an energy peak at periods between 2 and 3
seconds during the open water season with a significant wave height
of 20 to 30 cm. They note that although the waves approach the
shoreline at a steep angle, their relatively low height produces
only moderate longshore currents and sediment transport. During
storm periods Wiseman et al. (1974) reported waves of 2.5 m amplitude and 9-10 sec period. In the Beaufort Sea the waves are fetch
limited due to the proximity of the pack ice.

OUTER SHELF CIRCULATION

Bering Sea water has been shown by several authors (Johnson,
1956; Hufford, 1975; Mountain, 1974; Paquette and Bourke, 1974)
to enter the Beaufort Sea and travel eastwards along the shelf
break. Figure 2 taken from Mountain (1974) shows the temperature
maximum associated with this Bering Sea water at a cross-shelf
section at 150° W longitude. Mountain (1974) identified two distinct water masses which he called Alaskan coastal water and Bering

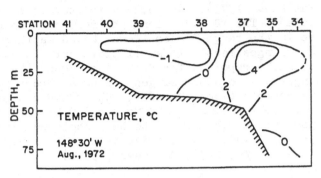

Fig. 2. Summer temperature section across the shelf from Mountain (1974).

Sea water. Figure 3 is adapted from Mountain (1974) by Aagaard (1981). Alaskan coastal water has temperatures off Barrow of 5–10° C and salinities less than 31.5 °/oo and is shown as temperature on a density anomaly surface σ_t = 25.00 (Figure 3). It mixes with ambient surface water as it moves eastwards and is not identifiable east of 148° W. The Bering Sea water is more saline and is shown as temperature on a density anomaly surface σ_t = 25.80 in Fig. 3. It has a deeper temperature maximum and can be traced as far as Barter Island at 143° W (Aagaard, 1981).

The temperature maximum on the shelf is primarily a summer phenomenon and is not identifiable after November. Temperatures are at freezing point down to 40 m with density anomaly of 26.5 by late November (Aagaard, 1981).

UPWELLING

Hufford (1974) reported a summer upwelling event, where water from 125 m appeared on the shelf, which he related to easterly winds during a period of open water. Aagaard (1981) showed (Fig.4) that water warmer than 0°C and more saline than 34.5 °/oo at section Oliktok East (OE) had origins deeper than 200 m in October–November, 1976. The Lonely West (LW) Section, 120 km west of the Oliktok East section, showed that even at the innermost station the bottom 10 m of the water column was warmer than –1°C and more saline than 34.5 °/oo. Aagaard (1981) notes that the apparent upwelling is not a local phenomena but that it does not appear to be related to wind-driven coastal upwelling in the examples given. At such times he notes that the geostrophic shear changes sign at the outermost stations at a depth of 50–80 m with a westward core speed

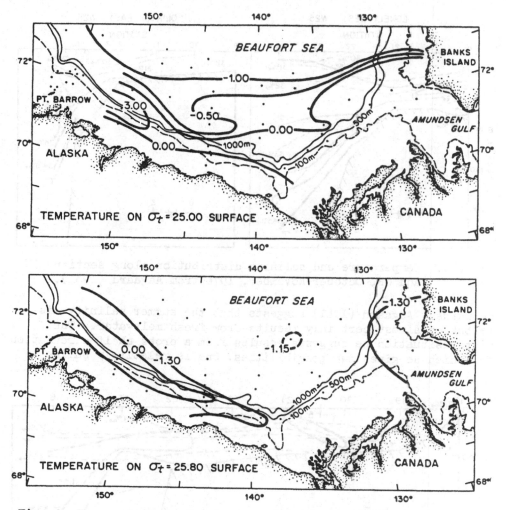

Fig. 3. Temperature on density surfaces corresponding to Alaskan
coastal water (σ_t = 25.00) and Bering Sea water
(σ_t = 25.80), August–September, 1951. Adapted from
Mountain (1974) by Aagaard (1981).

estimated at 55 cm sec^{-1} with an upslope component.

CROSS SHELF CIRCULATION

    Figure 5 taken from Aagaard (1981) shows a salinity section
in Harrison Bay in 1975. This has remnants of the summer situation
with a 2 °/oo salinity increase seaward. In contrast, Fig. 6 shows
the same section in winter where the salinity decreases by 0.8 °/oo

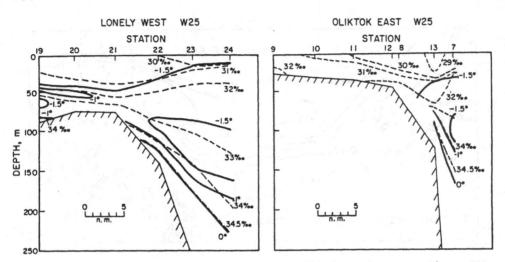

Fig. 4. Temperature and salinity distribution along sections LW
          and OE, October–November, 1976 from Aagaard (1981).

seawards. Aagaard (1981) suggests that the summer salinity distribution almost certainly results from fresh meltwater. The
winter situation he suggests results from a cross shelf circulation
for which he gives two possibilities. One is relatively saline

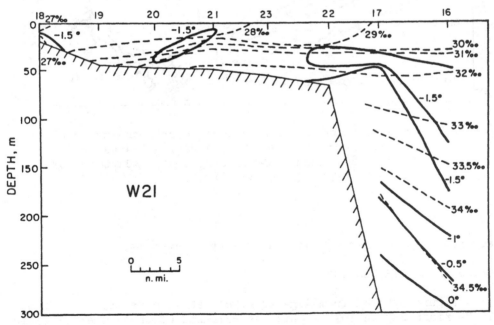

Fig. 5. Temperature and salinity section along section HB,
         November 1975, from Aagaard (1981).

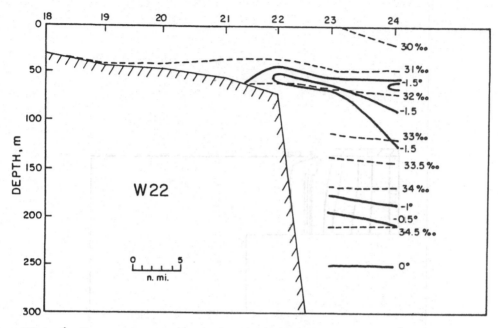

Fig. 6. Temperature and salinity distribution along section HB, February 1976, from Aagaard (1981).

Chukchi Sea water being advected onto the shelf and later sinking from the shelf to drive cross shelf circulation. The second alternative is the sinking of brine produced by the freezing process on the inner shelf and moving seaward in the lower layer. This would require the movement onshore of less saline water in the upper layer.

Such a mechanism has been postulated on the inner shelf from observations in Stefansson Sound from current meters (Matthews, 1981a) and under-ice drifters (Matthews, 1981b). Kowalik and Matthews (1982b) presented some numerical experiments in which a seaward moving salt flux under landfast ice could generate currents on a shallow lagoon of 1 cm sec^{-1} based on observed environmental parameters.

Figure 7, taken from Kowalik and Matthews (1982b) shows current streamlines derived in a brine driven numerical experiment in which growing ice covers part of a shelf. In addition to a brine drainage circulation there is also a suggestion of upwelling at the shelf break. Schell etal. (1982) has presented a transect across the Beaufort Sea Shelf in late winter (Fig. 8). Here in the lagoon section, Stefansson Sound, higher salinities associated with late winter conditions are clearly observed.

Beyond the barrier islands out to 60 m depths a bottom layer

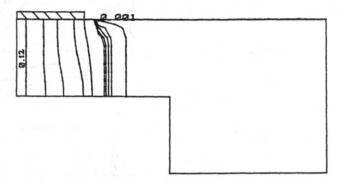

SALINITY (‰.) DAY 30

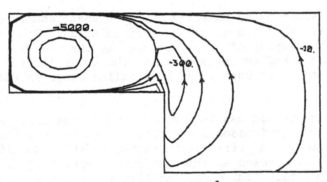

STREAM LINES(CM²/S) DAY 30

Fig. 7. Streamlines and salinity sections generated by a numerical
model from Kowalik and Matthews (1982b).

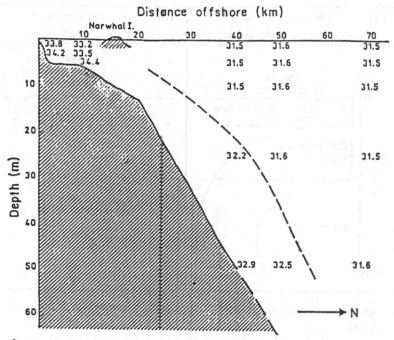

Fig. 8. Salinity profile through Stefansson Sound and Beaufort
Sea Shelf, 28 May 1982, from Schell et al. (1982).

of saline water is delineated with about 10 m thickness. Schell
(1982) relates this to the brine drainage process using other
nutrient tracers and a simple model. Aagaard et al., (1981) suggest
that this brine drainage is an important mechanism for maintaining
the halocline in the Arctic Ocean. They point out, however, that
for the volumes of saline water observed in the Arctic Ocean, open
water areas are required through much of the winter. They point
out that the Beaufort Sea does not typically show persistent ice-
free areas in winter.

        Schell, in a personal communication in 1977 to the author,
first suggested that such a brine drainage mechanism existed.
Moreover, the return flow of less saline water near the surface
appeared to be extremely important to explain the observed abundance of epontic algae. The return flow was deduced by Matthews
(1981a) from observations in Stefansson Sound and clearly shown
in the numerical experiments of Kowalik and Matthews (1982b). Schell
et al. (1982) has now used his own observations on nutrient dynamics with the data reported above to explain the observations of
surprisingly large populations of ice algae at the ice-water interface.

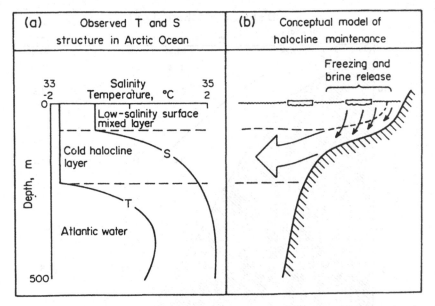

Fig. 9. Halocline maintenance model schematic from Aagaard et al.,
         (1981).

The mechanism has been clearly demonstrated and is a major
component of cross shelf circulation in the fall and winter. Figure
9 is taken from Aagaard et al.,(1981) and schematizes the halocline
model in relation to observed temperature and salinity distribution
in the Arctic Ocean. This form of the T and S distribution is most
clearly observed in the Eurasian Basin. In the Canadian Basin off
the Beaufort Sea coast the structure is obscured by a temperature
maximum at 75-100 m resulting from Bering Sea water as discussed
above.

The upper layer is a cold low salinity mixed layer to about
75 m depth. Below that layer, at 75-225 m depth, is a cold halocline layer which appears to derive from brine production on the
shelves. This overlies the warm saline Atlantic water. Full details of the contribution of the Beaufort Sea shelf to this process
are not yet available, but as Aagaard et al., (1981) suggest, it is
probably not as large as the contribution from wider shelves of the
Arctic rim where open leads are a persistent winter feature.

INNER SHELF

Shoreward of the 40 m contour very little data are available

except in the lagoons and protected embayments. The region is partially ice-covered in summer (Stringer et al.,1981) making it hazar-
dous for ship operations and is part of the shear zone between shore-
fast ice and the Arctic pack ice in winter. Consequently, moored
instruments are difficult to maintain. Surface drifters released
both in summer and winter suggest that the currents of the inner
shelf are predominantly eastward (Matthews, 1981b). This is to be
anticipated in summer during the ice-free season from the prevailing
easterly winds. However, there are periods of strong westerly wind
associated with storm systems which produce major positive surges
(Schafer, 1966) and which account for the greatest coastal erosion
rates (Reimnitz and Maurer, 1979). Detailed descriptions of the
dynamics of this marginal ice zone await the development of suitable equipment.

NEARSHORE CIRCULATION

    The nearshore region comprises lagoons and open embayments.
Matthews (1982b) has reviewed the seasonal circulation patterns
based mainly on data from the OCSEAP program. He used examples
from different lagoons to yield a seasonal picture. The seasonal
cycle will be reviewed here supplemented by additional data.

    Figure 10, showing data off Milne Point in Simpson lagoon,
typifies the open water characteristics in a coastal lagoon.
Salinity (^{o}/oo), temperature (^{o}C), current vectors (cm sec^{-1}) and
sea level (cm) are shown for the first two weeks in August. Salinity is generally less than 30 ^{o}/oo and fluctuates sharply as fresh
water runoff mixes with coastal water. The temperature is generally above 7°C reflecting the high insolation, 24 hour daylight
for two months, and the warming effect of the water of riverine
origin. Currents are wind-driven and variable with a distinct
semidiurnal tidal component. Currents are from 0-25 cm sec^{-1} until
periods of strong easterly winds. The second half of the diagram
shows the effect of strong easterly winds which are the dominant
feature driving the summer circulation. Salinity rises as temperature falls indicating that water of Beaufort Sea origin is passing
through the lagoon. The divergence produces a drop in sea level,
in this case of about 60 cm which is four times the tidal range of
15 cm. The net easterly flow, even during the first part of the
month when variable winds occur, appears throughout the water
column. Figure 11, taken from Matthews (1981b), shows the idealized tracks of bottom drifters released in Simpson lagoon at the
same time as the data in Figure 10. Drifters released off Beechy
Point at the end of July were washed ashore at Oliktok Point, east
of Milne Point by mid-August.

    The summer eastward surface flow is very marked in some years.
Figure 12, taken from Matthews (1981b), shows drifters which covered
distances of about 300 km eastwards at speeds of 7-10 cm sec^{-1},

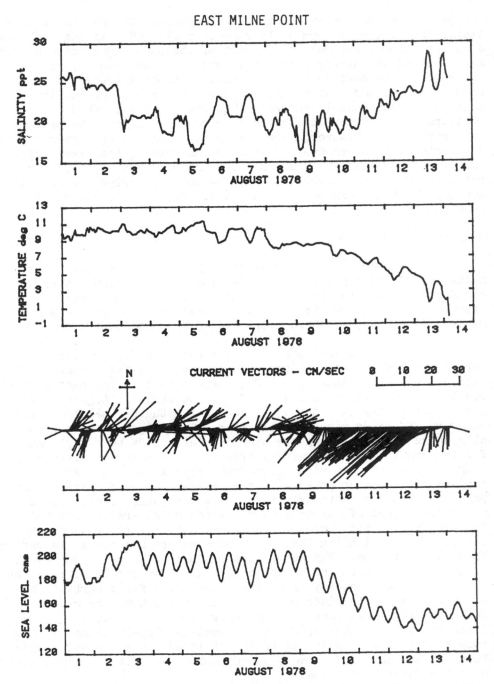

Fig. 10. Salinity, temperature, current vectors and sea level
in Simpson Lagoon, 1-14 August 1978, from Matthews
(1982).

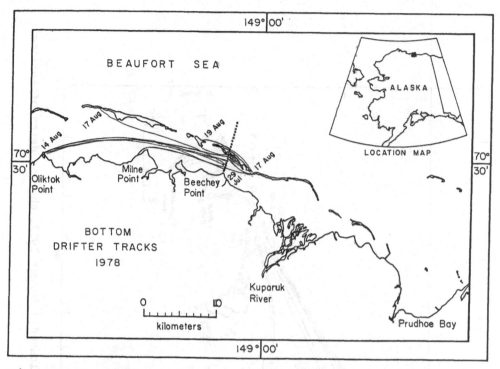

Fig. 11. Idealized tracks of bottom drifters released in Simpson
        Lagoon on 27 July 1978, with dates of first recoveries
        from Matthews (1981b).

which is 3.4% of the wind transport. However, the greatest coastal
erosion occurs with westerly winds and the effects of strong westerly winds in a coastal embayment are shown in Fig. 13. These data
were taken in Harrison Bay 10 km north of Atigaru Point at 6.25 m
depth in 8 8 m water. Harrison Bay is an open embayment unlike
the Simpson Lagoon or Stefansson Sound lagoon systems. The Colville River empties into its southern extremity. Under a variable
wind system water from the Colville River forms a two-layer system
with an upper layer 3-5 m thick. In the salinity record in Fig. 13,
penetration of this upper layer almost to the bottom is seen on
3-4 August as a period of low (<22 °/oo) salinity. Currents are
variable but the coastal mixed water of riverine origin slowly
exits the bay to the west.

From 15 August onwards a major storm developed with strong
persistent westerly winds. By late August a strong southeasterly
movement can be seen in the 30 cm sec^{-1} landward currents. The
water column is mixed from top to bottom and as the storm continued the salinity continued to fall. This suggests that Colville
River water was trapped in Harrison Bay. The storm continued into
September when freezeup begins. The strong currents, moreover,

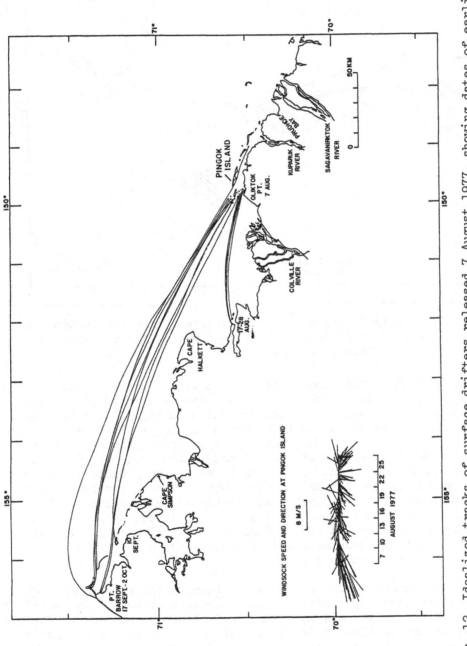

Fig. 12. Idealized tracks of surface drifters released 7 August 1977, showing dates of earliest recoveries from Matthews (1981b).

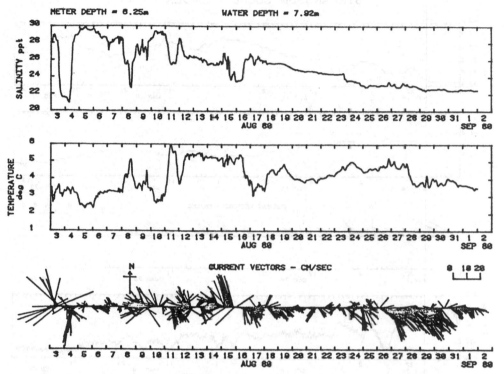

Fig. 13. Salinity, temperature and current vectors from Atigaru
Point in Harrison Bay, 3 August – 2 September 1980.

suggest that bottom sediments were resuspended at this time
(Naidu et al., 1982).

Freezeup begins in September and by early November 50 cm of
sea ice have formed. Matthews (1981a) described the freezeup
process and demonstrated a weak under-ice circulation due to brine
rejection. Figure 14 shows data from Stefansson Sound from November to February. The first 50 days' record shows salinity increas-
sing at 0.04 °/oo day^{-1}, which Matthews (1981a) showed to be
consistent with the formation of sea ice of salinity 6 °/oo at a
rate of 1 cm day^{-1}. Currents at 2 and 3 m above the sea floor in
6 m water depth were consistent with a sluggish (< 2 cm sec^{-1})
seaward drainage of brine. Kowalik and Matthews (1982b) have
modelled this lagoon brine-driven circulation and showed that it
is topographically controlled. Horizontal velocities of 3.5 cm
sec^{-1} were the minimum necessary to initiate and sustain drainage.
In the observations from Stefanson Sound, Matthews (1981a) reports mean currents of 1.9 cm sec^{-1} with peak currents of < 10 cm
sec^{-1}. Since brine drainage currents are superimposed on those

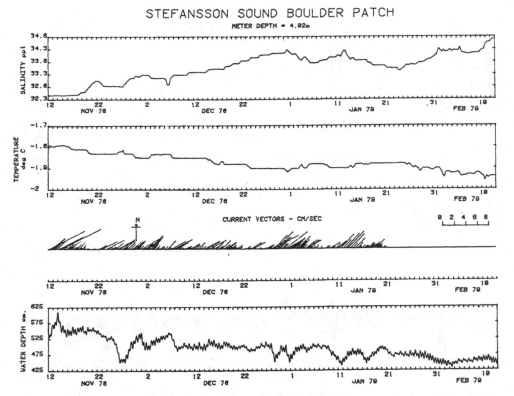

Fig. 14. Salinity, temperature and current vectors from an instru-
        ment 323 cm above the sea floor November 1978 -
        February 1979.

generated by sea level variations it seems likely that currents
large enough to initiate a drainage flow can occur. Figure 14
shows sea level variation in Stefansson Sound. A negative surge
of 1.6 m occurred in early November compared with the 15 cm tidal
range. This illustrates the occurrence of negative surges reported by Henry and Heaps (1976).

        The early part of the freezeup process is not fully under-
stood. However, late fall storms such as that shown in Fig. 13
are believed to stir up sediments which are incorporated into the
newly formed sea ice (Osterkamp and Gosink, 1982). Figure 15
shows an ice block 2 m thick taken from Egg Island channel off the
Kuparuk River delta in 5 m water depth in May 1978. Osterkamp and
Gosink (1982) suggest that such ice is formed when conditions
produce sufficient turbulence to suspend sediments at the same
time that frazil ice is present during the freezeup period.
Sediment-laden ice is not observed every year and appears to be
related to fall storm events. Naidu et al. (1982) reports that

Fig. 15. 2 m thick sediment laden ice block in Egg Island channel
May 1978, from Matthews (1982).

winds of 8 m sec^{-1} are needed to resuspend sediments and winds of
this strength are not uncommon along the Beaufort Sea coast.
Using Matthews' (1981b) criterion of currents 3.4% of the wind
speed, we expect sediment-laden ice to be formed with currents of
about 30 cm sec^{-1} in the lagoons.

By January the rate of ice formation has slowed considerably
due to the lower heat loss from thicker sea ice and especially the
snow cover (McPhee and Untersteiner, 1982). This is shown in very
small mean currents (< 1 cm sec^{-1}) in Stefansson Sound in January
and February. Brine however collects in isolated pockets and
channels and can reach values well in excess of 180 ^{o}/oo by late
spring (Schell, 1974). Currents under the landfast ice in late
spring are very small (< 1 cm sec^{-1}) due to the lowered ice growth
rate, the incidence of fewer surges, and the lowered free volume
under the ice cover.

Spring breakup in the landfast ice is initiated in early
June by the overflowing of coastal rivers (Matthews and Stringer,
1982). In the Egg Island Channel, the main channel through the
Simpson Lagoon barrier island chain, change from cold saline water
(43 ^{o}/oo) to fresh water occurs in a few hours even at the bottom
of the 50 m deep channel. This is clearly shown in Fig. 16.
Hypersaline water near its freezing point gently oscillates with

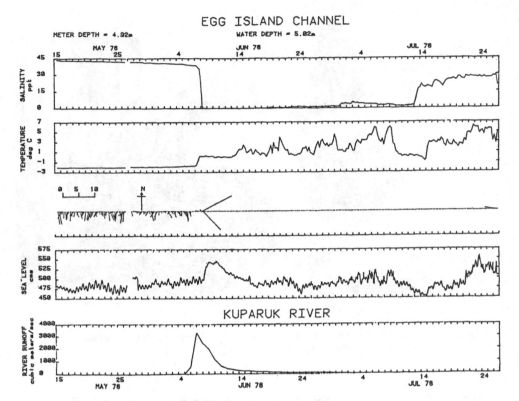

Fig. 16. Salinity, temperature, current vectors, sea level and
         river runoff for Egg Island Channel, Simpson Lagoon for
         15 May – 26 July 1978 from Matthews and Stringer (1982).

small tides before breakup. The breakup floodwaters appear in the
tidal signal as a 60 cm rise which occurs with strong currents and
fresh cold water at freezing point. Matthews and Stringer (1982)
showed that the spring flooding occurs each year on May 30 with a
standard deviation of 8 days.

They also showed, from examination of satellite data, that
the lagoon is icefree and saline oceanic water re-enters the lagoon
by 12 July with a standard deviation of 5 days. The six week period
from spring flood to reestablishment of coastal water is an annual
process to which the biological processes are synchronized (Johnson
and Richardson, 1981). The coastal lagoons are important breeding
and nursery areas to birds and fish during the short summer season.
The full complexity and dynamics of the arctic ecosystem are explored in detail by Johnson and Richardson (1981).

With the reestablishment of saline coastal water in mid-July,
we complete our description of the seasonal cycle of the barrier
islands and nearshore region of the arctic ocean.

CONCLUSIONS

Recent exploration of the dynamics of the Beaufort Sea shelf has revealed some of the physical processes upon which the active arctic ecosystem depends. In summer and fall Bering Sea and Alaska Coastal water travels eastwards along the shelf break. Upwelling events appear in the summer and are apparently related to wind events. A different type of upwelling has been reported to occur on the outer shelf for which an explanation has not been offered.

A major part of the cross shelf circulation appears to be related to brine drainage flow and the counter current of less saline surface waters. The brine drainage can be seen as far as 50 km offshore in 60 m water in late winter. The landward surface flow under the ice cover appears to be important to epontic algae which are a major component of the arctic ecosystem.

Summer circulation on the inner shelf is dominated by a westward wind-driven flow at about 3% of the wind speed. Coastal river water mixes with the cold saline water and as freezeup approaches in September coastal salinity is about 30 ^{o}/oo. If fall storms produce winds of 8 m sec^{-1} and currents of about 30 cm sec^{-1}, sediments will be entrained in the ice cover in the shallow (< 10 m) parts of the shelf. Ice grows at about 1 cm day^{-1} for about 50 days in fall producing ice at about 6 ^{o}/oo and increasing seawater salinity at 0.04 ^{o}/oo day^{-1}. A sluggish brine drainage, topographically controlled in lagoons, is initiated and superimposed on currents generated by tides and sea level variation. Surges are commonly > 2 m with negative surges occurring more frequently in winter while the tidal range is 15 cm.

Breakup of the landfast ice in lagoons occurs in late May of early June and is initiated by coastal rivers overflowing landfast ice. The lagoons contain freshwater derived from river runoff and meltwater from landfast ice. Pockets of very saline water (> 40 ^{o}/oo) are flushed from the lagoon and temperatures reach 7°C or more. Saline coastal water (~ 30 ^{o}/oo) re-enters the lagoons by mid July. Summer and fall storms during the open water season account for the mean annual coastal retreat of 1.1 m.

Recent work has filled in some of the details of Beaufort Sea shelf dynamics and forms a basis for future work.

The work reported here results from many discussions with colleagues, too numerous to mention, working in the Beaufort Sea. Their help, cooperation and stimulation is gratefully acknowledged. The financial support from the State of Alaska through the Federal Budget Impact Fund allowed the work to be brought to this present conclusion. The work would have been impossible without the support and encouragement of Nina Livingston.

REFERENCES

Aagaard, K., 1981, Current measurements in possible dispersal
        regions of the Beaufort Sea, in: "Final Report Research
        Unit 91, Ref. A81-02", Department of Oceanography, Uni-
        versity of Wasihington, Seattle, WA, 74 pp.

Aagaard, K., Coachman, L.K., and Carmack, E., 1981, On the
        halocline of the Arctic Ocean, Deep-Sea Res., 28(6A):
        529-546.

Bushuev, A.V., 1964, Plasticity and isostatic equilibrium of
        an ice cover,Trudy, Arktiki i Antarktiki Instituta, 267:
        105-109.

Carlson, R.F., and Kane, D.L., 1975, Hydrology of Alaska's
        Arctic, pp.367-373, in: "Climate of the Arctic", Gunter
        Weller and S.A. Bowling, eds., Geophysical Institute,
        University of Alaska, Fairbanks, Alaska.

Harrison, W.D., and Osterkamp, T.E., 1979, Heat and mass tran-
        sport processes in subsea permafrost, 1. An analysis of
        molecular diffusion and its consequences, J.Geophys.
        Res. 83(C9): 4707-4712.

Henry, R.F., and Heaps, N.S., 1976, Storm surges in the
        southern Beaufort Sea. J. Fisheries Res. Board of
        Canada, 33(10):2362-2376.

Hufford, G.L., 1973, Warm water advection in the southern
        Beaufort Sea, August-September, 1971. J.Geophys.Res.,
        78(15):2702-2707.

Hufford, G.L., 1974, On apparent upwelling in the southern
        Beaufort Sea, J. Geophys. Res., 79:1305-1306.

Hume, J.E., and Marshall Schalk, 1967, Shoreline processes
        near Barrow, Alaska: a comparison of the normal and the
        catastrophic. Arctic, 20(2):86-103.

Hunkins, K.L., 1962, Waves on the Arctic Ocean, J. Geophys.
        Res. 67:2477-2489.

Johnson, M.W., 1956, The plankton of the Beaufort and Chukchi
        Sea areas and its relation to the hydrography.
        Technical Paper Number 1, Arctic Institute of North-
        America, Arlington, VA. 32 pp.

Johnson, S.R., and Richardson, J.W., 1981, Beaufort Sea-
        Barrier Island- Lagoon ecological process studies;
        Final Report, Simpson Lagoon. Vols. 7 and 8, Final
        Reports of the Principal Investigators, Environmental
        Assessment of the Alaskan continental shelf, Arctic
        Project Office, Geophysical Institute, University of
        Alaska, Fairbanks, AK.

Kovacs, A., and Mellor, M., 1974, Sea ice morphology and ice
        as a geologic agent in the southern Beaufort Sea. p
        113-164, in: Reed, J.C. and Sater, J.E. (eds.) "The
        Coast and Shelf of the Beaufort Sea", Arctic Inst.
        North America, Arlington, VA, 749 pp.

Kowalik, Z., and Matthews, J.B., 1982a, The M_2 tide in the
    Beaufort and Chukchi Seas, J.Phys.Ocean., 12(7): 743-
    746.

Kowalik, Z., and Matthews, J.B., 1982b, Numerical study of
    the water movement driven by brine rejection from near-
    shore arctic ice. Accepted for publication, J.Geophys.
    Res., Oct.

Kozo, T.L., 1979, Evidence for sea breezes on the Alaskan
    Beaufort Sea Coast, Geophys. Res. Letters, 6(11):
    849-852.

Lewellen, R.I., 1974, Offshore permafrost of the Beaufort Sea,
    Alaska. p 417-426, in: Reed, J.C. and Sater, J.E. (eds.),
    "The Coast and Shelf of the Beaufort Sea",Arctic Inst.
    North America, Arlington, VA. 749 pp.

Matthews, J.B., 1971, Long period gravity waves and storm
    surges on the Arctic continental shelf. Proc. Joint
    Oceanogr. Assembly, Tokyo. p 332.
    Matthews, J.B., 1981a, Observations of under-ice circu-
    lation in a shallow lagoon in the Alaskan Beaufort Sea,
    Ocean Management, 6:223-234.

Matthews, J.B., 1981b, Observations of surface and bottom
    currents in the Beaufort Sea near Prudhoe Bay, Alaska,
    J. Geophys. Res., 86(C7): 6653-6660.

Matthews, J.B., 1982, Seasonal circulation in some Alaskan
    arctic lagoons. Accepted, Oceanologica Acta, ISCOL/
    UNESCO volume, Dec.

Matthews, J.B., and Stringer, W.J., 1982, Spring breakup and
    flushing of an arctic lagoon estuary. Submitted J.
    Geophys. Res. Nov.

McPhee, M.G., and Untersteiner, N., 1982, Using sea ice to
    measure vertical heat flux in the ocean, J.Geophys.Res.,
    87(C3):2071-2074.

Mountain, D.G., 1974, Bering Sea water on the north Alaskan
    Shelf. Ph.D. Thesis, University of Washington, Seattle,
    WA., 153 pp.

Naidu, A.S., Mowatt, T.C., Rawlinson, S.E., and Weiss, H.V.,
    1982, Barrier island-lagoon systems of northern Alaska:
    Lithologies, depositional processes, evolution, and
    stability, in: Barnes, Peter, Reimnitz, Erk and Schell,
    D.M. (eds.), "Coastal processes of the Beaufort Sea",
    (tentative title), Academic Press, N.Y. in preparation.
    Dec.

Osterkamp, T.E., and Gosink, J.P., 1982, Observations and
    analyses of sediment-laden sea ice, in: Barnes, Peter,
    Reimnitz, Erk, and Schell, D.M. (eds.), "Coastal Pro-
    cesses of the Beaufort Sea"(tentative title), Academic
    Press, N.Y. in preparation, Dec.

Osterkamp, T.E., and Payne, M.W., 1981, Estimates of perma-
    frost thickness from well logs in northern Alaska, Cold
    Regions Science and Technology, 5:13-27.

Paquette, R.G., and Bourke, R.H., 1974, Observations on the coastal current of arctic Alaska, J. Mar. Res. 32(2): 195-207.

Reimnitz, Erk, Barnes, P.W., Forgatsch, T.C., and Rodeick, C.S., 1972, Influence of grounding ice on the arctic shelf of Alaska, Mar. Geol., 13:323-334.

Reimnitz, Erk, and Maurer, D.K., 1979, Effects of storm surges on the Beaufort Sea coast, northern Alaska, Arctic, 32(4):329-344

Schafer, P.J., 1966, Computation of a storm surge at Barrow, Alaska. Archiv für Meteorologie, Geophysik und Bioklimatologie, Ser. A. Meteorologie und Geophysik, 15(3-4):372-393.

Schell, D.M., 1974, Regeneration of nitrogenous nutrients in arctic Alaskan estuarine waters. p649-664, in: Reed, J.C. and Sater, J.E. (eds.). "The Coast and Shelf of the Beaufort Sea". Arctic Inst. North America, Arlington, VA. 749 pp.

Schell, D.M., Ziemann, P.J., Parrish, D.M., Dunton, K.H., and Brown, E.D., 1982, Foodweb and nutrient dynamics in nearshore Alaska Beaufort Sea waters. Final report, Contract 3-5-022-56, Research Unit 537, Institute of Water Resources, University of Alaska, Fairbanks, AK. 185 pp.

Searby, H.W., and Hunter, M., 1971, Climate of the North Slope, Alaska. NOAA Technical Memorandum NW3 AR-4, National Weather Service, Anchorage, AK. 53 pp.

Short, A.D., Coleman, J.M., and Wright, L.D., 1974, Beach dynamics and nearshore morphology of the Beaufort Sea coast, Alaska. p 477-488, in: Reed, J.C. and Sater, J.E. (eds.), "The Coast and Shelf of the Beaufort Sea". Arctic Inst., North America, Arlington, VA. 749 pp.

Stringer, W.J., Bauman, M.E., and Roberts, L.J., 1981, Summertime ice concentrations in the Harrison Bay and Prudhoe Bay vicinities of the Beaufort Sea. Geophysical Inst. Report, NOAA OCEAP, RU 427, University of Alaska, Fairbanks, AK. 35 pp.

Thorndike, A.S., and Colony, R., 1980, Arctic ocean buoy program, data report, 19 January 1979 to 31 December 1980. Polar Science Center, University of Washington, Seattle, WA. 131 pp.

Thorndike, A.S., and Colony, R., 1981, Arctic ocean buoy program, data report, 1 January 1980 to 31 December 1980. Polar Science Center, University of Washington, Seattle, WA. 127 pp.

Thorndike, A.S., Colony, R., and Nunoz, E.A., 1982, Arctic ocean data buoy program, data report 1 January 1981 to 31 December 1981. Polar Science Center, University of Washington, Seattle, WA. 137 pp.

U.S.G.S., 1978, Water resources data for Alaska, Water data report AK 78-1, Water year 1978. U.S. Dept. Interior,

Geological Survey, Anchorage, AK. p 269-276.

Walker, H.J., 1973, Spring discharge of an arctic river deter-
    mined from salinity measurements beneath sea ice.
    Water Resources Research, 9(2):474-480.

Walker, H.J., 1974, The Colville River and the Beaufort Sea:
    Some interactions. p 513-540, in: Reed, J.C. and Sater,
    J.E. (eds.), "The Coast and Shelf of the Beaufort Sea".
    Arctic Inst. North America, Arlington, VA. 749 pp.

Wiseman, Jr, W.J., Suhayda, J.N., Hsu, S.A., and Walters, C.D.,
    1974, p 49-64, in: Reed, J.C. and Sater, J.E. (eds.),
    "The Coast and Shelf of the Beaufort Sea". Arctic Inst.
    North America, Arlington, VA. 749 pp.

LOW FREQUENCY FLUCTUATIONS IN THE SKAGERRAK

Gary Shaffer

Oceanographic Institute
Gothenburg, Sweden

ABSTRACT

Fluctuations in the Skagerrak of period 5 days or more are analysed. Two distinct frequency bands are identified by coherence and phase analysis. In the first, a 5 to 6 day period band, fluctuations in the eastern Skagerrak and in the fjord Gullmaren appeared to be driven by the wind over the northern North Sea. Topographic waves propagating eastward along the Norwegian Trench may communicate this wind forcing. In the second, a 2 week or more band, the major Empirical Orthogonal Function analysis current mode from the coastal zone in the eastern Skagerrak was more coherent with northern North Sea rather than local wind. A quasi-steady topographic gyre or topographic boundary layer model is proposed for the very low frequency barotropic circulation in the Norwegian Trench-Skagerrak region. It is unclear whether the 2 week or more period internal fluctuations in Gullmaren are more related to conditions in the Skagerrak or in the Kattegatt. Strong but infrequent local up-welling or downwelling forced by the wind parallel to the Swedish Skagerrak coast complete the picture.

INTRODUCTION

The fjord Gullmaren lies at the eastern end of the Skagerrak (Fig. 1). Internal oscillations of period longer than the inertial period have been known and investigated for most of this century and have been variously attributed to wind or atmospheric pressure and to local or remote forcing (Petterson, 1909; Petterson, 1920). Neutrally buoyant floats registered oscillations of the permanent halocline there as early as 1909 (Petterson, 1917). The pressure difference between two fixed depths was observed during the summer of 1918 at three positions in west Sweden - Böttö in the northern

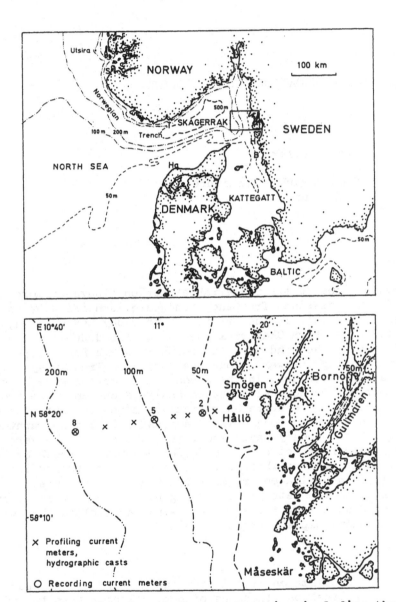

Fig. 1. The Norwegian Trench–Skagerrak region including the area
of study for the spring 1979 field program. H = Hättan,
B = Böttö, Ha = Hanstholm, T = Tregde and S = Stavanger.

Kattegatt (2 to 20 m), Hättan in the southermost Skagerrak (2 to 25 m)
and Bornö (2 to 25 m) (Petterson, 1919). The results of this experiment are redrawn in Fig. 2 together with the SW-NE wind at Hållö and
an approximate scale relating to halocline displacement.

All series of observations of halocline depth at Bornö found
in the literature exhibit the same character as in Fig. 2: the
lower frequency oscillations, often grouped around periods of 5-7
days and of about 2 weeks, dominate with amplitudes of up to 5 m or
more and up to 10 m respectively. This also holds for the cold
winters with ice cover over Gullmaren and the eastern Skagerrak
(Johnson, 1943). Johnson noted that halocline depth and atmospheric
pressure curves coincided well during parts of the severe 1940-42
winters and concluded that, as the wind could not directly force
water movements in the eastern Skagerrak owing to the ice cover, the
"barometric effect" was important. On the other hand, H. Petterson
(1920) found generally good correlation between local winds and internal displacements over most of the year.

It became clear that low frequency fluctuations with periods
of 5 days or more were coupled to fluctuations in the eastern
Skagerrak: Gullmaren was a convenient "probe" to monitor phenomena
with much larger scales. The Skagerrak and the northern North Sea
form a topographic unit characterized by the Norwegian Trench, which
reaches 700 m depth in the Skagerrak from a sill depth of about

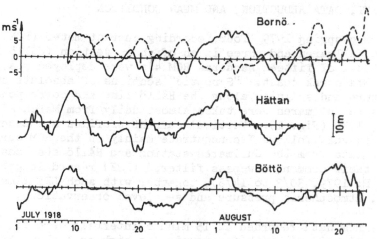

Fig. 2. Pressure difference observations (see text) from the
locations off the Swedish west coast (Fig. 1) during the
summer of 1918 as well as SW wind (dashed line) components
from Hållö. An approximate vertical scale for the observations in terms of vertical halocline displacements (up-
ward-positive) is given. Redrawn from Petterson (1919).

270 m off Utsira. The Kattegatt and North Sea south of the trench
are much shallower with mean depths of 23 m and 80 m. Salty North
Sea water (S > 34 °/oo) flows into the Skagerrak along the north
coast of Denmark and southern edge of the Norwegian Trench as the
Jutland current. Off Denmark's northeast tip this water meets
brackish Kattegatt surface layer water (S ~ 20-30 °/oo), which flows
northward as the Baltic current in a wedge along the Swedish coast
over the North Sea water. This current, part of an estuarine circulation with fresh water supply from the Baltic Sea, proceeds northward and then westward along the coast of the Skagerrak to become
the Norwegian coastal current. The stratification along the Swedish
Skagerrak coast is therefore characterized by a strong permanent
halocline ($\Delta\sigma_t \sim$ 4-8), which permeated Gullmaren and other inlets
along this coast. The halocline is centered at about 15 m depth
at the coast and intersects the sea surface within 10-20 km from
shore except for a thin layer of somewhat saltier, brackish water
usually covering most of the eastern and northern Skagerrak.

Internal oscillations in Gullmaren clearly have to be understood in terms of conditions in the Skagerrak or even further afield.
Earlier workers lacked necessary tools for this jobs like modern
instruments and statistical analysis methods as well as appropriate
mechanistic models. Work along these lines has recently been
started based on a 1979 field study (Shaffer and Djurfeldt, 1982,
henceforth called SDS). In what follows I concentrate on periods
of 5 days or more, using these and earlier results.

EXPERIMENT, DATA REDUCTION, AND MEAN CONDITION

During spring 1979, four recording current meter (RCM) moorings
- one in Gullmaren and three in an offshore section (Hållö line) -
with a total of fifteen Aanderaa RCM 4's were deployed near the
Swedish Skagerrak coast. "Synoptic" sections of absolute current,
temperature and salinity along the Hållö line and corresponding
profiles in Gullmaren were taken almost daily from March 27 to
April 6, 1979 (Fig. 1). Vertical current profiles are described in
SDS and in what follows I concentrate mainly on the RCM current and
salinity data from the Gullmaren station and Hållö stations H2 and
H5 over their common, low-pass filtered (LPF) record lengths
(Period P: 1979, 30/3 to 14/5), together with such LPF time series
of wind, atmospheric pressure and sea level observations.

The RCM data recorded at 15 minute intervals were filtered to
suppress all signals with frequencies as high or higher than the
local inertial frequency (0.07 cph). The symmetrical filter spanned
109 hours and passed 1% and 99% of the power at 0.069 and 0.042 cph
respectively with a half power point at 0.055 cph. The data of the
other parameters were reduced in similar manners.

Table 1 shows means, standard deviations and their ratios

Table 1. Means, standard deviations (σ) and their ratios of alongshore current (+V = 340°) for Hållö station 2 and 5 RCM's
and of inward current (+V = 46o) for the Gullmaren station
RCM's over the period 790330-790514).

RCM	$\overline{V}$	σ	$V\sigma^{-1}$
H2/27	6.5	8.1	0.8
H2/54	0.3	8.9	0.0
H5/10	45.0	15.8	2.8
H5/28	28.7	14.6	2.0
H5/47	20.3	15.4	1.3
H5/99	6.3	9.3	0.7
G/5	3.5	13.1	0.3
G/16	-0.4	9.5	0.0
G/27	-1.6	5.6	0.2
G/48	-0.6	3.0	0.2

based on the LPF alongshore (+V → 340°) and inward (+V → 46°) current
data from the Hållö line and Gullmaren respectively. The strong,
persistent, northward and baroclinic flow near the surface at H5
can be identified as the Baltic current. In Gullmaren the means
were much less than standard deviations. The mean stratification
during this period was as described earlier. Mean alongshore
currents based on all three moorings on the Hållö line, but for a
much shorter common record length, showed weak (6-7 cms^{-1}) northward, barotropic flow at H8 and results very similar to those in
Table 1 for H2 and H5 (SDS).

A DOWNWELLING EVENT AND ITS RELAXATION

Onshore winds are usually more frequent and stronger than
alongshore ones at the Swedish Skagerrak coast. This was the case
during spring 1979 (March-June) when onshore winds (W, SW, E, NE)
exceeded alongshore winds (S, SE, N, NW) by 60% to 40% and 7.9 to
5.9 m s^{-1} at Smögen (Fig. 1). For SW winds alone these values were
32% and 9.1 m s^{-1}. Strong, southeasterly winds did, however, occur
between March 25 and March 27, 1979 with speeds up to 16 m s^{-1}. This
proved to be the strongest alongshore wind event of spring 1979.

Figure 3 shows "synoptic" section salinity and alongshore
current results from the Hållö line and salinity and inward current
profiles in Gullmaren from March 27 near the end of this wind event
and from March 29 after its relaxation. On March 27 the brackish,
Kattegatt water wedge was observed compressed and deepened near
the coast. The Gullmaren halocline had followed the coastal zone
one downward (Fig. 3a). This was caused by downwelling at the
coast driven by the strong south-easterlies and the shoreward Ekman

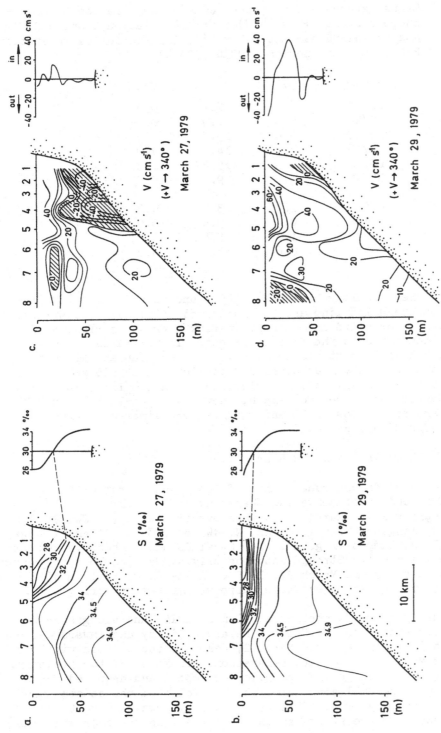

Fig. 3. "Synoptic" sections of salinity and alongshore current from the Hållö line (left) and inward current profiles from the Gullmaren station (right) before and after the relaxation of a strong downwelling event during March 1979.

transport associated with them. By March 29 the downwelling had
ceased, the brackish wedge had spread seaward, and the nearshore
halocline had risen "pulling" the Gullmaren halocline with it
(Fig. 3b).

      The alongshore currents at the Hållö line on March 27 were
dominated by a region of strong vertical shear near the coast which,
in the lack of strong barotropic flow, led to strong subsurface
flow against the wind. This, together with the stratification, fits
well with the classical picture of baroclinic response to a down
welling event albeit in the unusual setting of a brackish water
wedge. Only weak inflow-outflow was observed in Gullmaren on March
27 (Fig. 3c). Two days later the situation was dramatically different: The vertical shear was much weaker as could be anticipated
from the flatter isohalines in Fig. 3b. The northward, barotropic
flow centered around H5 had increased. The upward movement of the
Gullmaren halocline apparently had induced large internal pressure
gradients along the fjord and intense inflows and outflows. About
half of the water above the sill depth ($\sim$ 50 m) in Gullmaren was
exchanged within one day by this event.

      This example shows clearly how strong downwelling (or upwelling)
of at least several days duration can produce large, locally driven,
low frequency fluctuations near the Swedish Skagerrak coast.
Although infrequent, such events represent one of the mechanisms
responsible for the large fluctuations of halocline depth and thus
for great water exchange in Gullmaren and elsewhere along the
Swedish west coast. These mechanisms are probably even more important along the southern coast of Norway where frequent west- southwesterly wind events lead to seaward Ekman transport and upwelling
(Aure and Saetre, 1981).

THE LOW FREQUENCY FLUCTUATIONS: STATISTICAL ANALYSIS

      In this section it is attempted to identify organized types of
low frequency motions in the eastern Skagerrak and in Gullmaren
during spring 1979, to unravel the space and time structure associated with them and to search for relationships among these types
and other parameters, all with the help of statistical analysis
tools. Coherence squared and phase spectra were calculated for all
LPF alongshore current record pairs from H2 and H5 during period P.
The relative number of such pairs exhibiting statistically significant coherence as functions of frequency is plotted in Fig. 4.
Coherence squared significance levels were calculated according to
Julian (1975) for 10 degrees of freedom. The 95% curve indicates
that alongshore flow was best organized for well defined bands
corresponding to periods of slightly over 2 days, 5-6 days, and 2
weeks or more. Comparison with power spectra suggested that the
proportion of organized motions (currents and waves?) to "noise"
(turbulence?) is particularly great in these bands (SDS) so that

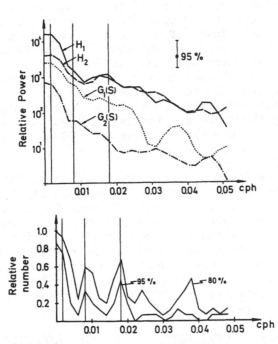

Figure 4. Power spectra of the first two Hållö line EOF current
          modes and the first two Gullmaren EOF salinity modes
          (top). Ratios of coherent alongshore current RCM pairs
          (to the 80% and 95% significance level), Julian (1975))
          from Hållö line stations 2 and 5 to the total number of
          possible pairs (15) as a function of frequency (bottom).

Table 2. Coherence squared and phase spectra estimates of alongshore current records from selected Hållö line RCM pairs
for frequency bands corresponding to the 5-6 day and 2
week or more periods. Only coh^2 values above .51 (95% significance level), are included. 95% confidence intervals
are given for the phase estimates (Jenkins and Watts, 1968).
The second record leads the first for positive phase.

RCM pairs	0.002 cph (504 hrs)		0.008 cph (126 hrs)	
	coh^2	phase (degrees)	coh^2	phase (degrees)
H5/47–H5/99	0.87	16 ± 7	0.70	20 ± 19
H5/28–H5/99	0.83	31 ± 12	0.53	15 ± 26
H5/28–H5/47	0.96	17 ± 2	0.79	6 ± 15
H5/10–H5/28	0.76	36 ± 16	–	–
H5/10–H5/47	0.68	55 ± 20	–	–
H5/10–H2/27	0.56	37 ± 25	–	–
H5/28–H2/27	0.74	7 ± 17	–	–
H5/47–H2/54	–	–	–	–
H5/99–H2/54	–	–	0.63	143 ± 22

the possibility of explaining the observations by deterministic
models should be best in these bands. In the following I concentrate on the 5-6 day and 2 week or period motions. The 2 day or
more period motions were considered in SDS.

Table 2 shows coherence squared and phase estimates for the
two spectra bands of interest for several selected RCM pairs with
$coh^2 > .51$ (95% significance level). Each of the bands exhibits
somewhat different coherence and phase structure. For the 504 hour
period, the signal was coherent throughout the H5 water column and
with the H2/27 but not with H2/54 RCM. Shallower currents lagged
deeper ones. For instance, H5/28 lagged H5/99 by 42 ± 17 hours.
For the 126 hour period the signal was coherent at H5 below 28 m
and along the bottom between H5/99 and H2/54; shallower currents
again lagged deeper ones.

Empirical Orthogonal Function (EOF) analysis was applied to
the P period RCM records from the Hållö line and Gullmaren. Unfortunately the common record length was too short and too much variance was concentrated at periods of 2 weeks or more to be able to
isolate each of the two bands of interest by band-pass filtering.
EOF analysis organizes the in-phase variance in a set of N time
series into as many modes which converge quicker than any other
orthogonal base set of functions (Davis, 1976). Eigenfunctions

(mode spatial structure) and eigenvalues (amplitude function time series) are the output from the analysis.

Figure 5 shows the first two Hållö EOF modes which are so-called "overall" modes (Huyer et al., 1978), using onshore and alongshore current component time series as input. The quasi-barotropic H_1 mode accounted for 46% of the total variance (over 70% for the H5/28 and 47 m records). The H_1 power spectra (Fig.4) shows that this mode was dominated by the 2 week or more period

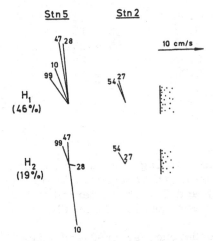

Figure 5. The first two EOF "overall" current modes for the Hållö
         line plotted as vectors in space. The numbers on each
         vector identify the depth of its RCM. The length of each
         vector is scaled by the modal standard deviation.

motions as was the baroclinic H_2 mode which accounted for 19% of the total variance ($\sim$ 50% for the H5/10 record). The strongest currents were found at H5 for both modes.

EOF analysis of the Gullmaren current and salinity data yielded the following results: the first two current modes, G_1 and G_2, accounted for 68 and 22% of the total variance respectively. Their coordinates for G5, 16, 27 and 48 m were 12.7, -6.0, -3.6 and -0.1

cm s^{-1} and 2.8, 7.3, -2.3 and 1.4 cm s^{-1} respectively. These have been scaled by the modal standard deviation and are the coordinates along the fjord axis. Crossfjord coordinates were negligible. The zero-crossover of the apparent two layer flow described by both modes was shallower for G_1 than for G_2. The first two salinity modes, $G_1(S)$ and $G_2(S)$, which accounted for 82 and 15% of the total variance respectively had the following coordinates (again scaled by the modal standard deviation): 0.53, 2.50, 0.27 and 0.03 o/oo and 1.06, -.21, -.18 and -.03 o/oo respectively. $G_1(S)$ reflected vertical displacements of the halocline past the G/16 m RCM whereas $G_2(S)$ reflected surface layer salinity changes. The power spectra of both modes (Fig. 4) were dominated by very low frequency fluctuations. The relative importance of the 5-6 day period was greater for $G_1(S)$.

Table 3 summarizes some of the results of extensive calculations of coherence squared and phase spectra among the period P

Table 3.   Coherence squared and phase spectra estimates for a number of combinations of Hållö line and Gullmaren EOF modes (see text), Smögen and Utsira wind stresses (τ_s and τ_u) and adjusted sea level (ASL) from Smögen for two frequency bands corresponding to the 5-6 day and 2 week or more periods. Only coh^2 values above .32 (80% significance level) are included. 95% confidence intervals are given for the phase estimates.

RCM pairs	0.002 cph (450 hrs)		0.008 cph (112.5 hrs)	
	coh^2	phase (degrees)	coh^2	phase (degrees)
H_1 - ASL	0.57	68 $\pm$ 25	-	-
H_1 - $\tau_s(40°)$	0.47	18 $\pm$ 29	-	-
H - $\tau_u(90°)$	0.62	27 $\pm$ 22	-	-
H_1 - H_2	0.65	86 $\pm$ 21	-	-
ASL - $\tau_s(40°)$	-	-	-	-
ASL - $\tau_u(110°)$	0.34	-43 $\pm$ 36	0.71	-36 $\pm$ 18
$G_1(S)$- ASL	0.50	147 $\pm$ 28	0.69	-47 $\pm$ 19
$G_1(S)$- H_1	-	-	-	-
$G_1(S)$- $\tau_s(40°)$	-	-	0.51	-47 $\pm$ 27
$G_1(S)$- $\tau_u(110°)$	-	-	0.76	-79 $\pm$ 16
$G_1(S)$- $G_2(S)$	-	-	0.67	-164 $\pm$ 20

time series of the EOF modes described above, LPF wind stress from
Smögen and Utsira (Fig. 1), τ_s and τ_u, and LPF adjusted sea-level
(ASL) from Smögen for the spectral estimates closest to the 5-6 day
(112.5 hrs) and 2 week or more (450 hrs) period bands. The combinations involving wind stress are for those wind directions which
yielded maximum coh^2. A certain pattern emerges for each of these
bands. For the 450 hour period, H_1 was more coherent with northern
North Sea (τ_u) than local (τ_s) wind stress. $G_1(S)$ was not coherent
with either of them nor with H_1. Adjusted sea level was significantly but rather weakly coherent with H_1, $G_1(S)$ and τ_u. Although
it led H_1 and τ_u, it was nearly out of phase with $G_1(S)$ as would be
expected from steric adjustment. Although H_1 and H_2 were coherent
with about 90° lag – consistent with the earlier coh^2 and phase
calculations and implying that H_1 and H_2 really represent one fluctuation type with vertical phase lag – $G_1(S)$ and $G_2(S)$ were unre-
lated.

For the 112.5 hour band, no combination including H_1 showed
significant coherence. It is doubtful if the 5-6 day fluctuation
type which emerged in the RCM pair coherence and phase analysis is
represented by H_1 at all. The structure of H_1 was clearly determined by the very energetic 2 week or more fluctuation type (Fig. 4).
On the other hand, $G_1(S)$ and ASL were highly coherent with the
northern North Sea wind stress and with each other. Negative $G_1(S)$
– deep halocline in Gullmaren – lagged ASL and τ_u by 143° and
101° respectively. Other combinations (including the G_1 and G_2
modes for instance) have been calculated and are considered in SDS.

THE 5-6 DAY PERIOD: TOPOGRAPHIC WAVES IN THE NORWEGIAN TRENCH –
SKAGERRAK?

As shown above, the fluctuations of halocline depth in Gullmaren
and adjusted sea level at Smögen were more coherent with the wind
stress at Utsira than with the local wind stress at Smögen in the
5-6 day band during spring 1979. A comparison of the spectra of
these wind stresses shows the variance at these positions to be of
the same order for periods greater than 5 days. $G_1(S)$, ASL and,
presumably the 5-6 day alongshore current fluctuation type found
earlier from coh^2 and phase spectra analysis of Hållö line RCM pairs
may have been remotely forced by the wind over the northern North
Sea.

The relationship between ASL and τ_u is probably explainable
as set-up of the whole north European shelf sea owing to west winds
in the adjacent North Atlantic. This set-up would be communicated
to the whole region within a matters of hours by long gravity waves,
explaining the observed phase lag of τ_u behind ASL (wind events
generally arrive later at Utsira than at the continental shelf break).
Subsequently, the west wind blowing over the Norwegian Trench –
Skagerrak region would force barotropic jets in the wind direction
along the Norwegian coast and along the western and southern edge

of the Norwegian Trench within some fraction of the inertial period
after the wind's onset. These develop as a consequence of relative
vorticity generation caused by initial barotropic cross-isobath
compensation flow to the Ekman divergence/convergence at the coasts.
The jet over the trench would propagate eastward into the eastern
Skagerrak as a free topographic wave. In that region of strong
stratification, baroclinic interaction with the barotropic wave may
be expected and, with that, fluctuations of the halocline near the
coast, the $G_1(S)$ mode.

Second class, topographic waves - in particular, so-called
continental shelf waves - have been the object of much study over
the last 10 to 15 years (Buchwald and Adams, 1968; Gill and Schumann,
1974 and others). They have been proposed to explain low frequency
fluctuations, particularly of adjusted sea level, along many of the
world's coasts (Hamon, 1966; Cutchin and Smith, 1973 and others).
For a given piece of sloping bottom, four different types of topographic waves can be considered which correspond to four vertical
wall - flat bottom combinations for bounding this piece. In SDS
the propagation characteristics of these four types are considered
for an exponential bottom profile. Whereas the bottom profile
across the Norwegian Trench (off Utsira for instance) suggests a
trench wave model, that across the western Skagerrak suggests a
wedge wave one. The steep slope off the Norwegian coast is modelled
as a vertical wall in each case. The dispersion relations for the
first modes of these two waves for an exponential bottom profile
(a good fit to the actual topography) and the appropriate local topographic parameters are shown in Fig. 6. The first mode continental
shelf wave dispersion relation with Skagerrak parameters is also
shown for comparison. Given a 5-day period ($\omega f^{-1} = .12$), the dispersion relations yield the following wave lengths (phase speeds)
for the wedge wave and trench wave respectively: 600 km (1.4 ms^{-1})
and 1310 km (3.0 ms^{-1}). The scales along the axis of the Norwegian
Trench for the Skagerrak and the rest of this trench which extends
north along the Norwegian coast to about $62°N$ - ~ 250 km and ~ 600
km respectively - compare well with the associated half wave lengths.
Thus, resonant forcing of these waves by a wind over the northern
North Sea with a period of 5-6 days is possible. A trench wave
propagating into the Skagerrak would be expected to retain its
frequency, slow down, shorten and proceed eastward as a wedge wave.

Unfortunately, few data are available from the Skagerrak
suitable for detecting topographic waves propagating along a coast
by coherence and phase spectra analysis of data from different
alongshore positions. The adjusted sea level fluctuations associated with such waves would, in general, be less than those due
to wind set-up in the Skagerrak. Analyses of ASL pairs from several locations along the Danish, Swedish and Norwegian coasts around
the Skagerrak showed very high coherence in the 5-6 day band but
very small phase lags, consistent with gravity wave propagation (SDS).

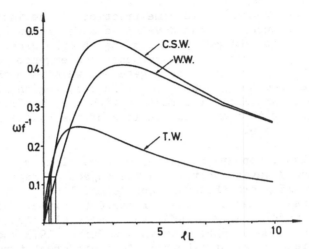

Figure 6. Dispersion curves for the first modes of different free
topographic waves over exponential bottom profiles "fitted"
to actual bottom topographies in the Norwegian Trench-
Skagerrak region. C.S.W. = Continental shelf Wave, W.W. =
Wedge Wave and T.W. = Trench Wave. ωf^{-1} = .12 corresponds
to a period of 5 days, the width of the exponential bottom
is 50 km in all cases, 1 is the wave number.

Simultaneous measurements of current, the dominant signature in
such waves, at different alongshore points are lacking. This points
out the need for future field work. Simultaneous observations of
alongshore current at the surface off Denmark's northeast tip and
at 50 m depth at 100 m water depth off Hållö from June and July
1967 were reported by Möller and Svansson (1978). These showed
clearly that during a period of repeated 4 to 5 day fluctuations,
the Hållö RCM lagged the other record by an amount corresponding
to an eastward phase speed of the order of 1 ms^{-1}.

The phase lag of a deep halocline in Gullmaren behind the
Utsira wind stress of 101 ± 16° in the 5-6 day band and the assumption of about 300 km separation between the region of wind forcing
and the Swedish Skagerrak coast yields an eastward phase speed of
2.5 ms^{-1}. This is consistent with the above hypothesis of eastward
propagating trench, and then wedge, waves and their interaction
with the stratification near the Swedish coast. The details of
such an interaction are beyond the scope of this paper. The complexity of the Eastern Skagerrak as a wave guide – abruptly turning
topography, frontal zone off Denmark's northeast tip, brackish
water wedge with its strong, shallow halocline and large mean
current shears, quite strong stratification over the shelf below

this halocline - precludes a simplified analysis. It is interesting
to note, however, that the results of the RCM pair coherence and
phase analysis for the 5-6 day band - significant coherence only
below the surface layer and along the bottom with slight upward
phase propagation - are reminiscent of bottom trapped waves (cf.
Wang, 1975). Topographic waves tend to become bottom trapped on
propagation into a region of strong stratification. Note that the
sense of the vertical shear associated with bottom trapping after
a west wind event would be opposite to that of the mean shear off
the Swedish Skagerrak coast.

     Information about the propagation of low frequency halocline
depth fluctuations along the Swedish Skagerrak coast can be derived
from H. Petterson's (1919) data (Fig. 2). These curves were read
at 6 hour intervals corresponding approximately to the original
data collection frequency. Power, coherence squared and phase spectra were then calculated for the Bornö, Hättan and Böttö longest
common record length (Fig. 7). Particularly high coherence was
found between the Bornö and Hättan records for periods of 5 days or
more. For the spectral estimates at 120, 180 and 360 hours, Bornö
was found to lag Hättan by $51 \pm 21°$, $25 \pm 12°$ and $16 \pm 3°$ which,
with their separation distance of 75 km, yields northward phase
speeds of $1.2 \pm .5$, $1.65 \pm .8$ and $1.3 \pm .2$ ms^{-1} respectively. These
are not unlike the first mode wedge wave phase speed nor the first
internal (Kelvin) wave mode phase speed of 1.0 ms^{-1} which can be
calculated from the observed stratification during the spring of
1979. Hättan is also highly coherent with Böttö over periods of
several days or more. However Hättan does not lag but actually
leads Böttö slightly for periods exceeding 5 days. The northward
phase speeds in this case for the 120, 180 and 360 hour spectral
estimates were $.3 \pm .5$, $- .1 \pm .4$, $-1.0 \pm .8$ ms^{-1}. This is further
evidence that the low frequency fluctuations near the coast in the
eastern Skagerrak do not originate in the Kattegatt but rather in
the northern North Sea - Skagerrak. During the summer of 1918,
however, the local wind was more coherent with low frequency halocline fluctuations in Gullmaren than during the spring of 1979.
Petterson (1919) stated: "A rough calculation shows that the Bornö
curve is on the average 3 hours ($\pm$ 1 hour) behind that of Hättan and
4 hours ($\pm$ 2 hours) after Böttö.... (it is) extremely improbable
that these movements can be due to any progressive boundary wave
motion proceeding either parallel or at an angle to the coast line".
However the re-examination above implied such "boundary wave motion".

THE 2 WEEK OR MORE PERIOD: TOPOGRAPHIC GYRES IN THE
NORWEGIAN TRENCH - SKAGERRAK?

     As was shown in Table 3, the first EOF current mode from the
Hållö line was considerably more coherent with the northern North
Sea wind stress than with the local one in the 2 week or more period
band where most of the energy of this mode was concentrated. This,

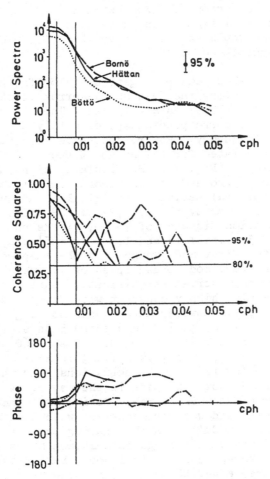

Figure 7. Power, coherence squared and phase spectra based on the
longest common record length of the three records in
Figure 2 and 10 degrees of freedom. The coh^2 and phase
pairs are Bornö–Hållö wind (———), Bornö–Hättan (————),
Hättan–Böttö (–·–·–·) and Bornö–Böttö (·····).

the phase lag of H_1 behind τ_u, and other evidence to be presented
below support the interpretation that the H_1 mode was indeed forced
by the west (east) wind over the northern North Sea. Upon onset of
west (east) winds over the trench not only would topographic waves
be generated as described above but also a pressure gradient would
be set up which would force a deep return flow along the axis of
the trench. The resulting "double cell" circulation pattern would
rotate initially counter clockwise at the eastern end of the Skagerrak along with the advance of the topographic wave there until bottom
friction would become important and stop this rotation. Under the
assumption that quasi-steady barotropic circulation models are
appropriate to describe the circulation in the Norwegian Trench –
Skagerrak for periods of 2 weeks or more, we would expect to find
one of two types of circulation there:

1. a quasi-steady topographic gyre characterized by local, cross
    trench compensation to Ekman divergence (2–D continuity),
    currents in the wind direction near each shore and a deep re-
    turn flow driven by a pressure gradient along the trench axis
    independent of the cross-trench co-ordinate (Csandy, 1974)

    or,

2. quasi-steady topographic boundary layers near each coast
    characterized by divergent boundary currents which accept the
    Ekman divergence and redistribute it around the trench's end
    (3–D continuity). The return currents would be driven by
    alongshore pressure gradients with offshore scales less than
    the trench width (Pedlosky, 1974; Csanady, 1978).

    The dominance of one of these circulation types over the other
would depend, for one thing, on the ratio of trench length to width
as shown below. The H_1 mode off Hållö would be expected to be
connected with the nearshore current off Denmark's northern coast
in both models.

    In the following I sketch the topographic gyre model and apply
it to the Norwegian Trench – Skagerrak. I then present a scale
analysis to determine which of the above two models would be favoured in this region. Finally I draw upon some field observations
to help settle this point.

    The vertically integrated equations for relatively slow,
steady, large scale motions in a basin of variable depth filled
with homogeneous water are:

$$- fv = - g\eta_x + \left[(\tau_w^x - \tau_B^x)(\rho h)^{-1} \right] \qquad (1)$$

$$\left[fu \right] = -g\eta_y + (\tau_w^y - \tau_B^y)(\rho h)^{-1} \qquad (2)$$

$$(uh)_x + (vh)_y = 0 \tag{3}$$

Here u and v are vertically averaged velocities in the x and y directions respectively, η is the surface elevation above equilibrium, τ_w and τ_s are the surface wind and bottom stresses respectively, $h(x,y)$ is the water depth and the subscripts denote partial differentiation.

The special case of a long, straight channel whose axis is in the y-direction is considered. No variations are allowed in this direction but a non-zero η_y exists, supported by some distant channel end ($\eta_{yy} = 0$). The motions are driven by a constant wind in the y direction. The boundary conditions at the channel sides are

$$uh = 0, \quad x = 0,B \tag{4}$$

Since $(vh)_y = 0$, Eqns. (3) and (4) imply that u = 0 everywhere. For these reasons all terms in brackets in Eqns. (1)-(3) vanish. Since $\eta_{xy} = 0$, η_y is found to be a constant called G in the following. Note that no assumptions of "narrow" channel width nor closeness to shore have been made. Ekman layers and geostrophically-balanced cross-channel flow are "included" in the model but their cross-channel transports add up to zero everywhere. Now, if the bottom friction is expressed by the quadratic law,

$$\tau_B^y = \rho C_d |v| v \tag{5}$$

C_d is the drag coefficient – this definition "explains" the neglect of τ_B^x in Eqn.(1) – the equations solved by Csanady (1974) appear. In the above notation

$$v = \pm (\tau_w^y)^{\frac{1}{2}} (\rho C_d)^{-\frac{1}{2}} | 1 - hh_o^{-1} |^{\frac{1}{2}} \tag{6}$$

where h_o is the depth at which the wind stress term balances the alongshore pressure gradient and, consequently, where v = 0. At depths less than h_o, the wind stress term exceeds the pressure gradient term and flow is in the wind direction. The opposite is true for $h > h_o$. Note that v is scaled by a balance of wind and bottom stress terms near the coast. As the channel is bounded at (at least) one distant end, the net alongshore transport must vanish in the steady state,

$$\int_o^B (vh) \, dx = 0 \tag{7}$$

which together with Eqn. 6 gives a recipe for calculating h_o (and subsequently v) for a given bottom profile

$$\int\limits_{h<h_o} h(h_o-h)^{\frac{1}{2}} \, dx = \int\limits_{h>h_o} h(h-h_o)^{\frac{1}{2}} \, dx \tag{8}$$

I applied this model to a cross section in the western Skagerrak (Hanstholm–Tregde, Fig. 1). A value of 372 m was obtained for h_o. For realistic choices of a westerly wind stress of 10^{-1} Nm^{-2}, and a bottom drag coefficient of 5×10^{-3} in- and out-transports of ± .79 SV result (1 Sverdrup (SV) = 10^6 m^{3}s^{-1}). The asymmetric bottom profile of the section in effect "forces" a cyclonic circulation over most of this cross-section for a westerly wind (+ .60 sv on the Danish side). The largest in-transport and current shears are predicted over steep bottom slopes and the largest out-transports (and currents) are predicted over the greatest depths, all located toward the Norwegian side of the Skagerrak.

In the topographic boundary layer model, the interior is quiescent except for the surface Ekman transport. This scales u in Eqn. (2). Also 3-D continuity implies that both terms in Eqn. (3) are comparable. A longshore scale results from combining these two

$$M \sim \rho^{\frac{1}{2}} \, f \, \overline{B} \, \overline{H} \, (C_d \tau_w^y)^{-\frac{1}{2}} \tag{9}$$

where $\overline{B}$ is a characteristic width of the topography and $\overline{H}$ is a mean depth. The alongshore scale of the channel should be greater than this for the topographic gyre type circulation to be favoured. With our earlier choices for τ_w^y and C_d, appropriate ρ and f, $\overline{B}$ = 20 km and $\overline{H}$ = 200 m, one obtains M $\sim$ 700 km. Since M is comparable to the length of the Norwegian Trench, a choice of the one model over the other based on these scaling arguments is difficult.

Note that M decreases with increasing wind velocity, favouring the topographic gyre type circulation (a wind velocity of about 8 ms^{-1} corresponds to the choice of τ_w^y above). Larger effective bottom friction, owing to higher frequency barotropic currents, would have the same effect. Bottom friction calculated from depth averaged currents may be suspect near the Swedish and Norwegian coasts where highly baroclinic, brackish coastal currents are found. The topographic gyre results above showed however that this nearshore region is of secondary importance in that model since the overwhelming amount of total bottom friction essential for the quasi-steady state was associated with currents outside this nearshore region.

The current structures associated with each of the two models sketched above are quite different (see Csanady 1978 for details of the topography boundary layer model, his "arrested topographic wave"). Thus field observations with sufficient time and space resolution might help decide which, if any, of the two models

better describes the 2 week or more period barotropic fluctuations
in the Norwegian Trench-Skagerrak region. The observations off
Hållö dealt with above are not suitable for this. However, during
the Joint North Sea Data Acquisition Program (JONSDAP) of the spring
of 1976, a number of RCM moorings were deployed across the Norwegian
Trench to the west and soutwest of Utsira. Although only preliminary results of this program have been reported (Furnes and Saelen,
1977; Riepma, 1980), some of them are quite interesting. The mean
currents over the month and a half long study period are very suggestive of a topographic gyre type circulation over the Norwegian
Trench upon which a baroclinic current near the Norwegian coast is
superimposed: near bottom currents "into" the trench, towards the
south and southeast there, were found on both sides of it and the
strongest flow "out" of it was found along its deep center. Owing
to the mean vertical shear near the coast of order $10^{-3}s^{-1}$, mean
northward flow was observed near the surface there.

The 2 week or more period fluctuations of halocline depth near
the Swedish Skagerrak coast remain a puzzle. The $G_1(S)$ mode was
coherent with neither local nor northern North Sea forcing nor with
alongshore flow during the spring of 1979. During the summer of
1918 however, vertical fluctuations of the Gullmaren halocline were
coherent with the local wind (Fig. 7). Hättan led positions to both
the north and south, indicating that the fluctuations originated in
the Skagerrak rather than Kattegatt. Stigebrandt (1983) proposed
a Kattegatt-Baltic Sea model which predicted halocline depth and
surface salinity in the Kattegatt. A quasi-steady, rotating hydraulic control was proposed for the brackish water outflow to the
Skagerrak. Were this "passive" Skagerrak assumption correct at
least for periods of 2 weeks or more, the halocline depth along the
Swedish Skagerrak coast would be essentially controlled by that in
the Kattegatt. Unfortunately no halocline depth time series of
sufficient quality seems to be available from the Kattegatt during
the spring of 1979 to test this prediction. One can only speculate
at the present that both Skagerrak and Kattegatt dynamics can be
important for 2 weeks or more band halocline fluctuations but to
different degrees under different meteorological conditions. Obviously future work is needed here.

The $G_2(S)$ mode, on the other hand, which described very low
frequency changes in surface layer salinity in Gullmaren and which
was not coherent with $G_1(S)$ at these frequencies is likely to be
controlled by surface layer salinities in the Kattegatt for the
usual conditions of low runoff to the fjord. Svansson (1975) shows
that, for time scales of months or more, surface layer salinity
in Gullmaren correlates well with Baltic sea level, i.e. with in- and
outflow from the Baltic.

SOME CLOSING COMMENTS

    In this paper I have ascribed the role of driving the low fre-
quency fluctuations in the Skagerrak to the wind. But what about
atmospheric pressure, which was also significantly coherent with
many of the EOF modes and parameters considered here and which,
through the inverse barometer effect, indeed was responsible for
the largest sea level changes in the eastern Skagerrak during the
spring of 1979? The reason for neglecting the direct effect of
atmospheric pressure lies in its inefficiency in driving alongshore
flow. Adams and Buchwald (1969) and Gill and Schumann (1974) demonstrated this for forced continental shelf waves but the reasoning
holds for motions over sloping bottoms near a coast in barotropic
rotating systems in general: the efficiency of any driving mechanism for generating alongshore flow in such a system is related to
the rate at which it can generate relative vorticity. Near a coast,
Ekman convergences and divergences driven by the alongshore wind
cause initial cross isobath flow with relative vorticity generation
(wind curl effects are less important over the scale of the Skagerrak). Atmospheric pressure generates relative vorticity through
the inverse barometer effect. In SDS, rates of vorticity generation
owing to these two effects were compared for typical Skagerrak parameters. The wind proved to be at least 20 times more efficient
than atmospheric pressure in driving barotropic alongshore flow
there.

    Slow along isobath flow of the order of 1 cm s^{-1} results by
continuity in the Skagerrak from atmospheric pressure changes.
However, such changes are certainly important for the exchange
through the narrow Danish straits between the Baltic Sea and Kattegatt. In this indirect way, flow in the baroclinic Baltic current
may be influenced by atmospheric pressure fluctuations (Stigebrandt
1983).

    Another major question concerned the relative importance of
local or remote wind forcing of low frequency fluctuations in the
eastern Skagerrak. The most detailed answer to date has been given
here and in SDS. The 5-6 day and 2 week or more current fluctuations in the eastern Skagerrak were found, in general, to be remotely forced by the wind over the northern North Sea (or in part
even farther to the west?). Wind driven topographic waves and
topographic gyres/topographic boundary layers for the Norwegian
Trench-Skagerrak appeared to be the mechanisms. The internal fluctuations appeared likewise to follow suit (with some reservations
for Kattegatt influence on the 2 week or more period fluctuations).
I suggest that the good visual correlations between local atmospheric pressure and internal fluctuations at Bornö reported by

Johnson (1943) for the "ice" winters in the early 1940's were not
a result of the "barometric" effect. Alternatively, the winds
associated with these pressure systems probably had forced topographic waves over the ice-free Norwegian Trench-western Skagerrak.
These had then interacted with the stratification in the eastern
Skagerrak to produce the observed internal fluctuations in Gullmaren.

For certain frequency bands and meteorological situations,
local forcing in the eastern Skagerrak was most important. This
was shown for the 2 day or more band in the coastal zone (cf. Fig.
4) and in Gullmaren in SDS. Also, strong but infrequent upwelling/
downwelling events driven by the local wind lead to large amplitude
fluctuations with periods of up to about one week. These are of
great importance for water exchange in the fjords and embayments
along the Swedish Skagerrak coast.

West wind events dominate the wind field over most of the year
in the Norwegian Trench-Skagerrak region. Low pressure systems often pass over southern Norway such that winds associated with them
tend to be oriented parallel to the trench over all its length, to
have a northerly component off west Norway. It is likely that the
mean circulation over time scales of months is to some degree a sum
of topographic wave/topographic gyre type response to these wind
events. This appeared to be the case for the JONSDAP-Utsira data
mentioned above. The topographic gyre example calculated above for
the western Skagerrak resulted in reasonable quasi-steady transports
for reasonable choices of west wind strength and bottom friction.
A smaller baroclinic transport of about .2 SV leaves the Skagerrak
in the Norwegian coastal current. A comparable transport into the
Skagerrak is required. Where this transport might occur and the
details of its entrainment into the coastal current are beyond the
scope of this paper. I note only that considerable topographic
wave energy may be dissipated in the eastern Skagerrak much of
which may be made available for mixing there.

ACKNOWLEDGEMENTS

I wish to thank the staff of the Oceanographic Institute,
Gothenburg for their help in the field work and with the prepara-
tion of this manuscript. Special thanks go to L. Djurfeldt and
O. Åkerlund for their invaluable assistance with the statistical
analyses. This work was supported by grants from the Swedish
Natural Research Council and the Nordic Committee for Physical
Oceanography.

REFERENCES

Adams, J.K., and Buchwald, V.T., 1969, The generation of con-
    tinental shelf waves, J.Fluid Mech., 35:815-826.

Aure, J., and Saetre, R., 1981, Wind effects on the Skagerrak
    outflow, The Norwegian Coastal Current, Proc.Symp. on
    the Norwegian Coastal Current, Geilo, Norway, 9-12 Sept.
    1980. Roald Saetre and Martin Mork (Ed) Univ. of Bergen,
    1981, Vol. I, 263-293.

Buchwald, V.T., and Adams, J.K., 1968, The propagation of con-
    tinental shelf waves, Proc. Roy. Soc. London, A305:
    235-250.

Csanady, G.T., 1974, Barotropic currents over the continental
    shelf, J.Phys.Oceanogr, 4:357-371.

Csanady, G.T., 1978, The arrested topographic wave,
    J. Phys. Oceanogr., 8:47-62.

Cutchin, D.L., and Smith, R.L., 1973, Continental shelf waves:
    low frequency variations in sea level and currents over
    the Oregon continental shelf, J. Phys. Oceanogr., 3:
    73-82.

Davies, R.E., 1976, Predictability of sea surface temperature
    and sea level pressure anomalies over the North Pacific
    Ocean, J. Phys. Oceanogr., 6:249-266.

Furnes, G.K., and Saelen, O.H., 1977, Currents and hydrography
    in the Norwegian coastal current off Utsira during
    JONSDAP 76, The Norwegian Coastal Current Project, 2/77.

Hamon, B.V., 1966, Continental shelf waves and the effects of
    atmospheric pressure and wind stress on sea level,
    J. Geophys. Res., 71:2883-2893.

Huyer, A., Smith, R.L., and Sobey, E.J.C., 1978, Seasonal
    differences in low-frequency current fluctuations over
    the Oregon shelf, J. Geophys. Res., 83:5077-5089.

Jenkins, G.M., and Watts, D.G., 1968, Spectral Analysis and
    Its Applications, San Francisco, Holden-Day, 525 pp.

Johnson, N., 1943, Studier av isen i Gullmarsfjorden, Svenska
    Hydr.-Biol.Komm.Skrifter, Hydrografi XVII.

Julian, P.R., 1975, Comments on the determination of signifi-
    cance levels of the coherence statistic, J.Atmos.Sci.,
    32:836-837.

Möller, P., and Svansson, A., Investigations in the Northern
    Kattegatt during the International JONSDAP -76 Period
    INOUT, March-April 1976, Fish. Bd. Sweden, S.Hydrography
    Rep., 15:6 pp.

Pedlomsky, J., 1974, Longshore currents, upwelling and bottom
    topography, J. Phys. Oceanogr., 4:214-226.

Petterson, H., 1917, Some new instruments for oceanographical
    research, Monthly Weather Review, 47(2):100-105.

Petterson, H., 1920, Internal movements in coastal waters and
    meteorological phenomena, Geografiska Annaler, Svenska
    Sällskapet för Antropologi och Geografi, 2:33-66.

Petterson, O., 1909, Gezeitenähnliche Bewegungen des Tiefen-
    wassers, Publ.Circ.Cons.Int.Explor.Mer., 47:1-21.

Riepma, H.W., 1980, Residual currents in the North Sea during
    the INOUT phase of JONSDAP -76, First results, Meteor.
    Forschungsergebnisse, Reihe A, 22:19-32.

522   G. SHAFFER

Shaffer, G., and Djurfeldt, L., 1982, On the low frequency
        fluctuations in the eastern Skagerrak and in Gullmaren,
        Submitted to J.Phys.Oceanogr.
Stigebrandt, A., 1983, A model for the exchange of water and
        salt between the Baltic and the Skagerrak. This volume
Svansson, A., 1975, Physical and chemical oceanography of the
        Skagerrak and the Kattegat I. Open sea conditions,
        Rep. Fish. Bd. Sweden., 1:88 pp.
Wang, D.P., 1975, Coastal trapped waves in a baroclinic ocean,
        J. Phys. Oceanogr., 5:326-333.

REDUCED GRAVITY MODELLING OF OUTER OSLOFJORDEN

Svein Arne Gjerp

Norwegian Hydrodynamic Laboratories
Trondheim, Norway

ABSTRACT

A reduced gravity, numerical model is used to simulate time dependent, vertically averaged brackish water flows in the outer reaches of Oslofjorden.

Sufficient measurements of wind and current velocities, fresh water inflow and stratification conditions were available for calibration and testing. Using density profiles, fresh water in-flow and wind velocity as input, simulated and observed currents were compared at three different locations for three different weather simulations. Current vectors and rms. deviations of speed are presented.

The model seems to be satisfactory for giving a crude esti-mate of the current pattern averaged over the surface layer.

INTRODUCTION

The numerical calculations presented in this paper were made as a part of a project whose purpose was to simulate spreading and transport of probable oil spills in coastal areas, taking the dyna-mics of the surface water masses and the wind drift of the oil-spills into account. A description of the complete model and some test examples were given by Emblem et al. (1980).

The model has been applied to oil spill simulations in the southern reaches of Oslofjorden (Fig.1)(Gjerp et al., 1981, 1981b). The drift simulations were based on current simulations, and in this paper we concentrate only on the current simulations and

their accuracy. Hence, comparisons between simulations and observations have been made for four specified data sets.

MODEL FORMULATION

A detailed description of the model was presented in Gjerp
et al. (1980). Therefore we only give a short review of the basic
assumptions and theories, and concentrate on the application of
the model.

River inflows in the northern and southern parts of the
modelled area – Drammenselva (Q_{DE}, Fig.1) and Glomma (Q_{GL}, Fig.1)
respectively – produce a more or less pronounced two layer structure, and in the present simulations we assumed no mixing between
the two layers. However, frictional forces were included as friction
between the two layers and as wind stress on the surface. The distribution of the fresh water inflows and the relatively short time
scale in the drift simulations implied the consideration of constant
densities in each layer. These assumptions were supported by
salinity and temperature recordings (Audunson et al., 1974) which
suggested that, for a simulation period of approximately one week,
the changes in the hydrographic conditions were minor. In addition,
instantaneous mixing between the fresh water and the surrounding
and underlying saltwater at the river mouths was assumed.

Recordings (Audunson et al., 1974) had shown that the currents
in the upper layer were definitely stronger than in the deeper parts
of the fjord. In the present work only currents in the upper layer
were of interest and pressure gradients in the lower layer were
neglected.

Persistent winds may cause storing of water in specific areas,
leading to elevation of the free surface and deepening of the interface. This will again involve pressure gradients in the water masses,
which may affect the currents. In addition, the stratification may
involve a concentration of momentum in the upper layer. These combined effects of wind and stratification were taken into account
by the two-dimensional and two layer model with moveable interface
and free surface. We thus simulated the momentum balance of the
upper layer within the validity of these crude assumptions.

Except near Filtvet (Fig.1), harmonic analysis indicated that
tidal currents usually were of minor significance in the area.
Therefore they were neglected.

Using these assumptions together with the conservation laws
of mass and momentum, we obtained a reduced gravity model for the
dynamics of the upper layer; it included nonstationarity, nonlinear
convective terms, pressure gradients in the upper layer due to
moveable free surface and layer interface, Coriolis force, wind-

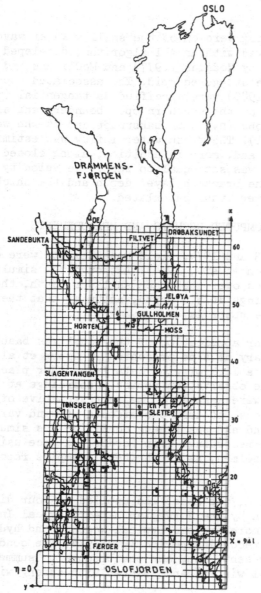

Fig. 1. Map of Oslofjorden. The modelled area is covered with the
        finite difference grid used for the simulations (grid-
        size 1 km, timestep 3600 s). The numbers show the posi-
        tions of the current meters: 6=Gullholmen, 8=Slagen,
        9=Sletter. Geometrical boundaries and the direction of
        horizontal axes are indicated (η is the estimated free
        surface elevation).

generated surface stresses, and interfacial shear stresses between the two layers.

The resulting version of the shallow water wave equations was numerically solved with an ADI algorithm, developed by Leendertse (1967), altered by Hodgins (1977) and McClimans and Gjerp (1978) who used a space staggered grid. The associated boundary conditions (Gjerp et al., 1980) were specified as tangential (v_t) and normal (v_n) velocities at the southern open boundary and as estimated free surface elevations (η) in Drøbaksundet and at the western open boundary (Fig. 1). The fresh water inflow was estimated as previously described and, more in addition, along closed boundaries the normal velocity was set equal to zero. The velocity components averaged over the brackish layer depth and the thickness of the brackish layer were thus calculated.

COMPARISON OF COMPUTED RESULTS AND MEASUREMENTS

During 1973 and 1974, field measurements were obtained in Oslofjorden (Audunson et al., 1974). Although the simulations represent integrated values over the brackish layer depth, they were compared with the recordings at 5 m depth, assuming that these were representative of the upper layer.

This further application of the model was based on the same input and boundary data as presented in Gjerp et al. (1980). However, due to a lack of wind data from other places in the area during the whole observation period, recordings at 10 m height at Ferder (Fig.1) were assumed to be representative of the modelling area. Nevertheless, to test the effects of wind variations along the fjord, a land and sea breeze condition was simulated twice - once using wind recordings from Ferder and once using wind recordings from Gullholmen, as in the test examples reported in Emblem et al. (1980).

As mentioned, the model was applied to four different weather situations given by the Norwegian Meteorological Institute to represent four more or less typical weather and hydrographic conditions in the recording period 1973-1974. The conditions were assumed to represent "land and sea breezes", "summer with strong winds", "typical winter" and "winter with strong winds".

Unfortunately there were no current meter recordings during the summer condition with strong winds. Comparisons between measurements and simulations were thus made for the other three conditions. Comparisons were made at the three locations marked "6", "8" and "9" (Fig.1) for the last two days of the simulation periods. Three days of initial simulations were made to get rid of initial effects, starting with zero velocity and free surface elevation in the whole area.

Attempting to quantify the deviations between the calculations and the observations at the three recording stations, the
root mean square values (σ) for the deviations independent of
directions have been given on each figure. σ_F represents the deviations for the cases based on wind observations at Ferder, while σ_G
represents the calculations based on Gullholmen wind.

The first case is the so-called typical winter. The salinity
difference between the two layers was 9 o/oo and the mean brackish
layer depth was 12 m. The freshwater discharges were 450 m^3/s and
210 m^3/s for Glomma and Drammenselva respectively. The beginning
of the period was characterized by south-easterly winds, causing
an inward transport and accordingly a deepening of the interface.

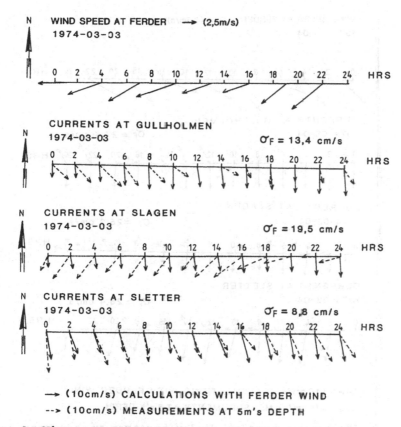

Fig. 2. Wind recordings at Ferder together with comparisons
        between current velocity simulations and measure-
        ments at 5 m depth for the first day of a typical
        winter condition.

  Comparisons were made for the last two days of the period
(Fig. 2, Fig. 3) when the dominant wind direction was from the
north-east. Looking at the comparisons, we see the agreement is
not too good, especially on the last day. The turning of the windfield, together with the established pressure gradients, obviously
resulted in an outward current in the model, as shown in the figures.
Therefore it was impossible to simulate the observed inward currents
at the end of the period with this kind of model, using our kind
of boundary conditions. Of the many reasons for these observed
inward currents, the most acceptable seem to be interactions with
the Skagerrak basin south of the southern open boundary and local
shear currents in the upper layer. However, we were not able to
find the exact explanation.

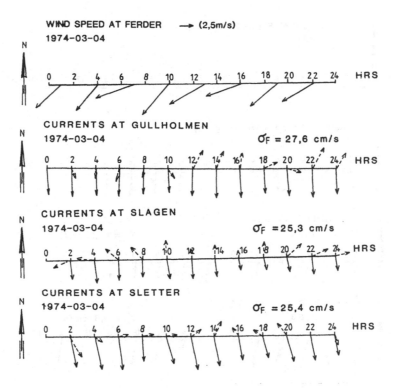

Fig. 3. Wind recordings at Ferder together with comparisons
        between current velocity simulations and measure-
        ments at 5 m depth for the second day of a typical
        winter condition.

At Gullholmen ("6") and Slagen ("8") the calculated currents were mainly southerly directed, deviating at times considerably from the measured directions. However, Fig. 1 shows that the measuring sites at Gullholmen and Slagen were only about 1.5 gridlengths from the western boundary which went in the north-south direction for 4 to 5 gridlengths. It was therefore to be expected that the simulated currents would be parallel to the boundary at these places, whereas at Sletter the recording instrument was farther from the boundary and the agreement was therefore generally better.

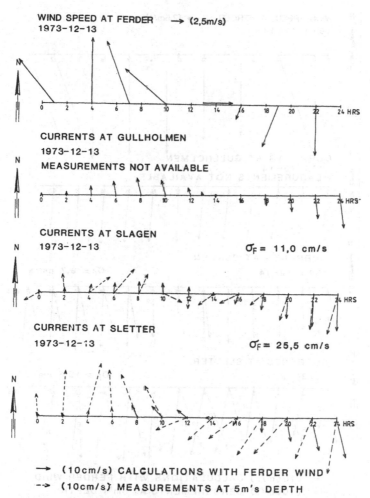

Fig. 4. Wind recordings and comparisons of computed currents with measurements at 5 m depth for the first day of a winter condition with strong winds.

During the winter condition with strong winds (Fig. 4, Fig. 5) the salinity difference between the two layers was 5 o/oo and the mean brackish layer depth 20 m. Fresh water discharges were 170 m^3/s from Drammenselva and 360 m^3/s from Glomma.

The windfield at Ferder changed direction from southwest to north during the simulation period. There were no recordings at Gullholmen during this period but as shown in Fig. 4 and Fig. 5 the currents were mainly wind dominated both in the observations and in the simulations. On the first day (Fig. 4) the simulated currents were too small, especially at Sletter ("9"), whereas they

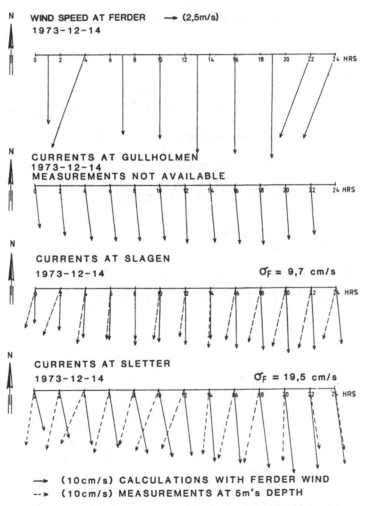

Fig. 5. Wind recordings and comparisons of computed currents with measurements at 5 m depth for the second day of winter conditions with strong winds.

were of the same order of magnitude as the observations on the last
day (Fig. 5). However, the observed direction was a little more
westerly than in the simulations, mainly because of boundary effects.

   The windfield variations along a fjord may be considerable.
To study these effects we simulated summer conditions with dominating
land and sea breezes, using wind data from Ferder and Gullholmen
(Fig. 1) respectively. The mean brackish layer thickness was 10 m,
the salinity difference 11 o/oo and the fresh water inflows 210 m^3/s
and 1330 m^3/s from Drammenselva and Glomma respectively. Both the
comparisons between the mean wind fields and the calculated currents
based on these observations are shown in Fig. 6 and Fig. 7. There
seem to be no systematic deviations between the wind fields at the

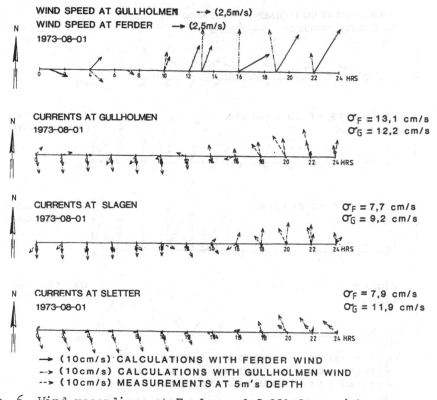

Fig. 6. Wind recordings at Ferder and Gullholmen with compa-
        risons of calculated currents based on wind data from
        Ferder and Gullholmen and measurements at 5 m depth for
        the first day of summer conditions with land and sea
        breeze.

two stations although we often find the strongest and most southerly
winds at Gullholmen, probably because of topographic effects.

   Looking at the simulations we see that there were some deviations between the two cases but the agreement can be improved by
using local winds. This statement is based on the fact that Gullholmen winds seemed to give best agreement between observations and
measurements at Gullholmen while Ferder wind gave best agreement
at Sletter which was the outmost observation station. Therefore,
if possible, it is important to take care of the geographical
variations in the wind field. This may be done by interpolation if
more than one recording station is available.

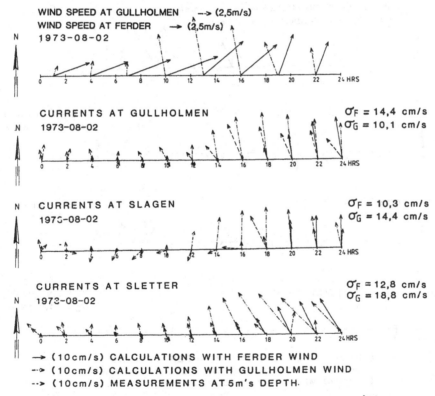

Fig. 7. Wind recordings at Ferder and Gullholmen with compa-
        risons of calculated currents based on wind data from
        Ferder and Gullholmen and measurements at 5 m depth
        for the second day of summer conditions with land and
        sea breeze.

CONCLUSIONS

Although there were some deviations between calculations and observations, this rather simple model seems to be satisfactory for giving a crude estimate of the current pattern averaged over the surface layer. This success probably owes to the well-defined two layer structure caused by the dominating freshwater discharges at each end of the area and it may even be improved by the availability of some local wind measurements.

A complete two layer model or more complex multilayer models would probably give better results, but it is doubtful that such a model is worth the investment. In particular the treatment of the open boundaries would be far more cumbersome.

ACKNOWLEDGEMENTS

The author would like to express his gratitude towards O. Oldervik, who participated in the calculations, and T.A. McClimans, H. Rye and other colleagues at NHL, who took part in the discussions about this work and gave valuable suggestions.

REFERENCES

Audunson, T., Thendrup, A., and Billfalk, L., 1974, Kjerne-kraftverk i Oslofjordsområdet. Rapport nr 14, Report STF60 F5030, Norwegian Hydrodynamic Laboratories, Trondheim (In Norwegian).

Emblem, K., Gjerp, S.A., Lervik, E., Næser, H., Oldervik, O., Rye, H., and Selseth, I., 1980, Spredning av olje i kyst-farvann. Rapport nr 2. Presentasjon av beregningsmodellene. Report STF60 A80070, Norwegian Hydrodynamic Laboratories, Trondheim (In Norwegian).

Gjerp, S.A., Lervik, E., Oldervik, O., and Emblem, K., 1981a, Spredning av olje i kystfarvann. Rapport nr 3. Strøm- og oljespredningsberegninger for Ytre Oslofjord, Report STF60 A81122, Norwegian Hydrodynamic Laboratories, Trondheim (In Norwegian).

Gjerp, S.A., Lervik, E., Oldervik, O., and Emblem, K., 1981b, Spredning av olje i kystfarvann. Vedlegg til rapport nr 3. Strøm- og oljespredningsberegninger for Ytre Oslofjord. Report STF60 A81125, Norwegian Hydrodynamic Laboratories, Trondheim (In Norwegian).

Gjerp, S.A., Oldervik, O., Næser, H., and Rye, H., 1980, a two-dimensional, two-layer numerical model for the dyna-mics of stratified coastal water masses, and its appli-cation to Oslofjorden, in "The Norwegian Coastal Current Symposium. Geilo, 9-12 september 1980", R. Sætre, M. Mork, ed., Reklametrykk A.S., Bergen.

Hodgins, S., 1977, An improved computational method for the
    shallow water wave equations based on the Leendertse
    (1967) finite difference scheme, Report STF60 A77058,
    Norwegian Hydrodynamic Laboratories, Trondheim.

Leendertse, J.J., 1967, Aspects of the computational model for
    long period water wave propagation. Memo. RM-5294-PR, The
    RAND Corporation, Santa Monica.

McClimans, T.A., and Gjerp, S.A., 1978, Numerical study of
    distortion in a Froude model, in "Proceedings 16th Coastal
    Engineering Conference, Hamburg 1978", Am. Soc. of Civ.
    Eng., New York.

ON ENTRAINMENT IN TWO-LAYER STRATIFIED FLOW

WITH SPECIAL FOCUS ON AN ARCTIC SILL-FJORD

Fl. Bo Pedersen

Institute of Hydrodynamics and
Hydraulic Engineering
Technical University of Denmark

INTRODUCTION

In coastal regions of the ocean the flow pattern is often influenced by density stratification due to temperature variations (heating, cooling, freezing) or fresh water inflows or both. This internal density structure has a great bearing on the mean flow as well as on the turbulence, which in turn governs mixing and hence the density stratification. This coupling of flow and mixing is considered from an energy-point of view and it is demonstrated that a rational hypothesis about the efficiency of the mixing process leads to a diagnostic entrainment function.

The utility of the theory is illustrated by the mixing processes occurring in a small ice-covered sill fjord an Greenland, where serial measurements are made by our Institute. Our primary concern is the free penetrative convection generated by salt rejection during the period of ice growth; but, as demonstrated by the sparse up to now measurements, phenomena such as spill over the sill and internal tidally generated resonant seiches must be taken into account for correct interpretation of the observations.

ENTRAINMENT

General View

If we confine ourselves to flow situations where a density interface separates an active turbulent flow from a passive ambient fluid, pure entrainment of non-turbulent fluid into the turbulent fluid occurs or, conversely, the turbulent fluid entrains the non-turbulent.

Basically, entrainment stems from interfacial instabilities which produce either cusps or vortices, agents for the mixing of the two fluids. Lack of turbulence in the ambient fluid prohibits transport away from the interface there, whereas turbulence in the active fluid may take part of the entrained fluid away from the interfacial region. During this mixing process the entrained water is subject to an increase in potential energy, due to the work done against gravity, and to an increase in turbulent kinetic energy. As the energy supply to this gain stems from the energy transferred from the mean flow to the turbulence, it is physically reasonable to relate the total gain in turbulent mechanical energy to the energy available for the production of turbulence or, which leads to the same, to look at the efficiency of the mixing. By doing so for a large class of physical processes, it looks probable (Bo Pedersen, 1980a) that the efficiency $\mathbb{R}_f^T$ of the mixing process - named the bulk flux Richardson number - is a constant, namely

$$\mathbb{R}_f^T = 0.045 \text{ for } \mathbb{F}_\Delta < \mathbb{F}_{\Delta,cr} \text{ or } \mathbb{R}i > \mathbb{R}i,cr \text{ (cusp)} \qquad (1a)$$

$$\mathbb{R}_f^T = 0.18 \text{ for } \mathbb{F}_\Delta > \mathbb{F}_{\Delta,cr} \text{ or } \mathbb{R}i < \mathbb{R}i,cr \text{ (vortices)} \qquad (1b)$$

where $\mathbb{F}_\Delta$ is the densimetric Froude number

    $\mathbb{R}i$ is the Richardson number

    index cr denotes critical

$$\mathbb{R}i = \frac{1}{\mathbb{F}_\Delta^2} = \frac{(\Delta\rho/\rho)gy}{v^2}$$

where $\Delta\rho g$ = buoyancy difference

    y = depth of active layer

The hypothesis of a constant bulk flux Richardson number leads directly to a diagnostic entrainment function. This will be illustrated for free penetrative convection.

Free Penetrative Convection (FPC)

FPC plays a crucial role in polar lakes, fjords and coastal regions due to the extreme cold there. FPC is the turbulent movement created by a source of buoyancy flux into an ambient fluid, illustrated in Fig. 1. The unstable buoyancy flux may stem from heating, cooling, salt-rejection or airbubble generation. It creates a molecular diffusion layer next to the wall. In the boundary transition layer the buoyancy flux converges along lines as vertical plumes, which form the well mixed, highly turbulent convective layer. The depth y of the convective layer grows with

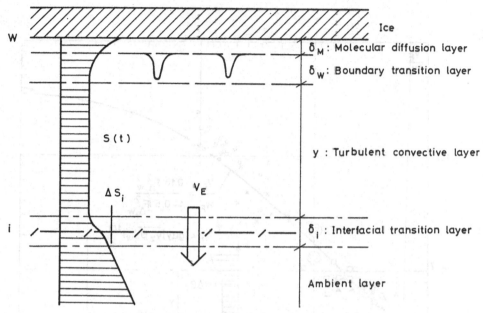

Fig. 1. Sketch (not to scale) of the different zones in free
         penetrative convection beneath growing sea ice. S(t) is
         salinity as a function of time. w: wall; i: interface.

         Buoyancy fluxes:

         w: $\rho g \beta \, \overline{(S'v_z')}_w$ downwards directed

         i: $\rho g \beta \, \overline{(\Delta S_i)} V_E$ upwards directed

         β = coefficient of saline expansion

time due to the penetration of turbulence into the ambient layer.

    The main objective of our fjord measuring program is to measure
this entrainment velocity as a function of the buoyancy flux and the
stability of the interface.

    In Fig. 2 (from Bo Pedersen (1980a), Fig. 6.7) are plotted
some laboratory and field data for FPC. The only field data shown
stem from Babine Lake, where Farmer (1975) measured FPC caused by
solar heating beneath lake ice. Measurements of FPC under a growing
sea ice sheet have been performed by Lewis and Walker (1970) in
Cambridge Bay, and further analysed by Gade et al. (1974) and

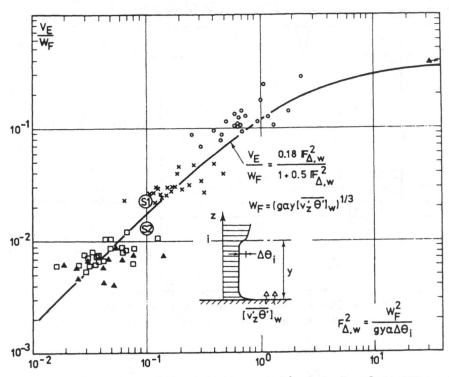

Fig. 2. Laboratory and field data on entrainment for free penetrative convection compared with theory. The points are
based on data referred by:

o, x, ◘ Heidt (1975) (laboratory experiments)

S1, S2 Willis and Deardorff (1974) (laboratory experiments)

▲   Farmer (1975) (field data from solar heating
       beneath lake ice)

Θ = potential temperature ($^\circ$C)

α = coefficient of thermal expamsion (per $^\circ$C)

v'_z = vertical component of turbulent velocity fluctuation

$\alpha\Theta$ in atmospheric inversion rise has the same physical
meaning as βS in halocline erosion below sea ice, (from
Bo Pedersen, 1980).

Perkin and Lewis (1978). As pointed out by these authors, Cambridge Bay is rather shallow and short, which makes it difficult to extract the mixing caused by FPC from the mixing caused by internal seiches and dense bottom currents along the shallows. Hence further field measurements are urgently needed to throw some light on the mixing mechanism in the polar regions.

In order to establish an entrainment function for FPC, the bulk flux Richardson number concept is applied:

$$\mathbb{R}_f^T = \frac{POT + KIN}{PROD} \tag{3}$$

where POT = gain in potential energy
      KIN = gain in turbulent kinetic energy
      PROD = production of turbulent kinetic energy

The PROD stems from the unstable buoyancy flux generated at the wall. As a consequence of the high rate of mixing in the convective layer the salinity S (or equivalent property) is constant with depth, which implies that the flux must be nearly linearly distributed over the depth. Therefore the PROD is simply the released potential energy and, if salinity expansion is the buoyancy agent,

$$PROD = 0.5 \; \rho g \beta y \; \{\overline{v_z' \, S'}\}_w \tag{4}$$

where β = coefficient of saline expansion.

Accordingly, if we denote the entrainment salt flux as $\overline{\Delta S_i} \times V_E$ where ΔS_i is the salinity jump across the interface, we obtain an analogous expression for the gain in potential energy, the work done against gravity:

$$POT = 0.5 \; \rho g \beta y \; (\overline{\Delta S_i}) V_E \tag{5}$$

For convenience we introduce the convective velocity scale

$$W_F = \sqrt[3]{0.5 \; \beta g y \; \{\overline{v_z' \, S'}\}_w} \tag{6}$$

which is proportional to the average turbulent kinetic energy in the well mixed layer, $\bar{e} \simeq 0.22 W_F^2$, (Bo Pedersen, 1980a, chapters 6 and 13). Hence the gain in turbulent kinetic energy is

$$KIN = 0.22 \; \rho W_F^2 V_E \tag{7}$$

Now solving the $\mathbb{R}_f^T$ relation with respect to the entrainment velocity V_E, we obtain the entrainment function (Fig. 2)

$$\frac{V_E}{W_F} = \frac{0.18 \quad \mathbb{F}_{\Delta,w}^2}{1+0.5 \quad \mathbb{F}_{\Delta,w}^2} \tag{8}$$

where

$$\mathbb{F}_{\Delta,w}^2 = \frac{W_F^2}{g\gamma\beta(\overline{\Delta S_i})} \tag{9}$$

The general shape of this function is representative of all types of entrainment function. Let us therefore briefly analyse the physical implications:

Low values of $\mathbb{F}_{\Delta,w}^2$: This is the most common condition in nature and corresponds to a negligible kinetic energy compared to the potential energy. In this case the rate of energy of the entrained mass is a constant fraction of the rate of turbulent energy production, which implies a dimensionless entrainment velocity proportional to the densimetric Froude number squared.

This general relationship has been found for other geophysical processes by many authors, see for instance: Rouse & Dodu (1955), Kato & Phillips (1969), Nihoul (1973), Edwards and Darbyshire (1973), Turner (1973), Kullenberg (1977), Stigebrandt (1976). Especially in connection with FPC caused by ice growth, the equality of POT and $\mathbb{R}_i^T$ PROD gives the simple relationship (see equation (4) amd (5))

$$(\overline{S_i})V_E = \{\overline{v_z'S'}\}_w \cdot \mathbb{R}_f^T \tag{10}$$

or that the rate of erosion of the interface is directly proportional to the ice growth rate

$$V_E = \frac{S_{ice}}{\Delta S_i} \quad \mathbb{R}_f^T \times \frac{\rho_{ice}}{\rho_w} \times \text{(ice growth rate)} \tag{11}$$

High values of $\mathbb{F}_{\Delta,w}^2$: The increasing importance of the turbulent kinetic energy relative to the potential energy has a dramatic influence on the entrainment function, which increases at a falling rate with the densimetric Froude number and eventually becomes a constant when the potential energy goes to zero, for free convection into a homogeneous layer. For further discussion see Bo Pedersen (1980a).

In summary on FPC: The knowledge of an entrainment function for FPC allows us to estimate the deepening of the upper layer beneath growing sea ice, provided we know the rate of growth of the ice sheet and its salt content (i.e. the salt-buoyancy flux).

MIXING PROCESSES IN AN ARCTIC SILL FJORD

Background for the Project

Increasing economic interest and environmental concern for the
Arctic have intensified research activity in the arctic seas. As
part of an interdisciplinary project for investigation of the coastal
zone of western Greenland, our institute is carrying out a series
of field measurements, whose goal is to clarify the mixing processes
taking place in an ice-covered sill-fjord.

The Agfardlikavsâ Fjord, Mârmorilik (71°N, 51°W), Greenland
was chosen for some obvious reasons:

1. The mining company GREENEX A/S has kindly placed its camp
   and laboratory facilities at our disposal and furthermore
   supplied transportation and lodging together with practical
   aid at Mârmorilik.

2. Regular monthly oceanographic measurements have been made
   by GREENEX A/S since 1978, Pedersen (1980, 1981).

3. The dissolved metals and the excess temperature of the
   tailings - which are discharged into the bottom layer -
   are both outstanding tracers, especially during periods of
   weak density stratification.

4. The Greenland Fishery Investigation Authority has a ship -
   R/V "Adolf Jensen" - operating in Disko Bay. It was placed
   at our disposal for installation (Sept. 81) and recovery
   (Sept. 82) of the automatically recording current meters
   and TC-chains in the fjord.

5. Pollution control in the fjord calls for detailed knowledge
   of mixing conditions in Agfardlikavsâ Fjord.

6. It is the most intensively investigated fjord in Greenland
   and is of a moderate size.

The Fjord Dynamics

Agfardlikavsâ is the innermost fjord in a fjord system,
separated from Qaûmarujuk by a sill at 22m depth, see Fig. 3 (from
Møller et al (1982)). The total run-off of about 10^8 m^3/year
is totally dominated by the summer melt water, see Fig. 5 (from
Møller et al (1982)). The air temperature varies from a monthly
average of about 7°C in July to -20°C in January. During the last

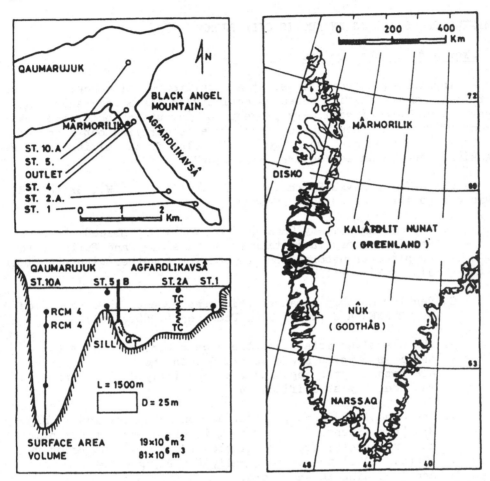

Fig. 3. Location of Mârmorilik, location of measurement stations
         and longitudinal section of Agfardlikavsâ Fjord. (from
         Møller et al. (1982)).

winter a maximum weekly ice growth of 0.15 m was observed and the
maximum ice-thickness measured was 0.9 m. Other relevant oceanographic conditions for the winter are the tide (amplitude $\leqslant$ 0.75 m)
and the tailing discharge to the bottom layer (0.05 m^3/s, 1200 kg/m^3,
15°C).

        In the following we give a brief outline of the phenomena to be
studied, referring to Fig. 5 and Table 1.

Free Penetrative Convection (FPC)

        Ice growth commences in December, when a pronounced two-layer
system exists. Salt rejection from the growing sea ice has two

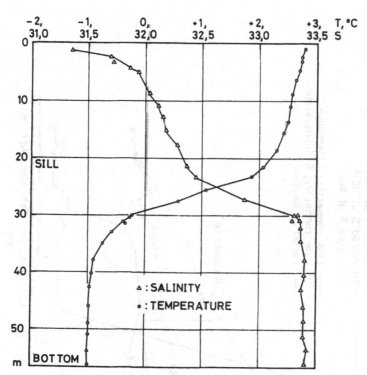

Fig. 4.   An example of salinity and temperature profiles.   Based on
CTD-measurements 13th Sept. 1981. St. 2a.   Agfardlikavsâ,
Greenland (from Møller et al. (1982)).

effects.   First, it creates a turbulent upper layer which erodes –
entrains – the lower layer and thereby increases the upper layer
depth.   Second, it increases the upper layer salinity and thereby
decreases the stability of the system.   Hence an accelerating erosion
takes place , which can be observed during December and January.
Our instrumentation for observing this FPC is illustrated in Table 1
(from Møller et al. (1982)].

In the shallows, entraining dense bottom currents are formed
which – contrary to the FPC – may transport salty water downwards
into the bottom layer, see Table 1.   The theory for entraining dense
bottom currents is further discussed by Bo Pederson (1980a,b).

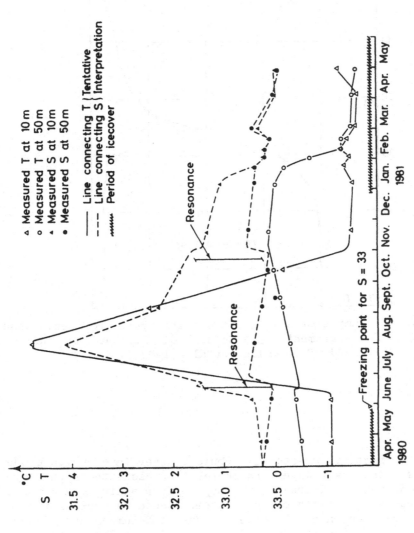

Fig 5. An example of annual variation of salinity and temperature for Agfardlikavså, Greenland. Measurements from 1980/81 (from Pedersen (1980,81)).

Table 1. The planned measurements and instruments used in Agfardlikavsâ, Greenland, during 1981/82. (From Møller et al. [1982]).

SUBJECT	INSTRUMENTS	DATA	PRECISION	STABILITY	ST.	DEPTH	PERIOD	INTERVAL
WINTER CONVECTION	RCM-4 AND TC-CHAIN	T	0,05°C	FAIR	2A	5 m	JAN.-APR.	60 min.
		C	0,1 $\frac{mmho}{cm}$			18 23 28 33 38 43 48 53 m	SEPT. 81 TO SEPT. 82	120 min.
EXCHANGE WITH Q-FIORD	RCM-4	T	0,05°C	FAIR	5	5 m	JAN.-APR.	60 min.
		C	0,1 $\frac{mmho}{cm}$			25 m	SEPT. 81 TO SEPT. 82	60 min.
		V	1 cm/s					
BOTTOMCURRENTS	RCM-4	T	DO	FAIR	1	38 m	SEPT. 81 TO SEPT. 82	60 min.
		C	DO					
		V	DO					
ICEFORMATION	CTD	C	0,02 $\frac{mmho}{cm}$	FINE	4 2A	PROFILE	JAN. 82 TO APR. 82	PROFILE
		T	0,01°C					
		D	0,2 m					
CALIBRATION	CTD	DO	DO	DO	1 2A 4 5 10 A	AT INSTRU- MENT DEPTHS	SEPT. 81 AND JAN. 81 APR. 82 AND SEPT. 82	
	HG-TERM WATER- SAMPLER	T	0,02°C	EXTRA FINE				
		S	0,01‰					

Ice-Formation

Ice formation is governed by the upper boundary conditions, the meteorology - and by the lower boundary conditions, the oceanography. Both were measured during the one year test period.

The heat content in the lower layer is seen to increase steadily during the spring and autumn due to the over-temperature (~15°C) of the tailing. Besides being an excellent tracer for the lower layer, it bears on the heat balance of the fjord and hence on the ice-formation. This is due to the FPC, which brings up the relatively warm lower layer water to the surface layer. Another important effect of the FPC in sea water is its ability to create a turbulent heat exchange between the ice and the water.

One may notice, Fig. 5, the significant difference in the measured temperature of the upper layer water and the theoretical freezing temperature at the actual salinity, a phenomenon we study further.

Internal Waves

The continuous increase in meltwater run-off during the spring and the continuous decrease during the autumn make a tuning of the system into an internal resonance condition inevitable, as pointed out by Lewis in an internal report, March 1980. For a density difference of about 0.4 (kg/m^3) the uninodal internal seiche is in resonance with the major tidal frequency (12.5h), see Fig. 5. During these events a pronounced two-way entrainment takes place and, furthermore, the interfacial position is lowered due to spill over the sill, see Fig. 6. The instrumentation in stations 1, 2A, and 5 (see Table 1) may throw some light on these phenomena.

The measurements performed by Lewis on the lee side of the sill in Agfardlikavsâ fjord (Lewis and Perkin (1982)) illustrate the "inverse" spill over the sill (see Fig. 7). Up to the 10th of November (1978) the meters were located in the lower relatively salt and warm layer (36.5 m depth). During the resonance event - which is clearly illustrated by the current meter record - the water masses surrounding the meters steadily changed towards the characteristics of the upper less salt and colder water, because of lowering of the interface, due to the "inverse" spill over the sill.

Observations by Others

The phenomena mentioned above have been recognized in Cambridge Bay as well and the reader is referred to Lewis and Walker (1970), Gade et al. (1974), Perkin and Lewis (1978), and Lewis (1980) for further discussion of the oceanography of arctic fjords. Recently

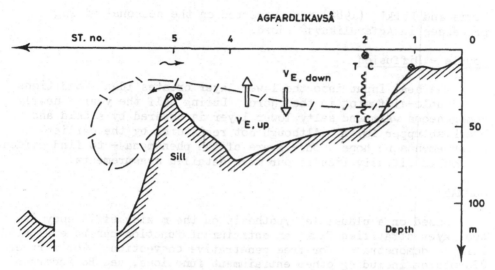

Fig. 6 Internal resonance seiche creates intense mixing
(two-ways entrainment) and outwards directed spill
over the sill.

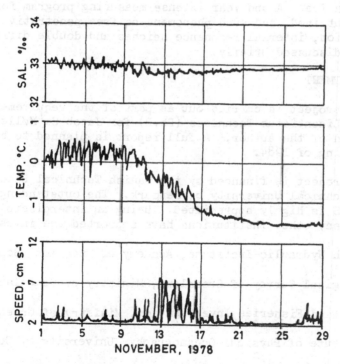

Fig. 7 Section of current-meter record commencing 1st November 1978. (From Lewis and Perkin [1982]).

Lewis and Perkin (1982) have reported on the seasonal mixing processes in Agfardlikavsâ fjord.

Double - Diffusion

The heat input into the lower layer creates ideal conditions for double-diffusion in the fjord. During half the year a nearly homogeneous warm and salty lower layer is covered by a cold and brackish upper layer. Although not registered by the earlier measurements we hope - once aware of the phenomena - to find evidence of double-diffusivities in our more detailed measurements.

SUMMARY

Based on a plausible hypothesis on the mixing efficiency in two-layer stratified flow, an entrainment function can be evaluated. This is demonstrated for free penetrative convection. For further discussion including other entrainment functions, see Bo Pedersen (1980a).

Field data on free penetrative convection in the ocean are sparse. In the arctic regions it plays a crucial role, especially during the time of ice-formation, where salt is rejected from the growing sea ice. A one year intense measuring program for an arctic fjord is outlined, and such phenomena as free penetrative convection, ice-formation, internal resonance seiches and double diffusivity have been discussed briefly.

ACKNOWLEDGEMENT

This project is carried out as part of the requirements for the Degree of Licentiatus Technices (Ph.D) for Jacob S. Møller under the supervision of the author. A full report is planned to be published in the spring of 1984.

The project is financed by the Danish Technical Research Fund and the Technical University of Denmark. The outstanding help of GREENEX A/S is highly appreciated. Being an interdisciplinary project, many other institutions have supported the investigations:

Danish Hydraulic Institute, Academy of Technical Science.

Geological Survey of Greenland, Ministry of Greenland.

Greenland Fisheries Investigation, Ministry of Greenland.

Institute of Physical Oceanography, University of Copenhagen.

Marine Pollution Laboratory, National Agency of Environmental Protection.

REFERENCES

Bo Pedersen, Fl., 1980a, A monograph on turbulent entrainment and friction in two-layer stratified flow. Technical University of Denmark, Institute of Hydrodynamics and Hydraulic Engineering, Series Paper No. 25: 397.

Bo Pedersen, Fl., 1980b, Dense bottom currents in rotating ocean. Journal of the Hydraulics Division, ASCE, Vol. 106. No HY8, Proc. Paper 15629, Aug. 1291-1308.

Edwards, A., and Darbyshire, J., 1973, Models of a Lacustrine thermocline. Fourth Liege Colloquium on Ocean Hydrodynamics, Liege 1972. Ed. by J. Nihoul. Mémoires de la Societé Royale des Sciences de Liege, 6e série, tome 4, 81-101.

Farmer, D.M., 1975, Penetrative convection in the absence of mean shear. Quarterly Journal of the Royal Meteorological Society, Vol. 101, 869-891.

Gade, H.G., Lake, R.A., Lewis, E.L., and Walker, E.R., 1974, Oceanography of an Arctic bay. Deep Sea Research, Vol. 21, 547-571.

Heidt, F.D., 1975, Zeitlicher Abbau der Stabilen Schichtung eines Fluids durch freie Konvektion. Sonderforschnungsbereich 80. Ausbreitungs- und Transportvorgänge in Strömungen. Universität Karlsruhe, Report SF80/ET/61, 133.

Kato, H., and Phillips, O.M., 1969, On the penetration of a turbulent layer into stratified fluid. Journal of Fluid Mechanics, Vol. 37, 643-655.

Kullenberg, G.E.B., 1977, Entrainment velocity in natural stratified shear flow. Estuarine and Coastal Marine Science, Vol. 5, 329-338.

Lewis, E.L., 1980, Water movement in A and Q Fjords near Mârmorilik and the processes for pollutant transport. Institute of Ocean Sciences, Box 6000, Sidney B.C., Canada, March. Report to the Ministry of Greenland, Copenhagen.

Lewis, E.L. and Perkin, R.G., 1982, Seasonal Mixing Processes in an arctic fjord system. Journal of Physical Oceanography, Vol. 12.

Lewis, E.L. and Walker, E.R., 1970, The water structure under a growing sea ice sheet. Journal of Geophysical Research, Vol. 75, No. 33, 6836-6845.

Møller, J.S., Bo Pedersen, Fl., and Nielsen, Tue Kell, 1982, Measurements of the convective mixed layer under growing sea ice. Technical University of Denmark, Institute of Hydrodynamics and Hydraulic Engineering, Progress Report No. 56, 13-23.

Nihoul, J.C.J., 1973, The effect of wind blowing on turbulent mixing and entrainment in the upper layer of the ocean. Fourth Liege Colloquium on Ocean Hydrodynamics, Liege 1972. Ed. by J.C.J. Nihoul, Memoires de la Societé Royale des Sciences de Liege, 6e Serie, tome 4, 115-124.

Pedersen, K., 1980, 1981, Recipient kontroldata for havvand, A-fjord
        st. 4 og Q-fjord st. 10. (Sea water quality data for A-fjord
        st. 4 and Q-fjord st. 10). Internal report in Danish,
        Greenex A/S, Copenhagen.
Perkin, R.G., and Lewis, E.L., 1978, Mixing in an arctic fjord.
        Journal of Physical Oceanography, Vol. 8, No. 5, 873-880.
Rouse, H., and Dodu, J., 1955, Turbulent diffusion across a density
        discontinuity. La Houille Blanche, Vol. 10, 522-532.
Stigebrandt, A., 1976, Vertical diffusion driven by internal waves
        in a sill fjord. Journal of Physical Oceanography, Vol. 6,
        486-495.
Turner, J.S., 1973, Buoyancy effects in fluids. Cambridge University
        Press, 367.
Willis, G.E., and Deardorff, J.W., 1974, A laboratory model of the
        unstable planetary boundary layer. Journal of the Atmospheric
        Sciences, Vol. 31, 1297-1307.

SALT ENTRAINMENT AND MIXING PROCESSES IN AN UNDER-ICE RIVER PLUME

R. Grant Ingram

Institute of Oceanography, McGill University
Montreal, Canada

ABSTRACT

Comparative studies of open water and ice-covered conditions
of the Great Whale River plume in Hudson Bay show a marked difference in areal and vertical extent of the fresh water volume.
Detailed winter observations in zones of weak and moderate tidal
mixing demonstrate the importance of current magnitude and vertical
density gradient in controlling upward salt entrainment and
vertical mixing. The thickness of the frictional boundary adjoining
the ice cover varied as a function of the available tidal kinetic
energy and the freshwater input. Speculation as to the nature of
the transitions occurring at the time of the ice break-up is also
discussed.

INTRODUCTION

The way in which freshwater flows to the sea and mixes with
it ranges from gradual to abrupt. Transition occurs either within
the river or offshore depending on the relative strength of the
mean and tidal flow. The Great Whale River, a major Canadian river,
enters Hudson Bay on its south-eastern perimeter with a natural
discharge sufficiently high to form a large offshore plume throughout the year. During open water conditions (July to December), the
plume is of order 100 km^2 in surface area and some 2m in thickness
(Ingram, 1981). Because of the generally weak tidal current regime
in the nearshore, little mixing takes place, which allows for a
noticeable dilution of the ambient waters over some distance. Under
the ice, the plume influence is felt over a much broader area than
in summer, in spite of the reduced river runoff at that time. The
vertical extent of the plume near the river mouth is also altered,

being a factor of two to three times thicker in the winter-spring
period.

There has been a large number of laboratory studies in recent
years dealing with entrainment and mixing processes in stably stratified flows. The characteristics of the Great Whale River plume
described above may result from, at various times in the year and
geographical locations, a combination of several mechanisms which
have been investigated in the laboratory. These include studies of
wind induced mixing processes by Kato and Phillips (1969), Long
(1975) and Kundu (1981) and entrainment induced across the pycnocline as studied by Thorpe (1973) and Chu and Baddour (1980). In
comparison, few experiments have been completed in nature. The work
by Cordes et al. (1980) and by Partch and Smith (1978) were carried
out in open water under strong tidal conditions where turbulent
levels were very high and experimental design complicated by vessel
movement and other constraints. More recently, Lewis and Perkin
(1982) have completed a study of mixing processes in a Greenland
fjord both under the ice and in open water. A general review of
entrainment processes in two layer stratified flow can be found in
Bo Pedersen (1980).

The Great Whale River plume was chosen for the present study
on entrainment and mixing processes because of its location and
suitable hydrodynamical characteristics during the winter and spring.
These include (i) a weak tidal current regime in the adjacent coastal
waters, (ii) the presence of a stable 1-1.5 m thick ice cover from
which to deploy instrumentation for a 4-5 month period each year,
(iii) a nearby area (Manitounuk Sound) with much larger tidal
currents which is also influenced by the brackish plume throughout
its length, and (iv) the existence of comparative data for open
water conditions. The mean annual discharge of the Great Whale is
about 700 m^3/s, ranging from a minimum monthly average of less than
180 m^3/s in March to values in excess of 1320 m^3/s in June.

In regard to open water plume characteristics in Hudson Bay
and Manitounuk Sound reference may be made to Ingram (1981) and
Legendre et al. (1982). Furthermore, Legendre et al. (1981) have
examined the relationship of under-ice phytoplankton production to
water column stability and input solar radiation in this area. One
of the findings discussed by Ingram (1981) was the existence of a
much larger plume both in horizontal extent and thickness under
the ice than in open water. In accounting for the characteristics
of the plume, it was suggested that a marked reduction of turbulence in the brackish upper layer adjoining the ice was responsible
for the reduced salt entrainment.

The aim of the present work is to characterize the mixing and
entrainment processes across a sharp pycnocline under a complete

ice cover. Turbulence induced by tidal friction on the underside of ice, shear induced mixing across the pycnocline and other processes are discussed as possible explanations for the phenomena observed. The relationship of fresh (brackish) water input to pycnocline depth and speculation as to the changes occurring during break up are also presented.

OBSERVATIONS

Fig. 1 shows the station positions on a map of the study area. CTD sampling was done through holes in the ice using a Guildline instrument. Current meter data were obtained using Aanderaa instruments moored from the ice with large lead weight anchors at the base of each mooring line. Sampling rate was 10 minutes. In addition to oceanographical observations, river flow was monitored at an upstream location. Meteorological information (hourly) was obtained from the Canadian Atmospheric Environment Service Station in Great Whale located near the river mouth. The only major source of fresh water in the area is the Great Whale River. The river enters Hudson Bay at a location characterized by offshore tidal currents of less than 10 cm/s. In Manitounuk Sound, a nearby region also influenced

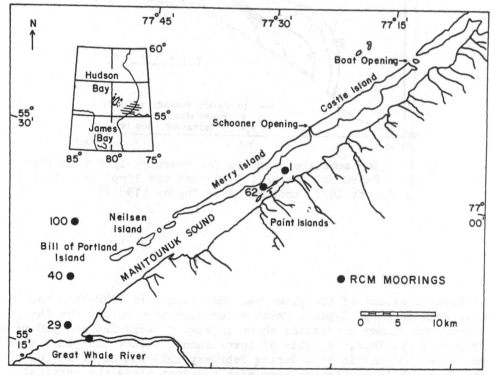

Fig. 1. Place map of study area showing station location.

by plume waters, much stronger tidal flows are encountered, with
values up to 100 cm/s in some of the more constricted regions.

In the winter of 1982, further field work confirmed the
presence of a much larger plume volume under the ice than in open
water as had been found earlier (Fig. 2). The axis along which the
greatest extension was noted paralleled the coast. However, the

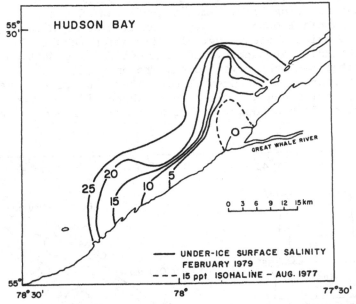

Fig. 2. Surface isohaline contours for February 1979 (modified
     from Baird and Anning (1979)) and the 15ppt isohaline
     in August 1977 (modified from Ingram (1981)).

horizontal extent of the plume was much larger in 1982 than had
been seen in 1979. Typical fresh water discharge values for the
winter and summer conditions shown in Fig. 2 were 225 and 900 m^3/s
respectively. Thus, in spite of lower runoff values the plume area
was some 1000 km^2 in area during February 1979 and at least 2000
km^2 in March 1982. Changes were also apparent along the vertical
axis. In Fig. 3, a cross-section extending from the river mouth to

the north shows distinct isohaline shoaling at 15-20 km offshore
and a near shore pycnocline some 5 m in depth under the presence
of an ice cover. Also shown on this figure are the mean current
directions computed from two month (March-May 1982) moorings at 3
and 8 m. Flow in the fresh layer was in a north-easterly direction,
parallel to the coast, while the ambient flow was in an opposite
sense. Profiling data taken in 1978 showed the current vector in
the layer adjoining the ice was predominantly offshore, as indicated by the arrow in Fig. 3. The upper 10 m of the water column
at the most northerly CTD station was homogeneous with a salinity
of 28.5 °/oo. Observations (not shown) taken in late April and
early May showed both a deepening of the pycnocline and an offshore
extension of the plume influence. Runoff values rose rapidly at
this time.

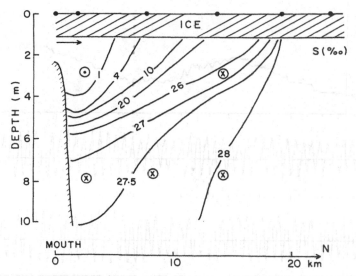

Fig. 3. Vertical section showing March 1982 isohaline distribution
        in upper layer of Hudson Bay from the mouth of the Great
        Whale River and extending in a northerly direction. Dots
        indicate CTD station positions. Circles designate mean
        flow over a two month period. Into the page is west, out
        of the page is east. (Unpublished data of S. Peck).

   In examining Fig. 2, the intrusion of brackish water into
Manitounuk Sound can be seen. Because of the winnowing of the cross-
sectional area as a function of distance into the Sound, tidal
currents in this region are notably stronger than those found in
nearby Hudson Bay. As a result of this tidal action, water mass
characteristics differ between the two regions. In order to understand the process occurring there, data at two stations, 1 and 62,
will be presented.

   At Station 1, current meter time series at a depth of 1.5 m or
0.5 m below the underside of the ice are shown in Fig. 4. Tidal
currents were semi-siurnal in nature, oriented along the channel
axis, with amplitudes of 10 cm/s or less. The salinity time series
showed little semi-diurnal variability but did show a positive slope
for salinity over periods of generally increasing tidal flow (neap
to spring). Salinity decreased with time during the complementary
part of the fortnightly cycle. Because of the unequal lengths of
the salinity trends a net increase in salinity occurred over the
length of the record. Temperatures were at or very near the freezing
point throughout the record (within the instrumental accuracy). In
examining the original data set, one finds that steep rises embedded
within the general upward trend of the salinity curve occurred
within the 10 min. sampling period. The same data were also filtered

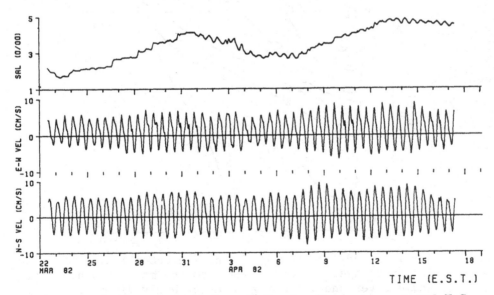

Fig. 4. Time series of salinity, E-W velocity component and N-S
        velocity component at station 1. (depth: 1.5 m).
        Decimated to hourly. Station depth: 39 m.

to remove all variability less than 25 hours in period. In addition
to the salinity trends mentioned above, the filtered velocity components showed a 2-3 day fluctuation of the along-channel current
with an amplitude of about 1 cm/s superimposed on a 2 cm/s mean
current directed into the Sound. Oscillations of water level, with
atmospheric pressure fluctuations removed, at a similar frequency
were also found on the tide gauge record taken at station 29. The
signals were approximately 180° out of phase.

Further down in the water column, at a depth of 3 m, much
stronger tidal currents were present, with peak values up to 25 cm/s
over the semi-diurnal cycle. The difference in current amplitude
over the spring-neap cycle was more accentuated than at 1.5 m. Also
apparent were large quarter and semi-diurnal fluctuations of the
salinity field. CTD profiles take at this location imply that
internal waves at this depth are less than 1 m in peak-to-peak
amplitude because of the sharp pycnocline. The "low-pass" time
series of this data set showed a similarly positive trend as at
1.5 m in salinity from neap to just after spring tides (Fig. 5).
Mean flow was also directed into the Sound but near zero in magnitude. Flow during periods of higher salinity was often directed out
of the Sound. Similar to the 1.5 m depth velocity field, a 2-3 day
variation of the mean flow was apparent throughout the record. There
was no significant correlation between fluctuations of the salinity
and velocity fields in this frequency range.

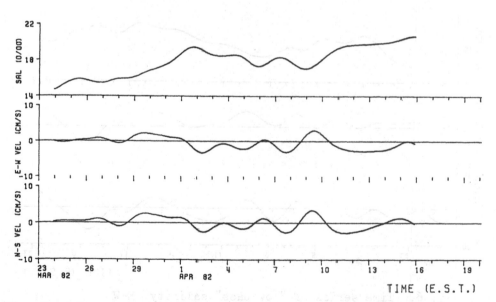

Fig. 5.   Time series of "low-pass" salinity, E-W velocity component
          and N-S velocity component at station 1 (depth: 3.0 m).

Data were also taken in a narrower section of Manitounuk Sound, at Station 62. Current meter time series at a depth of 3 m showed maximum semi-diurnal speeds of 60 cm/s and large salinity fluctuations. The "low-pass" time series showed salinity trends similar to that seen at station 1 and 2–3 day current variations superimposed on a small but inward mean flow (Fig. 6). At a depth of 8 m, the hourly time series showed similar tidal flow but smaller salinity fluctuations. The "low-pass" results indicated a near zero mean flow (Fig. 7). In comparing Figs. 6 and 7, the two "low-pass" salinity time series show opposite trends at 3 and 8 m depth with periods of increasing/decreasing salinity at the upper level paralleling decreasing/increasing salinity values at the lower level.

In comparing CTD profiles taken before and after the mean salinity rise between 7 and 16 April, one finds a net increase of salinity under the ice accompanied by a thinning of the upper boundary layer (Fig. 8). Although the instataneous profiles include internal wave effects, observations from the fixed current meters showed a distinct broadening of the pycnocline region over the period of interest. At a depth of 10 m, salinity values decreased over the same period.

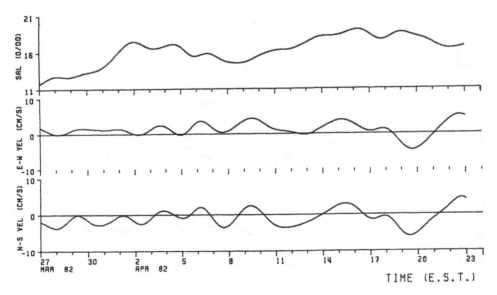

Fig. 6.   Time series of "low-pass" salinity, E-W
          velocity component and N–S velocity component
          at station 62 (depth: 3.0 m). Station depth: 45 m.

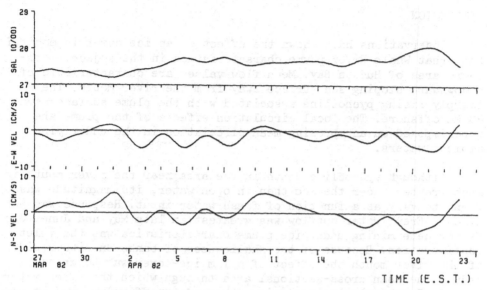

Fig. 7. Time series of "low-pass" salinity, E-W velocity component
and N-S velocity component at station 62 (depth: 8.0 m).

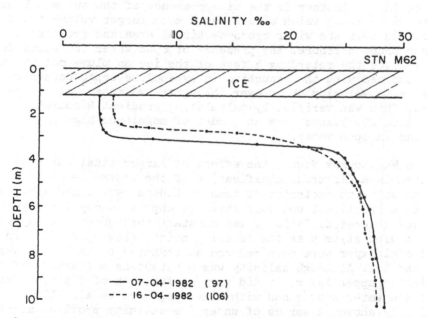

Fig. 8. CTD profiles taken at station 62 in April 1982 (only upper
10 m shown).

DISCUSSION

    Observations have shown the effect of an ice cover in modifying
the Great Whale River plume characteristics in the adjacent offshore area of Hudson Bay. Mean flow values are characteristic of
a downward sloping sea surface away from the river mouth. The
sharply rising pycnocline associated with the plume surfaces some
20 km offshore. The local circulation effects of the plume are
superimposed on an opposing mean circulation in the underlying
ambient waters.

    Although pycnocline depth in the area near the river mouth is
much greater under the ice than in open water, its magnitude was
found to vary as a function of fresh water input. Response was not
linear. Fresh water outflow was greatest in late May and June. One
factor determining under-ice plume characteristics was the inlet
configuration. Because of the shallowness of banks on either side
of the river mouth the effect of a 1 m ice-cover was to greatly
reduce the mean cross-sectional area through which the river water
passed. Rough calculations using observed runoff values and local
bathymetry over the shallow sand bar at the river mouth suggest a
Froude number of about 1 for both open and ice-covered conditions.

    In comparing the plume characteristics over a two to three
week period centred on ice breakup it is obvious that major changes
occur. One factor is the introduction of wind and wave mixing in the
surface layer. Another is the disappearance of the surface blocking
at the river mouth which allows for a much larger volume to be
introduced over the wider cross-sectional area and produces a
thinner plume offshore. The presence of open water in Hudson Bay
also removes the retarding effect of the ice on plume motion which
when combined with the reduction in pycnocline depth causes an
increase in velocity shear over the upper two meters of the water
column. This was verified by calculating gradient Richardson numbers
at Station 29. Values were an order of magnitude higher under the
ice than in open water.

    In Manitounuk Sound, the effect of larger tidal currents
resulted in a different modification of the under-ice plume and
ambient water characteristics than in Hudson Bay. Immediately under
the ice a frictional boundary layer of approximately 1.5 m in thickness was generated. Salinity was constant throughout the layer with
temperatures at/or near the freezing point. Flow speeds within the
frictional layer were much reduced as compared to values at depths
of 3 and 8 m. Although salinity was constant as a function of depth
within the upper layer, it did vary as function of distance from
the fresh water source and with time. Legendre et al. (1981)
previously showed a series of under-ice salinity profiles along
Manitounuk Sound which demonstrates a gradually shoaling pycnoc-

line and increasing salinity in the upper layer as one progresses
into the Sound.

Local modification of the salinity values in the brackish
upper layer at Station 1 was shown to vary as a function of the
tidal current magnitude. Because salinity increased over 9 day
periods while decreases occurred for about 6 days there was a net
increase over the total mooring period. Input of brackish water
was relatively constant over this period as shown by the "low-pass"
velocity plots and the monitored river runoff, which was also
approximately constant. Lewis and Perkin (1982) have also demonstrated the relationship of vertical mixing to tidal energy in a
study of Agfardlikavsa fjord. In their case a much weaker stratification was present than in Manitounuk Sound. Salinity trends
occurred over longer periods (two weeks) in conjunction with above-
average tidal current pulses.

The relationship of salinity increase to stability was investigated
using the 1.5 and 3.0 m depth current meter records at Station 1.
When the minimum calculated gradient Richardson numbers were in
the range of 2.4 to 3.0 over any semi-diurnal cycle the salinity
generally rose. If minimum values were consistently over 3 then
salinity values decreased or were steady. Since the pycnocline was
thinner (0.5 m) than the instrument separation, recalculation of
the Richardson number using similar velocities and densities but
substituting 0.5 for the separation yielded critical values of
0.8 to 1.0. Also of interest was the association of salinity increases to periods of flow into the Sound. This may imply that the
source of denser water in the upper layer might have been to the
southwest of Station 1. Since the overall salinity gradient from
the river mouth to the station contradicts this hypothesis, vertical
exchanges must be considered.

One of the areas where increased vertical exchange may occur
is located near Paint Islands. At Station 62, tidal velocities were
much greater than in the adjacent areas of the Sound. Current meter
records showed a relationship between salinity values at 3 and 8 m.
As salinity increased in the upper layer values decreased below
the pycnocline. The vertical mixing which occurred was dependent
on the dynamic stability of the water column which in turn varied
as a function of tidal energy. For semi-diurnal tidal cycles in
which the minimum gradient Richardson number (calculated using the
pycnocline thickness of 0.5 m) was in the range of 0.6 to 0.8,
salinity values rose in the upper layer. Similar to observations
at Station 1, there was an asymmetric pattern of salinity change
with time which led to a net increase of salinity over the mooring
period. As previously mentioned, the flux of brackish water was
relatively constant over the period of observation. Later in the
spring when runoff values rise dramatically there should be a net
decrease in the upper layer independent of tidal energy.

By using mean velocity values in the upper layer, a typical
value of 40 m³/s for fresh water input into the Sound was calculated
over the mooring period. This represents about 15-20% of the Great
Whale River output at this time. Taking the mean salinity value at
Station 1 and calculating the net upward salt flux over the entire
surface area between the entrance to the Sound and Station 1, one
obtains vertical velocity of 10^{-5} cm/s. If it is assumed that most
of the exchange takes place in the narrow region near Paint Islands
(Station 62) which is about 10% of the total surface area, then the
upward velocity is closer to 10^{-4} cm/s.

Some general points can be made about the processes occurring
in Manitounuk Sound. When the turbulence level at the pycnocline
is greater than that generated by the frictional boundary layer
adjoining the ice, salinity in the upper layer rises while it falls
in the lower layer. Furthermore, the pycnocline broadens such that
the thickness of the under-ice layer is less. An opposite response
is found during periods of weaker tidal flow when the stability
across the pycnocline inhibits entrainment and mixing processes.

New fluxes of fresh or brackish water in the upper layer were
quickly mixed in the layer under the ice by turbulence generated
there. No thin layers (< 1 m) appeared on any CTD traces taken in
the winter program. If this occurred during a period of weaker
tidal flow salinity decreased in the upper layer and the region
became thicker. Turbulence levels generated by friction on the ice
were seemingly higher than those resulting from interfacial stress
across the pycnocline at periods of lower than average tides and
vice-versa during tides greater than average. The asymmetry over
the fortnightly tide cycle results from the lowering of pycnocline
stability during stronger tidal periods which allows for an extension of the mixing for 1-2 days past the presence of average tides.

What happens to the water masses in Manitounuk Sound during
late spring and early summer? The fresh water input increases
dramatically in May and June which initially allows for the generation of a much thicker upper layer and decreasing salinity values
throughout the water column at all stations. In late May, when the
ice cover is destroyed, an increase of tidal currents at the surface
and wave activity quickly modifies the upper layer, resulting in a
mixed layer of 7-8 m in depth. Within a few weeks the Sound is
partially mixed with intermittent intrusions of 17-20 o/oo water
in the surface layer resulting from favourable wind stress. At
greater depth, much lower salinities occur than in winter.

In summary, the major changes resulting from the imposition
of an ice-cover on a fresh water plume overlying sea water have
been described. The relative importance of frictional stresses at
a solid upper boundary to interfacial stresses over the pycnocline
have been shown to modify the upper and lower layer by variable

mixing over the fortnightly tide cycle. Asymmetries result from the modification of the pycnocline during periods of above average tidal flow. Some speculation as to what occurs during ice break-up and transition to typical open water conditions were also presented.

ACKNOWLEDGEMENTS

This work is part of the program of the Groupe Interuniversitaire de Recherches Océanographiques du Québec. Research work supported by the Natural Sciences and Engineering Research Council of Canada, Hydro-Québec, Société de l'Energie de Baie James and Fisheries and Oceans Canada.
Jean-Claude Deguise was in charge of both field operations and data reduction.

REFERENCES

Baird, S.D., and Anning, J.L., 1979, LaGrande/Great Whale Winter Oceanographic Survey. Canada Centre for Inland Waters, Burlington, Canada. Report Series 79-2. "Unpublished Manuscript", 43 pp.

Bo, Pedersen, F., 1980, A monograph on turbulent entrainment and friction in two-layer stratified flow. Institute of Hydrodynamic Technical University of Denmark, Lyngby, 397 pp.

Chu, V.H., and Baddour, R.E., 1980, Stabilization of turbulence in plane shear layers, Proceedings of 2nd International Symposium on Stratified Flows, Vol. 1, pp. 367-377, Trondheim.

Ingram, R.G., 1981, Characteristics of the Great Whale River Plume, J.Geophys.Res.,86(C3):2017-2023.

Kato, H., and Phillips, O.M., 1969, On the penetration of a turbulent layer into a stratified fluid, J.Fluid Mech., 37:643-655.

Kundu, P.K., 1981, Self-similarity in stress-driven entrainment experiments, J. Geophys. Res.,86(C3):1979-1988.

Legendre, L., Ingram, R.G., and Poulin, M., 1981, Physical control of phytoplankton under the sea ice (Manitounuk Sound, Hudson Bay), Can. J. Fish. Aquatic Sci.,38(11): 1385-1392.

Legendre, L., Ingram, R.G., and Simard, Y., (in press), Aperiodic changes of water column stability and phytoplankton production in an Arctic Coastal Environment, Nat. Can.

Lewis, E.L., and Perkin, R.G., 1982, Seasonal mixing processes in an Arctic fjord system, J. Phys. Oceanogr., 12:74-83.

Long, R.R., 1975, The influence of shear on mixing across density interfaces, J. Fluid Mech., 70:305-320.

Partch, E.N., and Smith, J.D., 1978, Time dependent mixing
     in a salt wedge estuary, Est. Coast. Mar. Sci., 6:3-19.
Thorpe, S.A., 1973, Turbulence in stably stratified fluids:
     a review of laboratory experiments, Boundary-Layer Met.,
     5:95-119.

ON ENTRAINMENT OBSERVED IN LABORATORY

AND FIELD EXPERIMENTS

Erik Buch

Institute of Physical Oceanography
University of Copenhagen
2200 Copenhagen Ø, Denmark

ABSTRACT

The consistency between results from mixing experiments in two-layered fjords and results in the laboratory is investigated. The field data are interpreted by using a model in which the interfacial shear stress is regarded as the source of turbulent energy. Using the definition of the interfacial shear stress $\tau_i = \rho_0 u_*^2 = \rho_0 c_i U^2$, where the drag coefficient c_i is experimentally determined, the field results are found to agree quite well with the results of the laboratory experiments characterized by values of the overall Richardson number Ri_* in the interval $90 < Ri_* < 400$.

INTRODUCTION

In the last ten to fifteen years much effort has been devoted to explaining the phenomenon of shear induced entrainment. Most of the work is based on laboratory experiments, for instance Kato and Phillips (1969)(called the KP-experiment) and Kantha, Phillips and Azad (1977)(called the KPA-experiment). These two sets of experiments were carried out using the same laboratory equipment.

The KP-experiment started with a linearly stratified fluid and the following relation was obtained

$$\frac{u_e}{u_*} = 2.5 \, Ri_*^{-1} \tag{1}$$

where u_e is the entrainment velocity, $u_* = (\frac{\tau_0}{\rho_0})^{\frac{1}{2}}$ is the friction velocity due to the surface stress τ_o, $Ri_* = \frac{g'D}{u_*^2}$ is the overall Richardson number, g' is the reduced gravity and D the depth of the upper layer.

The KPA-experiment started with a fluid consisting of two homogeneous layers separated by a jump in density. For the same value of Ri_* a definite 100% higher entrainment rate was found compared to the KP-experiment. This discrepancy has been the subject of a number of studies, e.g. Phillips (1976), Kantha (1978), Price (1979) and recently Kitaigorodskii (1981).

Since laboratory experiments and theoretical work usually are done to explain things happening in nature, it is of great interest to compare their results.

FIELD EXPERIMENTS

The circulation in two-layered fjords is very much dependent on the vertical mixing across the pycnocline. In situ observations of mixing across the pycnocline in two-layered fjords have been made by means of the dye diffusion technique developed by Kullenberg (1969, 1974).

The fjords investigated all belong to type 1 according to the classification given by Pickard (1961). These types of fjords consist of two nearly homogeneous layers separated by an abrupt change in density. The velocity is almost uniform in the upper layer with direction out of the fjord, while a weak compensation flow occurs in the lower layer.

Buch (1981) assumed that the vertical mixing was generated solely by the interfacial shear stress,

$$\tau_i = \rho_o c_i U^2 \tag{2}$$

where c_i is the interfacial drag coefficient, U is the mean velocity of the upper layer, and it was additionally assumed that the upper layer was kept homogeneous by wind action. A simple model for the interfacial shear stress induced mixing was developed by studying the turbulent transport in a thin layer of thickness dz at the pycnocline in a Cartesian coordinate system with the z-axis positive upward. The local value of the flux Richardson number Rf at the interface is

$$Rf = \frac{g/\rho_o \rho'w'}{-u'w'\frac{du}{dz}} = \frac{-g/\rho_o K\frac{d\rho}{dz}}{\frac{\tau_i}{\rho_o}\frac{du}{dz}} = \frac{KN^2}{\frac{\tau_i}{\rho_o}\frac{du}{dz}} \tag{3}$$

where it is assumed that the Reynolds stress equals the interfacial
shear stress and the K-approximation expresses the buoyancy flux,
K being the vertical turbulent diffusion coefficient and N the
Brunt-Väisälä frequency.

Combining Eqns.(2) and (3) yields

$$K = Rf \; c_i \; (\frac{U}{N})^2 \; \frac{du}{dz} \tag{4}$$

This relation is consistent with previous theoretical relations,
Welander (1968), which can be seen by the following reasoning. The
vertical momentum transfer coefficient K_m can be expressed

$$K_m = \frac{Ri}{Rf} \; K \tag{5}$$

where $Ri = N^2(\frac{du}{dz})^{-2}$ is the gradient Richardson number. Inserting
Eqn.(4) into Eqn.(5) reveals

$$K_m \frac{du}{dz} = \frac{\tau_i}{\rho_o} = c_i U^2 \tag{6}$$

It is also noted that Eqn.(4) is consistent with the relation

$$K = \beta \; \epsilon \; N^{-2} \tag{7}$$

where β is a constant and ϵ is the rate of dissipation. The latter
is given by the energy flux from the mean motion to the turbulent
motion, which is

$$\frac{\tau_i}{\rho_o} \frac{du}{dz}$$

Buch (1981) found the factor $Rf \; c_i$ to be $2.0 \cdot 10^{-4}$ by plotting
observed values of K and $(\frac{U}{N})^2 \frac{du}{dz}$ versus each other.

Various investigators have suggested that the value of Rf
integrated over the upper layer in subcritical fjord flow is 0.05,
Stigebrandt (1976) and Bo Pedersen (1980). In Eqn.(3) the local
flux Richardson number is defined at the interface, in the most
stable part of the water column. This implies that Rf here will
approach its critical value, which is higher than the integrated
one. The critical value of Rf has been found to vary in the range
0.10 - 0.20 (Turner, 1973; Kullenberg, 1974) and here the value
$Rf_c = 0.10$ is chosen, leading to

$$c_i = 2.0 \cdot 10^{-3} \tag{8}$$

The rate of production of turbulent energy per unit area in a steady flow driven by a constant pressure gradient in a channel wide enough to permit neglection of side wall friction is given by Bo Pedersen (1980) in the following form

$$E_{prod.} = \tau_i (U - u_i) \tag{9}$$

u_i being the velocity at the interface.

The rate of increase of potential energy due to the entrained mass in a water column of unit area can be expressed as (Turner, 1973)

$$\frac{d\,E_{pot.}}{dt} = \frac{1}{2} \rho_o\, g' u_e D \tag{10}$$

Eqns. (9) and (10) are applicable to the brackish flow in a fjord provided the pressure gradient arising from the changing density of the flow can be neglected. It follows that the flux Richardson number, as defined below, can be written

$$RF \equiv \frac{\dfrac{d\,E_{pot.}}{dt}}{E_{prod.}} = \frac{\frac{1}{2}\rho_o\, D\, g' u_e}{\tau_i (U - u_i)} \tag{11}$$

from which an entrainment function can be found

$$\frac{u_e}{U} = 2\, Rf\, c_i\, (1 - u_i/U)F_\Delta^2 = \lambda F_\Delta^2 \tag{12}$$

where

$$F_\Delta^2 = \frac{U^2}{g'D} \tag{13}$$

is the densimetric Froude number.

In order to use eq. (12) for interpretation of results from field experiments it is necessary to express u_e in terms of parameters observed during the dye experiments. This has been done by Kullenberg (1977)

$$u_e = \frac{KN^2}{g'} \tag{14}$$

Plotting $\dfrac{u_e}{U}$ versus F_Δ^2 (Fig. 1) it is found

$$\frac{u_e}{U} = 5.2 \cdot 10^{-4} F_\Delta^2 \tag{15}$$

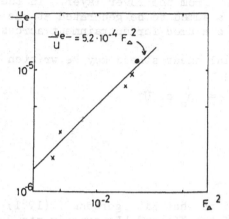

Fig. 1. The entrainment as a function of the densimetric Froude
number, determined from field experiments.

Although many dye experiments have been carried out in several
different areas only a few fulfil the conditions assumed in the
present study. Hence it is important to compare the results with
those given by appropriate laboratory experiments.

COMPARISON OF THE KPA-EXPERIMENT AND THE FIELD EXPERIMENTS

The KPA-experiments revealed no distinct power law dependence
on Ri_* and only in the range $90 < Ri_* < 400$, was $\frac{u_e}{u_*}$ found to be
proportional to Ri_*^{-1}, giving

$$E_{KPA} = \frac{u_e}{u_*} = 6\, Ri_*^{-1} \tag{16}$$

in qualitative agreement with the KP-experiment.

From Eqns. (15) and (16) it follows that

$$\frac{u_e}{u_*} = 5.2 \cdot 10^{-4} Ri_*^{-1} \cdot \left(\frac{U}{u_*}\right)^3 \tag{17}$$

To proceed, it is necessary to find an expression for $\frac{U}{u_*}$ in

terms of known parameters. In the laboratory experiments the turbulence was generated at the surface and the energy was partly dissipated and partly diffused down to the interface where it was used
for entraining water from the lower layer. In the field experiments
the turbulence was assumed to be generated at the interface and some
of this energy was consumed for entrainment across the interface.

The interfacial shear stress may be written

$$\tau_i = \rho_o u_*^2 = \rho_o c_i U^2 \tag{18}$$

giving

$$\frac{U}{u_*} = c_i^{-\frac{1}{2}} \tag{19}$$

Which is equivalent to what Kitaigorodskii (1981) found in his theoretical analysis of the KP- and KPA-experiments.

Inserting Eqn. (19) into Eqn. (17) and using the experimentally
obtained value of $c_i = 2.0 \cdot 10^{-3}$

$$\frac{u_e}{u_*} = 5.8 \cdot Ri_*^{-1} \tag{20}$$

- a result very near that obtained by KPA in the interval
$90 < Ri_* < 400$, see Fig. 2.

Eqn. (12) was derived assuming that the turbulence was shear-
generated and as seen in Fig. 1 the field experiments tend to corro-
borate this assumption.

The close resemblance between Eqn. (20) and Eqn. (15) consequently suggests that in the laboratory the turbulence was shear-
generated in the experiments characterized by values of Ri_* in the
interval $90 < Ri_* < 400$. The experiments with Ri_* values outside
this interval are seen to diverge from the relation given in Eqn.
(20).

For $Ri_* < 90$ the discrepancy is probably a real tendency since
it is to be expected that the exponent of Ri_* will approach zero
for $Ri_* \to 0$, also shown by the KP-experiments.

Bo Pedersen (1980) argued that with the experimental set-up
used in the KP- and KPA-experiments molecular effects would greatly
influence those experiments with values of $Ri_* \gtrsim 400$, a point also
discussed by KPA themselves.

With no extraneous effects present in the laboratory experi-

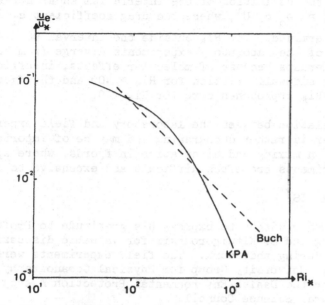

Fig. 2. The entrainment rate as a function of the overall
Richardson number.

ments the entrainment velocity $u_e = \frac{dD}{dt}$ is constant, the depth of
the mixed layer D is a linear function of time and departure from
linearity indicates the onset of unwanted sidewall – and other
– effects. As seen in Fig. 4 in the KPA-paper, the sidewall effect
becomes increasingly noticeable as Ri_* increases and the method
used by KPA to reduce it becomes crucial at high values of Ri_*
(see Figs. 7 and 8 in the KPA-paper). Therefore, insufficient
compensation of the friction at the walls may be an alternative
explanation of deviation from the Ri_*^{-1} power law for $Ri_* > 400$ in
the KPA-experiments.

SUMMARY AND CONCLUSIONS

The entrainment rate in some fjords was estimated by means of
in situ dye diffusion experiments and interpreted by means of an
earlier model (Buch, 1981).

The purpose of this paper is to investigate the relation between the results of the field experiments and the laboratory

investigations of Kantha et al. (1977). Such a relation is found
by applying the definition of the interfacial shear stress
$\tau_i = \rho_o u_*^2 = \rho_o c_i U^2$, where the drag coefficient c_i is experimentally determined. For Ri_* outside the interval $90 < Ri_* < 400$
the results of the laboratory experiments diverge from the Ri_*^{-1}
power law, perhaps because of molecular effects, insufficient
reduction of sidewall friction for $Ri_* > 400$ and the fact that the
exponent of Ri_* approaches zero for $Ri_* \to 0$.

The relation between the laboratory and field experiments found
in this paper is rather encouraging and may be of importance to
future work on mixing and circulation in fjords, where appropriate
mixing experiments are often difficult and expensive to carry out.

ACKNOWLEDGEMENTS

The author wishes to express his gratitude to Professor
G. Kullenberg and S. Kitaigorodskii for valuable discussions and
suggestions during this work. The field experiments were supported
by the Nordic University Group for Physical Oceanography, the Belt
Project under the Danish Environmental Protection Agency and the
Swedish Natural Science Council.

REFERENCES

Buch, E., 1981, On entrainment and vertical mixing in stably
        stratified fjords, Estuarine, Coastal and Shelf Science,
        12:461-469.
Bo Pedersen, F., 1980, A monograph on turbulent entrainment
        and friction in two-layer stratified flow, Inst. of
        Hydrodynamics and Hydraulic Engineering, Tech. Univ.
        of Denmark, Ser. Paper No 25.
Kantha, L.H., 1978, On surface-stress induced entrainment at
        a buoyancy interface. The Johns Hopkins University,
        Department of Earth and Planetary Sciences. Geophysical
        Fluid Dynamics Laboratory, Tech. Report TR 78-1.
Kantha, L.H., Phillips, O.M., Azas, R.S., 1977, On turbulent
        Entrainment at a stable Density Interface, J. of Fluid
        Mech., 79:753-767.
Kato, H., and Phillips, O.M., 1969, On the penetration of a
        Turbulent Layer into Stratified Fluid, J. of Fluid Mech.,
        37:643-655.
Kitaigorodskii, S., 1981, On the theory of the surface-stress
        induced entrainment at a buoyancy interface, Tellus,
        33(1):89-101.
Kullenberg, G., 1969, Measurements of horizontal and vertical
        diffusion in Coastal Waters, Kungl. Vetenskaps- og
        Vitterhets-Samhället, Gøteborg, Ser. Geophysica 2, 52 pp.

Kullenberg, G., 1974, An experimental and theoretical investigation of the turbulent diffusion in the upper layer of
the sea, Inst. Phys. Oceanogr., Univ. of Copenhagen,
rep. No 25.

Kullenberg, G., 1977, Note on entrainment in natural stratified vertical shear flow, Estuarine and Coastal Marine
Science, 5:329-338.

Phillips, O.M., 1976, Entrainment, in:"Modelling and Prediction of the Upper Layer of the Ocean", ed., E.B. Kraus,
Pergamon Press, New York.

Pickard, G.L., 1961, Oceanographic features of inlets in
British Columbia, Mainland Coast, J.Fish.Res.Bd.Canada,
18:907-999.

Price, J.F., 1979, On the scaling of stress-driven entrainment
experiments, J. of Fluid Mech., 90:509-529.

Stigebrandt, A., 1976, Vertical diffusion driven by internal
waves in a sill fjord, J. Phys. Oceanogr., 6:486-495.

Turner, J.D., 1973, Buoyancy effects in fluids, Cambridge Univ.
Press.

Welander, P., 1968, Theoretical forms for the vertical exchange
coefficients in a stratified fluid with application to
lakes and seas, Kungl. Vetenskaps- og Vitterhets-
Samhället,Göteborg, Ser. Geophysica 1, 27 pp.

GEOGRAPHICAL INDEX

SUBJECT INDEX